*Kostya (Ken) Ostrikov
and Shuyan Xu*
Plasma-Aided Nanofabrication

1807–2007 Knowledge for Generations

Each generation has its unique needs and aspirations. When Charles Wiley first opened his small printing shop in lower Manhattan in 1807, it was a generation of boundless potential searching for an identity. And we were there, helping to define a new American literary tradition. Over half a century later, in the midst of the Second Industrial Revolution, it was a generation focused on building the future. Once again, we were there, supplying the critical scientific, technical, and engineering knowledge that helped frame the world. Throughout the 20th Century, and into the new millennium, nations began to reach out beyond their own borders and a new international community was born. Wiley was there, expanding its operations around the world to enable a global exchange of ideas, opinions, and know-how.

For 200 years, Wiley has been an integral part of each generation's journey, enabling the flow of information and understanding necessary to meet their needs and fulfill their aspirations. Today, bold new technologies are changing the way we live and learn. Wiley will be there, providing you the must-have knowledge you need to imagine new worlds, new possibilities, and new opportunities.

Generations come and go, but you can always count on Wiley to provide you the knowledge you need, when and where you need it!

William J. Pesce
President and Chief Executive Officer

Peter Booth Wiley
Chairman of the Board

Kostya (Ken) Ostrikov and Shuyan Xu

Plasma-Aided Nanofabrication

From Plasma Sources to Nanoassembly

WILEY-VCH Verlag GmbH & Co. KGaA

The Authors

Prof. Kostya (Ken) Ostrikov
The University of Sydney
School of Physics A29
Sydney NSW 2006
Australien

Prof. Shuyan Xu
Nanyang Technological Univ.
Plasma Sources & Applications
1, Nanyang Walk
Singapore 637617
Singapur

Cover
Integrated Plasma-Aided Nanofabrication Facility (top left) and nanostructures of various dimensionality assembled by using plasma-aided nanofabrication: ultra-high-aspect-ratio silicon-based nanowires (top right), carbon nanoneedle-like structures (bottom left), and self-organized nanopatterns of silicon carbide quantum dots (bottom right).

All books published by Wiley-VCH are carefully produced. Nevertheless, authors, editors, and publisher do not warrant the information contained in these books, including this book, to be free of errors. Readers are advised to keep in mind that statements, data, illustrations, procedural details or other items may inadvertently be inaccurate.

Library of Congress Card No.:
applied for

British Library Cataloguing-in-Publication Data
A catalogue record for this book is available from the British Library.

Bibliographic information published by the Deutsche Nationalbibliothek
Die Deutsche Nationalbibliothek lists this publication in the Deutsche Nationalbibliografie; detailed bibliographic data are available in the Internet at <http://dnb.d-nb.de>.

© 2007 WILEY-VCH Verlag GmbH & Co. KGaA, Weinheim

All rights reserved (including those of translation into other languages). No part of this book may be reproduced in any form – by photoprinting, microfilm, or any other means – nor transmitted or translated into a machine language without written permission from the publishers. Registered names, trademarks, etc. used in this book, even when not specifically marked as such, are not to be considered unprotected by law.

Typesetting   Uwe Krieg, Berlin
Printing   betz-Druck GmbH, Darmstadt
Binding   Litges & Dopf GmbH, Heppenheim
Wiley Bicentennial Logo   Richard J. Pacifico

Printed in the Federal Republic of Germany
Printed on acid-free paper

ISBN: 978-3-527-40633-3

Contents

Preface *IX*

1 **Introduction** *1*
1.1 What is a Plasma? *3*
1.2 Relevant Issues of Nanoscience and Nanotechnology *12*
1.3 Plasma-Assisted Synthesis of Nanomaterials *19*
1.4 How to Choose the Right Plasma for Applications in Nanotechnology? *31*
1.5 Structure of the Monograph and Advice to the Reader *37*

2 **Generation of Highly Uniform, High-Density Inductively Coupled Plasma** *41*
2.1 Low-Frequency ICP with a Flat External Spiral Coil: Plasma Source and Diagnostic Equipment *42*
2.1.1 Plasma Source *44*
2.1.2 Diagnostics of Inductively Coupled Plasmas *45*
2.2 Discharge Operation Regimes, Plasma Parameters, and Optical Emission Spectra *50*
2.2.1 Electromagnetic Properties and Mode Transitions *50*
2.2.2 Plasma Parameters *52*
2.2.3 Discharge Hysteresis *54*
2.3 Electromagnetic Field Distribution and Nonlinear Effects *56*
2.4 Optical Emission Spectroscopy of Complex Gas Mixtures *62*
2.4.1 Optical Emission Spectra and Hysteresis *63*
2.4.2 $E \rightarrow H$ Transition Thresholds *66*
2.5 Modeling of Low-Frequency Inductively Coupled Plasmas *72*
2.5.1 Basic Assumptions *73*
2.5.2 Electromagnetic Fields *73*
2.5.3 Particle and Power Balance *74*
2.5.4 Numerical Results *76*
2.6 Concluding Remarks *81*

3	**Plasma Sources: Meeting the Demands of Nanotechnology** *85*	
3.1	Inductively Coupled Plasma Source with Internal Oscillating Currents: Concept and Experimental Verification *86*	
3.1.1	Configuration of the IOCPS *87*	
3.1.2	RF Power Deposition *89*	
3.1.3	Plasma Parameters *94*	
3.2	IOCPS: Stability and Mode Transitions *98*	
3.2.1	Optical Emission *99*	
3.2.2	Self-Transitions of the IOCPS Discharge Modes *101*	
3.3	ICP-Assisted DC Magnetron Sputtering Device *106*	
3.3.1	Enhancement of DC Magnetron Sputtering by an Inductively Coupled Plasma Source *109*	
3.3.2	Mode Transitions in ICP-Assisted Magnetron Sputtering Device *111*	
3.4	Integrated Plasma-Aided Nanofabrication Facility *115*	
3.5	Concluding Remarks *119*	
4	**Carbon-Based Nanostructures** *121*	
4.1	Growth of Carbon Nanostructures on Unheated Substrates *123*	
4.1.1	Process Details *124*	
4.1.2	Synthesis, Characterization, and Growth Kinetics *124*	
4.2	Temperature-Controlled Regime *130*	
4.3	Single-Crystalline Carbon Nanotips: Experiment *137*	
4.4	Single-Crystalline Carbon Nanotips: *ab initio* Simulations *141*	
4.4.1	Theoretical Background and Numerical Code *142*	
4.4.2	Geometrical Stability of Carbon Nanotip Structures *143*	
4.4.3	Electronic Properties of Carbon Nanotips *146*	
4.5	Plasma-Assisted Doping and Functionalization of Carbon Nanostructures *149*	
4.5.1	Doping of Carbon-Based Nanostructures: Density Functional Theory Considerations *149*	
4.5.2	Postprocessing of Carbon-Based Nanostructures: Experiments *152*	
4.6	Synthesis of Carbon Nanowall-Like Structures *156*	
5	**Quantum Confinement Structures** *159*	
5.1	Plasma-Assisted Fabrication of AlN Quantum Dots *161*	
5.2	Nanofabrication of $Al_xIn_{1-x}N$ Quantum Dots: Plasma-Aided Bandgap Control *167*	
5.3	Plasma-Aided Nanofabrication of SiC Quantum Dot Arrays *172*	
5.3.1	SiC Properties and Applications *172*	
5.3.2	SiC Growth Modes: With and Without AlN Interlayer *173*	
5.3.3	Quest for Crystallinity and Nanopattern Uniformity *181*	

5.4	Plasma-Aided Fabrication of Very Large-Aspect Ratio Si-Based Nanowires *188*	
5.5	Quasi-Two-Dimensional Semiconductor Superlattices Synthesized by Plasma-Assisted Sputtering Deposition *191*	
5.6	Other Low-Dimensional Quantum Confinement Structures and Concluding Remarks *199*	

6 Hydroxyapatite Bioceramics *209*
- 6.1 Basic Requirements for the Synthesis of HA Bioceramics *209*
- 6.2 Plasma-Assisted RF Magnetron Sputtering Deposition Approach *212*
- 6.2.1 Comparative Advantage *212*
- 6.2.2 Experimental Details *213*
- 6.3 Synthesis and Growth Kinetics *217*
- 6.3.1 Optimization of the Plasma-Aided Coating Fabrication Process *217*
- 6.3.2 Film Growth Kinetics *222*
- 6.4 Mechanical Testing of HA Films *226*
- 6.5 *In vitro* Assessment of Performance of Biocompatible HA Coatings *229*
- 6.5.1 Simulated Body Fluid assessment *230*
- 6.5.2 Cell Culture Assessment *233*
- 6.6 Concluding Remarks *236*

7 Other Examples of Plasma-Aided Nanofabrication *237*
- 7.1 Plasma-Assisted Er Doping of SiC Nanoparticle Films: An Efficient Way to Control Photoluminescence Properties *238*
- 7.2 Polymorphous (poly)Nanocrystalline Ti–O–Si–N Films Synthesized by Reactive Plasma-Assisted Sputtering *241*
- 7.3 Fabrication of Nanostructured AlCN Films: A Building Unit Approach for Tailoring Film Composition *244*
- 7.4 Plasma-Assisted Growth of Highly Oriented Nanocrystalline AlN *251*
- 7.5 Plasma-Assisted Synthesis of Nanocrystalline Vanadium Oxide Films *258*
- 7.6 Plasma-Treated Nano/Microporous Materials *264*

8 Further Examples, Conclusions, and Outlook *269*
- 8.1 Further Examples of Plasma-Aided Nanofabrication *270*
- 8.2 On Benefits and Problems of Using Plasma Nanotools *276*
- 8.3 Outlook for the Future and Concluding Remarks *280*

References *283*

Index *297*

Preface

Applications of low-temperature plasmas for materials synthesis and processing at nanoscales and fabrication of nanodevices is a very new and quickly emerging area at the frontier of physics and chemistry of plasmas and gas discharges, nanoscience and nanotechnology, solid-state physics, and materials science. Such plasma systems contain a wide range of neutral and charged, reactive and nonreactive species with the chemical structure and other properties that make them indispensable for nanoscale fabrication of exotic architectures of different dimensionality and functional thin films and places uniquely among other existing nanofabrication tools. By nanoscales, we imply spatial scales in the range between ~ 1 nm and a few hundred nm (1 nm = 10^{-9} m).

In our decision to write this book we were motivated by the fact that even though basic properties and applications of low-temperature plasma systems had been widely discussed in the literature, there have been no systematic attempt to show the entire pathway from the development of suitable advanced plasma sources and plasma-based nanofabrication facilities with the required parameters and operation capabilities to the successful nanoscale synthesis.

We started our research in the area of the plasma-based synthesis of nanomaterials, in a sense, from behind, only in mid-2001 after the establishment of the Advanced Materials and Nanostructures Laboratory (AMNL) within the Plasma Sources and Applications Center (PSAC) of Nanyang Technological University in Singapore. At that time we had a few RF plasma sources and also RF diode and DC magnetron sputtering facilities and, inspired by a number of breakthrough works on plasma-based synthesis of carbon nanotubes, nanoparticles, and various nanostructured films, decided to follow this exciting and very hot direction of research. At that time the nanomaterials research was among the top research priorities and we managed to secure a couple of million dollars in competitive research and infrastructure grants, established a new Plasma Sources and Applications Center, secured space for the AMNL, purchased our own field emission scanning electron microscope, hired people, purchased catalyzed templates and other necessary stuff and jumped straight into running experiments on nanostructure synthesis by using whatever plasma sources were available at that time in the lab.

However, to our great disappointment, we did not manage to reproduce many of the successful experiments on plasma-assisted synthesis of carbon nanotube-like structures, although all the recipes had been strictly followed. On almost the same substrates, at exactly the same substrate temperatures and working gases the nanostructures either did not grow at all or grew quite differently; in many cases the quality of the nanopatterns was very far from what had previously been reported in the literature.

At that time, there existed a widely accepted opinion that in the growth of carbon nanotubes, the role of the plasma environment is merely in providing vertical alignment of the nanotubes in the electric field, which originates due to charge separation in the plasma sheath between the plasma bulk and the nanostructured surfaces. The role of the ionized gas component was in most cases simply sidestepped or disregarded. Our disappointment was getting even worse when we tried to use different plasma rigs and found out that the results were actually very different in different vacuum chambers, although the process conditions, including surface temperature, were very similar. That led us to an intuitive guess that the observed differences had something to do with the plasma rather than surface conditions.

A bit later, when we realized that the plasma ion bombardment can substantially modify the surface during pretreatment, and also increase the surface temperature during the actual deposition stage, we tried to maintain the substrate DC bias, the main control tool of the ionic fluxes crashing into the surface, the same during the experiments in different plasma chambers. But the results were not the same again. This led us to an intuitive conclusion that the plasma does play a prominent role in the nanostructure synthesis and it is not merely the commonly accepted vertical alignment.

Early in 2003, when trying again and again to synthesize carbon nanotubes from a mixture of methane, hydrogen, and argon, we inadvertently forgot to turn an external substrate heater on. This heating element was supposed to heat our samples to temperatures at least as high as 550–600 °C, commonly used in most of the successful experiments at that time. To our biggest surprise, carbon nanostructures emerged and covered the entire substrate, and most amazingly, in a fairly uniform fashion! This eventually led us to the discovery of what we later called the "floating temperature regime". In this growth mode, the Ni/Fe/Co-catalyzed silicon substrates do not need to be externally heated at all and the growth temperatures are about the same as that of the neutral gas in the plasma reactor (or higher if ion bombardment is significant). This very unexpected and fascinating result was reported at the Gaseous Electronics Conference in San Francisco in October 2003.

Unfortunately, due to significant delays with the reliable substrate temperature and temperature gradient measurements and also bad luck with the editors and referees of a few journals which publish the results quickly (again, Murphy's law!), we did not manage to publish this result until January 2005 (the corrected proof had been published online since March 2004). At about the same time, the literature was virtually flooded by numerous reports on "lower, even lower, lowest" substrate temperatures that enable the plasma-assisted synthesis of various carbon-based nanostructures (just to mention commonly known single- and multiwalled nanotubes, nanofibers, nanocones, nano-pyramids, etc.).

This plasma-related "nanotube growth without heating" paradox was eventually resolved and conclusively related to the additional heating of solid surfaces by the plasma ions; this heating on its own can add more than 100 degrees to the surface temperature.

This is just one example evidencing that "the plasma does matter"! Very fortunately, we realized that reasonably quickly and decided to develop a new and versatile plasma facility, which would enable a direct control of the nanoassembly processes by the independently created and manipulated, in a controlled fashion, low-temperature inductively coupled plasma.

Before that, we had moved a long and thorny way along the trail of plasma source development and knew which main issues needed to be resolved before a successful plasma source could be created. Some of relevant attempts are described in Chapter 2 of this book. Not surprisingly, our new plasma facility, which soon evolved into the very successful Integrated Plasma-Aided Nanofabrication Facility (IPANF), turned out extremely useful in synthesizing not only "trivial" carbon nanotube-like structures but also a large variety of semiconductor quantum confinement structures of different dimensionality discussed in Chapter 5 of this monograph.

We emphasize that despite an enormous number of existing research monographs, textbooks, and edited volumes related to nanoscience and nanotechnology on one hand and the physics and applications of low-temperature plasmas on the other hand, we are not aware of any research monographs showing a consistent and complete "success story," which begins from the development of original plasma sources and processes and ends up with the evidence of successful synthesis and/or processing of nano-materials.

This monograph is based on collaborative research of the authors and their teams, which they started in 1999 in Singapore and continue, via a number of international linkage projects even after Kostya (Ken) Ostrikov moved to the University of Sydney, Australia and established his own Plasma Nanoscience research team. Over the years, this collaboration also involved a large number of researchers from different countries, which eventually led to the establishment of the International Research Network for Deterministic Plasma-Aided Nanofabrication.

The scope of this monograph lies within the "Plasma Nanoscience" subfield at the cutting edge interdisciplinary research at the cross-roads where the physics and chemistry of plasmas and gas discharges meets nanoscience and materials physics and engineering. This work certainly does not aim at the entire coverage of the existing reports on the variety of nanostructures, nanomaterials, and nanodevices on one hand and on the plasma tools and techniques for materials synthesis and processing at nanoscales and plasma-aided nanofabrication on the other one (even though it provides a very long but certainly not exhaustive list of relevant publications). Neither does our monograph aim to introduce the physics of low-temperature plasmas for materials processing. We refer the interested reader to some of the many existing books that cover the relevant areas of knowledge [1–11]. This work has a clear practical perspective and aims to demonstrate to the wide multidisciplinary academic and research community how important is to properly select and develop suitable plasma facilities and processes. This work also poses a number of open questions, which are expected to stimulate the interest of researchers from different areas toward more extensive use of plasma nanofabrication tools.

To make this undoubtedly very "research-heavy" monograph easily understood, at least in very basic terms, to a person with a basic high-school knowledge of physics and chemistry, we decided to write an introductory Chapter 1, which, first of all, explains what is the plasma, what is the nanotechnology and how these two things link together. After a brief overview of the main issues of the nanotechnology and plasma applications in nanofabrication, we give some basic ideas how to choose the right plasma-based process for the envisaged nanoscale application. The structure of this monograph will be introduced immediately after the most important things are clarified.

The authors greatly acknowledge contributions and collaborations of the present and past members of their research teams Plasma Sources and Applications Center (NTU, Singapore) and Plasma Nanoscience @Complex Systems (The University of Sydney, Australia) J. D. Long (very special thanks for his major contribution to the conceptual design and practical implementation of the IPANF facility), Q. J. Cheng, S. Y. Huang, M. Xu, E. L. Tsakadze, Z. L. Tsakadze, N. Jiang, P. P. Rutkevych, C. Mirpuri, V. Ng, L. Sim, W. Luo, J. W. Chai, Y. C. Ee, Y. A. Li, M. Chan, H. L. Chua, I. Denysenko, I. Levchenko, Y. P. Ren, A. Rider, and E. Tam.

We also greatly appreciate all participants of our international research network, as well as fruitful collaborations, mind-puzzling discussions, and critical comments of M. Bilek, I. H. Cairns, L. Chan, U. Cvelbar, C. H. Diong, N. M. Hwang, B. James, M. Keidar, S. Kumar, S. Lee, V. Ligatchev, O. Louchev, D. R. McKenzie, X. Q. Pan, P. A. Robinson, P. Roca i Cabarrocas, L. Stenflo, R. Storer, H. Sugai, G. S. Tan, L. Tan, S. V. Vladimirov, T. Woo, M. Y. Yu, and many other colleagues, collaborators and industry partners. We also thank all authors of original figures for their kind permission to reproduce them.

Last but not the least, we thank our families for their support and encouragement. Kostya (Ken) Ostrikov extends very special thanks to his beloved wife Tina for her love, inspiration, motivation, patience, emotional support, and sacrifice of family time over weekends, evenings and public holidays that enabled him to work on this book and also to his most patient and diligent student and companion Grace The Golden Retriever, who attended all practice sessions of all his keynote and invited talks in the Plasma Nanoscience area, and especially for her outstanding guiding skills that enabled all his thoughts and bright ideas he got during long walks with her.

This work was partially supported by the Australian Research Council, the University of Sydney, the Agency for Science, Technology, and Research (Singapore), Lee Kuan Yew Foundation (Singapore), Nanyang Technological University (Singapore), National Institute of Education (Singapore), Institute of Advanced Studies (Singapore) and the International Research Network for Deterministic Plasma-Aided Nanofabrication.

Sydney and Singapore *Kostya (Ken) Ostrikov and Shuyan Xu*
May 2007

1
Introduction

For decades, low-temperature (frequently termed "cold") plasmas have been extensively used in many industrial applications, just to mention a few: microstructuring of semiconductor wafers in microelectronic manufacturing; deposition of various coatings, protective, and other functional layers on solid surfaces; surface treatment, which includes hardening and modification of surfaces (e.g., metal tools such as drill bits, cutting blades), welding, drilling, cutting, and functionalization; chemical synthesis of ceramic and powder materials; toxic waste and flue exhaust management; all sorts of imaginable light sources from low-pressure bulb globes and commercial halogen lamps to high-intensity discharge lamps; plasma engines (ion thrusters) for the next-century adventures of space rocketeers; water purification, sterilization of medical instruments; energy converters, plasma antennas, satellite communication; various tools for isotope separation, materials characterization, detection of radioactive materials; gas lasers, and, of course, plasma TVs and large-panel displays, which have already found their place in our everyday's lives.

Moreover, 99 % of all visible matter in the Universe finds itself in the plasma state. It is a common knowledge that stars are nothing else but light-emitting giant hot-plasma balls that generate enormous amounts of heat and electromagnetic radiation as a result of nuclear fusion reactions when lighter elements fuse together and release nuclear energy $\mathcal{E}$ according to the famous Einstein's formula $\mathcal{E} = \Delta m c^2$, where Δm is the so-called defect of mass, which is a difference between the masses of the reacting species and products of the nuclear fusion reaction. After years of research and development, incredibly cheap electricity generated in nuclear fusion reactors is also becoming reality. Indeed, the International Thermonuclear Experimental Reactor (ITER) (probably the largest ever R&D project) will allow, for the first time, a positive energy gain in nuclear fusion reactions, will be commissioned and is just a couple of tens of years away from its international commercial operation.

Plasmas also play a prominent role in a variety of physical phenomena in the atmosphere and space. Bright atmospheric glows, solar wind, radio-emissions, bursty waves, shocks and other space phenomena owe their origin

Plasma-Aided Nanofabrication. Kostya (Ken) Ostrikov and Shuyan Xu
Copyright © 2007 WILEY-VCH Verlag GmbH & Co. KGaA, Weinheim
ISBN: 978-3-527-40633-3

to the plasma. Above all, does everybody know that a few-minute-long blackout in radio-communications during the re-entry of a space shuttle is also a plasma-related phenomenon?

The reader probably already understood that this monograph is about the plasma, which can be "cold" and "hot" and has so many natural occurrences and technological applications, absolutely impossible to cover in a single book. However, as the title suggests, it is also about nanofabrication and nanoassembly, which intuitively leads to a guess that this is something to do with nanotechnology. We intentionally avoided mentioning this very recent application in the first paragraph, which begins with "for decades"! This cutting-edge application emerged, first of all, because of continuing shrinkage of microelectronic features in sizes. The present-day ulra-large-scale-integration (ULSI) techhnology already deals with plasma-etching created features (e.g., trenches) as small as 90 nm and is gradually moving, according to the predictions of the International Technology Roadmap for Semiconductors (ITRS) [12] toward even smaller ones (∼40–50 nm). It is remarkable that almost a half of all process steps in the fabrication of semiconductor wafers and microchips involve plasma processing! Wafer processing in plasma reactors nowadays enables one to deposit stacks of barrier interlayers, with each of them being only a few to a few tens of nanometers thick.

On the other hand, Iijima's discovery of carbon nanotubes [13] was made from graphitic soot also synthesized in a plasma! Following this landmark work, a large variety of nanostructured films, nanoparticles, nanocrystals, individual nanostructures and their patterns and arrays have been successfully synthesized by using plasma-based tools and processes. One can thus think that the plasma indeed has something to do with carbon nanotubes and other nanostructures, the building blocks of nanotechnology.

However, a proper understanding of the relation between the plasma environment and fabrication of nanostructures is absolutely impossible without the knowledge of the main terms used, their basic properties, and typical examples of using plasmas for materials synthesis and processing at nanoscales. This is one of the main aims of this chapter. In the following, we will introduce the basic concepts and terminology (plasma, nanofabrication, etc.) used in this monograph, illustrate many uses of plasma-based processes in nanofabrication, analyze some of the most important issues in the development of suitable plasma tools and processes, and also explain the logic structure of this monograph.

1.1
What is a Plasma?

In this section we will introduce the most essential things the reader needs to know about the plasma. Owing to an incredibly large amount of knowledge accumulated to date about all sorts of plasmas and presented in a number of textbooks [14–17], we will focus only on the plasmas and their properties particularly suitable for materials synthesis and processing at nanoscales. None of the discussions in this chapter is exhaustive; our primary aim here is to provide the minimum amount of knowledge for comfortable reading of the remaining chapters of this monograph. Some of the basic concepts of nanotechnology and its relation with the low-temperature plasmas will be discussed later in this chapter. So, what is a plasma?

A plasma is generally understood as a fully or partially ionized gas. The ionization is a process involved in the creation of a positively charged ion from a neutral atom or a molecule. In this process, one or more electrons are stripped from the atom/molecule giving rise to a nett positive charge. The ionization process requires the atom/molecule to transit from the original, lower energy, to a higher energy, state and thus requires a certain amount of energy to be transferred to the atom/molecule. This transfer can occur via a number of channels, such as the impact of an incident electron (these kinds of processes are termed the electron impact processes), ion, or any other charged or neutral specie; energy transfer from a photon of a sufficient energy (this process is called photoionization); and external heating of the neutral gas (thermal ionization). Various and the many ways to ionize neutral gases are described elsewhere [18].

Laboratory plasmas are in most cases created and sustained in gas discharges in solid containers (e.g., vacuum chambers or discharge tubes). By a gas discharge one usually implies a process wherein a significant electric current appears in the originally neutral gas following its ionization [19, 20]. As was mentioned above, a large amount of external energy is required to ionize neutral gas atoms/molecules. The specific elementary mechanism of ionization in fact determines the most appropriate way to deliver the energy to the neutral gas. For example, intense laser beams can be used for photoionization and DC electric fields for acceleration of ionizing electrons that take part in electron impact processes. On the other hand, the actual amount of energy strongly depends on the operating gas pressure: generally speaking, the higher the pressure, the more molecules/atoms one needs to ionize, and hence the larger amount of energy is required to sustain the discharge.

At this point it would be prudent to read the title of this book again and notice the string "plasma sources" and ask what most essential parts a plasma source should have. First of all, there should be a vessel to contain the plasma. Secondly, there should be some means of delivering the energy required for the ionization of neutral species and sustaining the discharge. As follows from the above arguments, higher pressure operation would require more powerful means (e.g., more powerful DC or RF generators, lasers, etc.) of the ionization. Thus, operation of atmospheric-pressure plasma discharges normally implies larger power inputs to the neutral gas as compared to low-pressure (e.g., in the mTorr range) discharges.

This is one of the reasons why many laboratory plasmas are sustained at low pressures, when lower amount of input power is needed to sustain plasmas with comparable parameters. Thus, the costs of input power (and also of the power supply units) at low pressures are lower; however, the lower the pressure of the working gas, the more expensive is the vacuum pump system. This is the main tradeoff between the operation of plasmas at high (e.g., atmospheric) and low (e.g., mTorr pressure range) gas pressures. Thus, we arrive at the conclusion that a plasma source should have a vacuum system appropriate for the specific range of operating pressures; or, alternatively, not to have any vacuum system if the discharge is maintained at atmospheric pressures. The last essential thing of most of the existing commercial and laboratory plasma facilities is a gas handling system, which enables one to let working gases in the vacuum vessel and control the gas dosing, which is commonly done, e.g., by using special controllers of gas mass flows and partial pressures.

Of course, simplest gas discharges, such as a spark between two sharp electrodes in an open air, do not require all of the above components. The only thing needed in this case is a DC power supply and two electrodes. However, real present-day plasma fabrication facilities include not only all of the four essential components listed above but also a wide range of control, diagnostic and other auxiliary instrumentation. Some of this equipment will be described in relevant sections of this book.

From now, the reader might realize that a fully operational plasma source with all major components as well as with some "blows and whistles" is central to any plasma applications. Nonetheless, the question about what specific sort of plasmas one needs to use for applications in nanotechnology, still remains. We will discuss this issue, as well as some specific requirements for plasma sources suitable for nanoscale applications in more detail in Section 1.4.

Another important issue is to limit the number of possible options and to identify which sorts of plasma are worthwhile to use in general. For instance, would it be wise to use hot and fully ionized fusion plasmas with ion temperatures of excess of 100 million degrees? Our commonsense tells us: of course not! Such an environment is too hot for the assembly or processing of any solid substance. At such temperatures, not only the matter will melt and evaporate but will also most likely get stripped of some electrons and turns into the ionized gas (plasma) state. Apart from the synthesis of light-element nuclei from fusing nuclei of even lighter elements, no nanoscale processes are feasible in this environment.

One can thus conclude that a partially ionized low-temperature plasma is what really is needed for nanofabrication. The degree of ionization of the plasma

$$\zeta_k = \Sigma_k n_i^k / n_0 \tag{1.1}$$

is defined as the combined fraction of all ionic species k among neutral gas atoms/molecules. Here, n_i^k and n_0 are the number densities of ionic species k and neutral species in the discharge, respectively. In hot nuclear fusion plasmas $\zeta_k = 1$, whereas in cold processing plasmas ζ_k usually does not exceed 10^{-3}. Figure 1.1 shows sketches of fully and partially ionized plasmas.

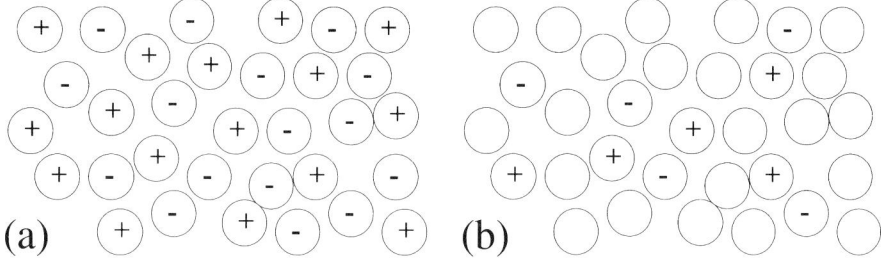

Fig. 1.1 Fully ionized plasmas (a) contain electrons and positive ions; partially ionized plasmas (b) contain electrons, ions, and neutral species.

We now arrive at the next obvious question: if the plasma needs to be cold, then how exactly cold should it be? Wait a minute! What exactly (plasma species) should be cold and how to define the plasma temperature? To this end, we note that partially ionized plasmas consist of at least three species (electrons, ions, and neutrals). Therefore, one needs to define three different temperatures for each of the three species: electron temperature T_e, ion temperature T_i, and temperature of neutrals T_n. Each of the temperatures T_j ($j = e, i, n$) can be introduced as $T_j = 2/3 \langle \mathcal{E}_k^j \rangle$ according to the kinetic theory of gases [14], where $\langle \mathcal{E}_k^j \rangle$ is the kinetic energy of species j averaged over their energy distribution function. For comprehensive review of different energy distribution functions (e.g., Maxwellian, bi-Maxwellian, Druyvesteyn-like, etc.) the reader can be referred to Lieberman's and Liechtenberg's textbook on plasma discharges and materials processing [15].

However, we still have not answered the question about how low the temperatures of the plasma should be to be able to contribute to nanoassembly (this term will be properly defined in the next section) processes. Again, our commonsense tells us that a room-temperature gas would be the best option.

For the neutrals and ions this is perfectly fine and they can, e.g., participate in the nanoscale synthesis processes (e.g., nucleate) in their adatom/adion form. However, an equivalent electron energy of 300 K would be just 0.026 eV, which is much lower than what is needed for electron-impact ionization, one of the main mechanisms that sustains the plasma. We should note that in an ionized gas with an average electron energy of 0.026 eV there is always a small but nonnegligible population of higher energy electrons (frequently termed "electron tail" in the physics of gas discharges) with the energies 10–30 times higher than T_e (thus, ranging from 0.26 to 0.78 eV). However, even this energy is far below the ionization potentials of most of the gases, e.g., 15.6 eV for argon, 13.6 eV for hydrogen, etc. We thus arrive at an important conclusion that the electron temperature should at least be of the order of 1 eV ($\sim$11,600 K), to maintain the plasma with room-temperature ions!

Such plasmas are termed nonequilibrium ($T_e \gg T_i \sim T_n$) low-temperature plasmas; in fact, this is one of the two major types of low-temperature plasmas used in materials processing. One might ask a question: would it be right to keep terming the plasma a low-temperature one if the temperature of its ionic and neutral species is raised up to the level comparable with the electron one? The answer is definitely yes, if one notes that temperatures of $\sim$1 eV are still a few orders of magnitude lower than required for nuclear fusion. Plasmas with $T_i \sim T_n \sim T_e$, wherein each of the temperatures is of the order of or higher than $\sim$1 eV, are commonly termed equilibrium or thermal plasmas.

In reality, it is certainly easier said "increase temperatures of ions and neutrals" than done. For instance, how exactly can one implement a transition from nonequilibrium ($T_e \gg T_i \sim T_n$) to thermal ($T_i \sim T_n \sim T_e$) plasmas? From the viewpoint of turning knobs of commonly available plasma sources, what should be changed to achieve such a transition? As we have already learned, there are two main controls of the plasma with the same composition: the input power and the gas feedstock pressure. So, let us try to keep the pressure low and increase the input power aiming to heat the ions and neutrals. A higher input power will definitely result in stronger electron-impact ionization and larger densities of electrons n_e and ions n_i.

At this point it would be instructive to point out that the plasma is always *charge neutral*, which means that the total electric charge residing on positively charged species (e.g., positive ions or cationic radicals) is balanced by the negative charge of the electrons and negative ions (atoms/molecules with one or more electrons attached) or negatively charged solid grains. In the simplest but most common case of plasma experiments when all (or at least the overwhelming majority of them) positive ions are singly charged (i.e., their charge is +1) and there are no negative ions or solid particles, the number densities of the electrons and ions are the same $n_e = n_i$ and are commonly termed the plasma density n_p. If there are more than one ionic species and/or the species are charged with multiples of electron charges, then $n_e = \sum_k Z_i^k n_i^k$, where Z_i^k is the (positive) charge on ionic (and/or cationic) species k. In this case (as well as in most other cases) it is safe to determine the plasma density as that of the electrons. However, if there are negative ions and/or other (e.g., dust grains, nanoclusters or nanoparticles) species in the discharge, a more accurate convention as to how to define the plasma density is required.

Let us turn our attention back to what happens when an input power is increased in a low-pressure gas discharge. Common experimental observations suggest that the plasma density (here $n_p = n_e = n_i$) usually increases linearly with input power but the electron temperature does not. Therefore, even though the overall number of ionizing electrons increases, the main electron population does not become hotter and remains at almost the same temperature. Moreover, the temperatures of ions and neutrals do not increase either (at least significantly)! One can try to increase the input power again and again but in practice a substantial care should be taken not to eventually damage the power generator. This is just one of the examples when even most advanced technology ($\equiv$ powerful generator) cannot beat the physics! Indeed, why do not we observe a substantial increase of the neutral gas temperature when the input power is doubled or tripled? The basic reason is that when the pressure is low, the rates of collisions between the multiple species are not high enough to thermalize (i.e., bring to thermal equilibrium) all the species in the discharge.

Thus, we are basically left with no other option but to try to manipulate the operating gas pressure to generate a thermal (but still low-temperature) plasma. When the pressure increases, what would most likely happen is that the rates of collisions between the discharge species will become high enough to substantially increase the ion and neutral temperatures and bring the partially ionized plasma to thermal equilibrium ($T_e = T_i = T_n$). It would now be wise to pose one obvious question: since we are talking about plasma applications in nanotechnology, and already learned that hot plasmas are not suitable for this purpose, then which low-temperature plasma (nonequilibrium or thermal) one should give preference to? This not-so-simple question will be answered in Section 1.4.

So far, we have introduced the very basic concepts related to the plasma and should now approach the issues related to plasma uses in nanofabrication more closely. However, we still do not know what the nanofabrication or even the nanotechnology is; the main issues of nanoscience and nanotechnology will be introduced in Section 1.2. For now, we just stress (again by picking another word "nanoassembly" from the book's title) that the plasma uses we are interested in will be primarily related to the synthesis of solid matter at nanoscales. Thus, the obvious question arises: our "cold" plasma should somewhere meet a solid, to contribute to some nanoscale growth on it.

In this regard, there could be two basic options: either some nanosized solid particles float in the plasma or some nanoscale objects are grown on a solid surface exposed to a plasma.

This is why it is essential to consider what actually happens when the plasma meets a solid surface. The most amazing thing is that when the plasma contacts a surface, a thin nonneutral layer of space charge, conventionally termed the plasma sheath, is formed. This phenomenon is intimately related to the plasma shielding (also called Debye shielding) of electric fields brought to the plasma by external objects and the notion of plasma confinement.

Figure 1.2 sketches some of the main physical phenomena that occur when a solid surface faces a plasma. Let us first consider what actually happens from the plasma confinement point of view. In nuclear fusion devices, hot plasmas are usually confined due to the balance between the magnetic pressure force (known in the physics of fusion plasmas as $\mathbf{j} \times \mathbf{B}$ force, where $\mathbf{j}$ is the current that flows through the plasma and $\mathbf{B}$ is the confining magnetic field) and kinetic pressure gradient force (this force is proportional to the gas density gradient). In low-temperature plasmas, it is common that the density of electrons and ions is higher in the central areas of the discharge chamber and decreases toward the surfaces of the vessel. However, very rarely such plasmas are confined by the magnetic fields. Then which force can counterbalance the kinetic pressure gradient force? The only viable possibility in this case would be to use an electric force. However, the plasma is always charge neutral, which means that no electric fields should exist in the plasma bulk area shown in Fig. 1.2. Thus, to create the electric field needed for the plasma

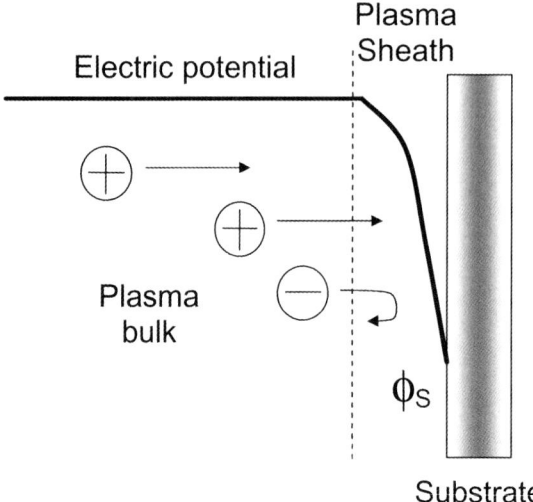

Fig. 1.2 Sketch of the near-substrate plasma area.

confinement, charge separation near the solid surface, is essential. Therefore, there is a need for a charge-nonneutral area (plasma sheath) in the vicinity of the surface.

At this point it would be wise to ask: how exactly the charge separation in the vicinity of the solid surface come about? Highly mobile plasma electrons reach the surface much faster than the ions and create a negative (with respect to the plasma bulk) electric potential on the surface, which repels the electrons and attract the ions as shown in Fig. 1.2. Let us consider the simplest case of a floating (disconnected from the chamber walls and the ground) substrate most widely covered in textbooks . As more electrons deposit on the surface, its potential further decreases, which results in a stronger ion flux to the surface and much reduced electron flux. When the surface potential reaches a certain value (called the floating wall potential), the ion and electron currents balance each other and the net current flowing through the substrate is zero. The electric field (directed from the plasma bulk toward the surface) thus arises and accelerates the positive ions toward the substrate. On the other hand, the electrons are repelled and their density in the plasma sheath is always less than that of the ions. This charge separation is the reason for the existence of the plasma-confining electric fields in low-temperature gas discharges.

However, as we have stressed above, overall charge neutrality of the plasma is equivalent to the nonexistence of electric fields (which, without external fields/actions can only appear as a result of charge separation) in the plasma bulk. We thus come to the obvious conclusion that the electric field should be confined in a relatively narrow area near the substrate. Moreover, the electron–ion separation should also take place in this area only. Hence, the electric potential should be nonuniform only in the vicinity of the solid surface as shown in Fig. 1.2.

We hope that our arguments have convinced the reader in the vital necessity of the plasma sheaths. However, some of the readers might still be curious how wide should the area of charge separation be and why is the sheath also commonly termed "Debye sheath." This is closely related to the unique ability of the plasma to shield any externally imposed electric fields; this phenomenon is called Debye shielding.

The essence of this phenomenon is that when a charged body is immersed in the plasma, a "coat" made of oppositely charged species shields the plasma bulk from this external electric field. In the case considered, a negatively charged surface is effectively screened from the plasma bulk by the "coat" mainly made of positively charged ions—the plasma sheath! If there were no thermal motions, this "coat" could be infinitely thin; in reality, it should have a finite thickness to accommodate for the effects of continuous ion recombination on the surface and mobile electrons that can still make it to the surface having overcome the potential barrier.

This finite thickness in the one-dimensional geometry of Fig. 1.2 can be quantified by using a balance of forces on an electron fluid element (here we do not consider the details of the plasma fluid theory and refer the interested reader to F. F. Chen's textbook [14])

$$-eE = \frac{k_B T_e}{n_e} \frac{\partial n_e}{\partial x} \quad (1.2)$$

and Poisson's equation

$$\frac{d^2 \phi}{dx^2} = -\frac{e(n_i - n_e)}{\varepsilon_0} \quad (1.3)$$

where $\phi(x)$ is the electrostatic potential, x is the coordinate normal to the surface, n_i and n_e are the densities of the ions and electrons, respectively, and k_B and ε_0 are Boltzmann's constant and the dielectric constant of vacuum, respectively. The solution of Eq. (1.2) gives

$$n_e(x) = n_0 \exp\left(\frac{|e|\phi(x)}{k_B T_e}\right) \quad (1.4)$$

which is Boltzmann's relation for electrons, one of the most important relations in the plasma physics. Here, n_0 is the electron density in the plasma bulk in Fig. 1.2.

From Eq. (1.4) one can note that in the plasma sheath, where $\phi(x) < 0$ (potential in the plasma bulk is chosen as the zero reference potential), the electron density is less than that in the plasma bulk and, moreover, exponentially decreases toward the substrate. This is perfectly aligned with our earlier conclusions. It is important to note that Boltzmann's relation has a simple and transparent physical interpretation. Since the electrons are very light, then should the forces onto them not balance, they would indefinitely accelerate. Thus, an electron density gradient instantaneously sets up a charge separation with ions, which in turn results in a balancing electric field.

After substitution of expression (1.4) into Eq. (1.3) and expansion into Taylor's series in the area of an efficient electrostatic shielding ($e\phi(x) \ll k_B T_e$), one obtains

$$\phi(x) = \phi_0 \exp\left(-\frac{|x|}{\lambda_D}\right) \quad (1.5)$$

where ϕ_0 is the (negative) potential at the surface and

$$\lambda_D = \sqrt{\frac{\varepsilon_0 k_B T_e}{n_e e^2}} \quad (1.6)$$

is the Debye length. More careful calculations suggest that the width of the plasma sheath λ_s is typically a few Debye lengths, i.e.,

$$\lambda_s = \gamma_s \lambda_D \quad (1.7)$$

where γ_s is a numerical coefficient typically ranging between 1 and 5 for unbiased substrates [15]. We emphasize that beyond a few Debye lengths, the plasma shielding is quite effective and the negative potential due to the substrate surface is negligible as sketched in Fig. 1.2. If an external bias U_s (in most cases it is also negative) is applied to the substrate, the width of the plasma sheath increases, as suggested by the following formula [15]:

$$\lambda_s = \frac{\sqrt{3}}{2} \lambda_D \left(\frac{2U_s}{k_B T_e} \right)^{3/4} \tag{1.8}$$

which is valid when $T_e \ll U_s$.

At this stage the readers' curiosity would probably peak and prompts their mind to question the need of so quite a lengthy discussion of the plasma sheath and its parameters. Indeed, why do we need these details in a "nano"-book? The answer is simple: it is the ion fluxes and electric fields in the near-surface areas as well as the electric charges (and also currents) on the surface what makes plasma-exposed solid surfaces so different!

One of the key things that one needs to know to proceed with the plasma-assisted synthesis of nanostructures is the energy and flux of the ions impinging on the surface. Apparently, the ions enter the plasma sheath area with a certain velocity and then accelerate under the action of the electric field. Amazingly, this velocity is intimately related to the whole existence of the plasma sheath! More specifically, in the near-surface area sketched Fig. 1.2 the plasma ions enter the plasma sheath with the velocity equal to or larger than the Bohm velocity

$$v_i \geq V_B = \sqrt{\frac{k_B T_e}{m_i}} \tag{1.9}$$

where m_i is the ion mass. Equation (1.9) expresses the commonly known Bohm sheath criterion [15]. Thus, the kinetic energy the plasma ions have at the edge of the (unbiased) plasma sheath $|x| = \lambda_s$ is $\mathcal{E}_i^s = T_e/2$ and is expressed in terms of the *electron* temperature rather than the ion one. In low-temperature nonequilibrium plasmas with $T_e \gg T_i$ the ion velocity at the sheath edge can be substantially larger than the ion thermal velocity $V_{Ti} = (k_B T_i / m_i)^{1/2}$.

The ion velocity at any point x within the plasma sheath $v_i(x)$ can be calculated by using the ion energy conservation

$$\frac{1}{2} m_i v_i(x)^2 = \frac{1}{2} m_i v_{i0}^2 - e\phi(x) \tag{1.10}$$

and ion continuity

$$n_0 v_{i0} = n_i(x) v_i(x) \tag{1.11}$$

equations, where n_0 and v_{i0} are the ion number density and velocity at the sheath edge. Combining Eqs. (1.10) and (1.11), one obtains

$$n_i(x) = n_0 \left(1 - \frac{2e\phi(x)}{m_i v_{i0}^2}\right) \qquad (1.12)$$

for the ion number density at any point within the plasma sheath. From Eq. (1.10), it follows that at the moment of impact at the nanostructured surface, the ion kinetic energy is approximately $T_e/2 + e\phi_s$. In the case of biased substrates ($\phi_s = U_s$), one usually has $T_e/2 \ll e\phi_s$ and it is quite accurate to assume that the ions are accelerated to the energy equal to the substrate bias. In this case the sheath width can be calculated by using Eq. (1.8). It is important to note, however, that the term $e\phi_s$ should also include contributions from micro- and nanoscaled morphology elements on the surface [21–24].

This is probably all of the most essential bits of knowledge the reader should have about low-temperature plasmas used in nanoscale applications. At this point we should map the most important issues of nanoscience and nanotechnology most relevant to this monograph.

1.2
Relevant Issues of Nanoscience and Nanotechnology

Presently, all high school children are probably aware that we live or at least entering the "nanoage" and that rapid advances in nanoscience and nanotechnology make their way into our everyday's lives. It is often said that the impact of nanotechnology over the next decade can be as significant as that of the commonly known groundbreaking inventions such as lasers, microchips, radio communications, and electricity. A number of countries have developed sophisticated, internetworked and well-coordinated national programs, with the most prominent one being the US National Nanotechnology Initiative, with almost US$1 billion spent in 2004 and additional US$3.7 billion allocated for 2005–2008 [25].

So, what is the nanoscience and nanotechnology in general? According to Richard Feynmann's vision given in his 1959 speech "There is plenty of room at the bottom" [26], the matter can be manipulated at atomic and molecular scales (just like Lego^{TM} or Tetris building blocks in popular children games) to create exotic, unusual assemblies, structures, patterns, etc. Expressed in a simple form, the main idea of nanoscience is to arrange atomic building blocks in an unusual, otherwise nonexistent, way, create something (e.g., nanostructures or materials), which is extremely small (with sizes in the nanometer range) and have the properties very different from the objects with "normal," macroscopic sizes, as well as to explore and develop suitable means of controlled assembly at nanoscales. Put in Feynmann's words, the ultimate goal of nanoscience is to arrange atoms one by one, the way we want them, just like bricklayers arrange bricks and other building blocks into pillars, walls, and eventually into architectural masterpieces.

The nanotechnology would then be some set of recipes and nanotools, which on one hand emerges from nanoscience via the process of innovation and commercialization, and on the other hand, has an outstanding potential to create new and transform existing industries (virtually any, from construction and agriculture to microelectronics, aerospace and IT) and eventually significantly improve the living standards. According to the recent report of the National Nanotechnology Strategy Taskforce (NNST) "Options for a National Nanotechnology Strategy" [27], global sales of products incorporating emerging nanotechnologies in 2014 could total US$2.6 trillion, which is as much as ICT and ten times more than biotechnology revenues.

As a synergy of a range of platform technologies with the most advanced nanoscience knowledge, the nanotechnology can become an underlying technology within a number of industrial sectors (including but not limited to manufacturing, health care, energy, electronics, and communications) and is capable to provide environmentally sustainable and cost-efficient manufacturing processes, cleaner and efficient energy sources, new exotic materials and coatings with nanoscale structure and features (termed nanostructured materials or simply nanomaterials below), extra-small microchips with enormous data processing and storage capacity, new ways of targeted delivery of drugs to individual cells, and purified water, just to mention a few. According to the NNST, in Australia (the country with a relatively lower presence of the manufacturing industry sector as compared with the US, Japan, and EU) only the nanotechnology will be used in up to 15 % of products within the next 10 years.

We hope that the reader has already appreciated the main global aims and benefits of the nanoscience and nanotechnology and will not expand on this matter any further. A more comprehensive coverage of the nanotechnology-related topics is given elsewhere [28–30].

Let us now be a bit more specific on the main issues we will be looking at in this monograph and revisit the working definition we gave for the nanoscience. The first notion we come across is some "small" things. How exactly small should they be to qualify to be considered in nanoscience? There are no clear boundaries as to the sizes of the objects involved; with the only one requirement that the nano-objects should not be as small as atoms/molecules nor as large as macroscopic things. This implies that anything with the sizes exceeding 1 nm, which is approximately 4–7 times larger than the sizes of most common atoms and molecules (excluding macromolecular matter), does qualify as a nano-object.

It is a bit more difficult to estimate the upper limit for the "nanosize," in part because of not-so-clear interpretation of how to define macroscopic things precisely. Indeed, if we start combining atoms following Feynmann's recipes, we will soon find that the properties of these assemblies will be very different depending on the number of atoms involved, and hence, the object size. A single atom exhibits atomic properties (e.g., will have a clear structure of atomic energy levels). Two atoms will have a different chemical organization including a bond between them and as a result a quite different structure of energy levels. The electronic structure of a cluster made of a few atoms will certainly be even more different, with the main feature in the appearance of the energy bands, which are quite narrow in the case of a few-atom clusters and broaden as the number of atoms increases. It is quite straightforward to estimate the number of atoms that can make the smallest nano-object. For example, there are approximately 28 atoms in a silicon crystal with a volume of 1 nm^3.

As the object size increases further, their properties become even different and eventually start resembling those of bulk crystals when they become sufficiently large. It is commonly accepted that nanoparticles of a size of ~1 nm feature properties no different from those of bulk materials. One more unusual thing that became reality owing to the recent advances in nanoscience is that the size dependence of electronic and other properties is much stronger when the nano-objects are real small. But exactly, how small? This is intimately related to the notion of electron (more precisely, the electron wave function keeping in mind the basics of the quantum mechanics) confinement, which turns out more efficient in the sub-10 nm size range. In low-dimensional semiconductor structures such as quantum dots, it is commonly accepted that the efficiency of the electron confinement is best when at least one of the sizes is less than the exciton's Bohr radius, which is also approximately equal to 10 nm.

The remarkable changes in the electronic structure lead to the prominent dependence of numerous properties of nanosized objects on their size. A "classic" example of such dependence is a dramatic change of gold spherical nanoparticles in color when their size is varied. Indeed, as the nanoparticle size is reduced from 30 down to 1 nm, their color changes from reddish-blue to orange and even become colorless [31]. On the other hand, when the gold particles are enlarged to the size of macroscopic crystals, their color reverts to the commonly expected yellowish golden. Therefore, the *size* does matter at nanoscales!

We emphasize that the above observation applies to spherical golden nanoparticles. In this case there is only one parameter that characterizes their size—obviously, the radius. On the other hand, ellipsoidal or cylindrical nanoparticles already have two different parameters that characterize their size. Let us take a nanocylinder as an example. The two parameters that determine its size are the length and the radius. If the length is much larger than the radius, the conditions for the electron confinement are much different along the cylinder axis and within parallel slice cut across the axis. However, since all the slices are identical, the electron confinement will be the same in every slice.

If we consider a cone, such slices will be larger near the cone's base and will be very small toward the cone's tip. Therefore, the electron confinement in the radial direction is more effective in the upper sections of the nanocone than near its base. But the most amazing thing is that there still only two parameters that characterize the nanocone's size and they are the same as in the case of the straight cylinder—the base radius and the height! So, what actually led to the remarkable change in the electron confinement? The answer is now obvious and it is the *shape*, which also means a lot at the nanoscales!

It goes without saying that it is also extremely important to make the nanoscale objects from suitable materials or any combination thereof. We thus work out another critical issue being the *elemental composition*. If it is a carbon nanotube or a carbon nanocrystal, then, apparently, they have to be made of carbon atoms and not any others. Then what is the difference between the nanotubes and the crystals if they are made of the same material? Of course, in their internal (also frequently referred to as the chemical) *structure*, which is yet another major issue in nanoscience and nanotechnology!

Let us turn our attention to Fig. 1.3, which shows two single-walled cylindrical carbon nanotubes. Structurally, they are hollow (there are no atoms inside) and are made of planar graphite (more precisely graphene) sheets wrapped in a specific way. Depending on the angle of wrapping up of the graphene sheet (which is made of a hexagonal network of carbon atoms), there are three possibilities of forming carbon nanotubes with the zigzag, armchair, or chiral structures. Put in a simple way, the main difference between these structures is in the way the hexagons with carbon atoms are tilted around the lateral surface of the nanotubes. There are other modifications of carbon nanotubes, with the two of them shown in Fig. 1.3. Panels (a) and (b) in Fig. 1.3 show sketches of a capped and open-ended (with no cap on top) single-walled nanotube, respectively. If a nanotube has more than one wall, it is called a multiwalled nanotube.

One remarkable structural feature of capped single-walled nanotubes in Fig. 1.3(a) is that all the atoms are intimately interlinked, with all their chemical bonds occupied by other carbon atoms. This is an example of a chemically pure structure, made of the same sort of atoms. For more details of the fascinating properties and applications of carbon nanotubes the reader can be referred to the landmark monograph [32].

Another nano-object with quite different structural properties is depicted in Fig. 1.4(a), where a nanocrystal with multiple unterminated dangling bonds

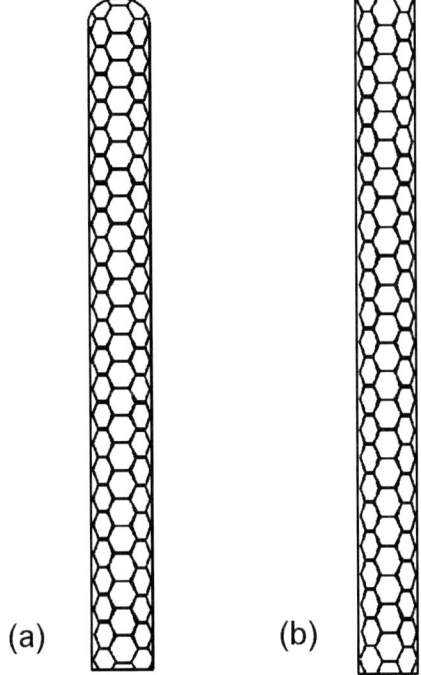

Fig. 1.3 Sketches of (a) capped and (b) open-ended single-walled carbon nanotubes.

is sketched. This structure has no internal voids and is fully filled with atoms of the same sort, which form an ordered three-dimensional crystalline lattice. This nano-object is also chemically pure; however, its multiple surface bonds can be easily terminated by highly reactive species such as atomic hydrogen (shown as black circles in Fig. 1.4(b)). Another nanocrystal in Fig. 1.4(b) is no longer chemically pure and is commonly known as a hydrogenated crystal X:H, where X denotes a chemical element.

For instance, if the silicon crystal is concerned, then Si:H would refer to a hydrogenated silicon nanocrystal. We note that amorphous nanoparticles can also be hydrogenated and denoted as a-Si:H nanoparticles. The structure sketched in Fig. 1.4(b) is an example of a nanostructure that lacks chemical purity. Thus, the *chemical purity* is yet another important issue to highlight.

It is relevant to note that if a nanoassembly is made of more than one sort of atoms, then there is one more essential requirement, which demands that the actual numbers of the atoms of all sorts are *stoichiometric*. In simple terms this means that the number of different atomic species is proportional to what should be according to the chemical formula of the material. For example, in SiC the numbers of silicon and carbon atoms are expected to be the same.

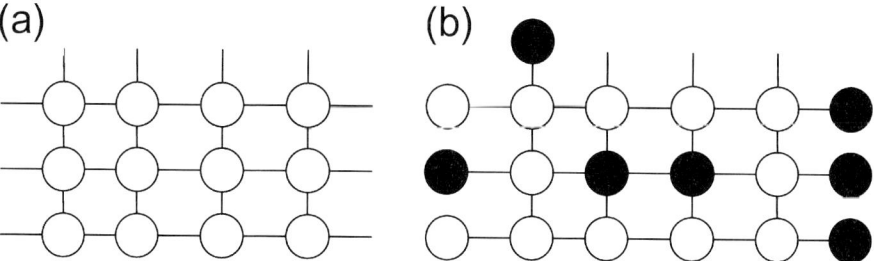

Fig. 1.4 Sketches of (a) pure and (b) hydrogenated nanocrystals.

This requirement is, however, very difficult to implement in many common nanofabrication techniques and is paid a significant attention in this monograph.

Some readers knowledgeable in materials science would immediately raise a reasonable question: a mere surface termination by hydrogen is not enough to consider the material hydrogenated and is usually neglected in the case of macroscopic solids! The most fascinating thing is that if an object is very small, then even a mere surface termination by hydrogen can lead to a substantial percentage of hydrogen atoms (hydrogenation) in the structure. The reason is that the ratio of the numbers of atoms located on the surface and in the interior (which is determined by the surface-to-volume ratio) strongly depends on the size of the nanoscale architecture concerned.

The smaller the size, the larger is the surface-to-volume ratio, and hence, the larger percentage of atoms is located on the surface. In this case, the relative population of dangling bonds available for bonding will be larger and more hydrogen atoms can stick to the surface of the nanoscaled object. What this means is that the chemical reactivity of a sufficiently small nanoparticle is determined by its surface-to-volume ratio and is much larger for sufficiently small nanoclusters. This is why a large variety of nanoparticles made of different materials is used by chemical, polymer, petrochemical, and other industries in numerous technologies utilizing various sorts of chemical catalysis. Thus, we can bookmark two more important issues being the *surface-to-volume ratio* and *reactivity* of the nanoparticles, which makes them invaluable building blocks of nanotechnology.

However, the most amazing thing in the nanoassembly from small clusters is that if one aims to create a new exotic material with very unusual properties, the reactivity of the clusters should not be too high. Our commonsense tells us that the most unusual properties cluster-made materials (materials composed of three-dimensional arrays of discrete, size-selected nanoparticles [33]) would have if they were made of nanoclusters of a small number of atoms but with just a few dangling bonds! In this case the nanoclusters would retain their own structure while combining in a larger assembly; in this case only the clusters would be interlinked. Otherwise, if the number of dangling bonds were as large as the number of the surface atoms, the small nanoclusters would combine together and form a large number of bonds between a large number of atoms from different clusters. Such multiple bonds would result in a much larger strain on the nanocluster structure, which will eventually break to form an object with a little difference from an atom-made assembly.

This simple intuitive assertion leads to a recent groundbreaking discovery that cadmium selenide nanoclusters $(CdSe)_{33}$ and $(CdSe)_{34}$ "self-passivate" their surface bonds to form much smoother surface structures without deleterious dangling bonds, which are drastically different from the common wurtzite structure of bulk CdSe [34]. It is quite typical for nanoclusters that the large surface-to-volume ratio controls the surface reconstruction and makes it stable. This unusual "self-passivation" of the surface (with the high surface ratio) effectively reduces the number of surface dangling bonds. In this case the clusters become less reactive and are less capable to merge together to form larger clusters and maintain their individuality, which is one of the critical requirements for nanofabrication of three-dimensional cluster arrays and cluster-assembled nanomaterials [33].

The above surface self-termination leads to much different structures of small nanoclusters, which in fact turn out to be a lot more strained compared to bulk forms of essentially the same materials. These strained chemical structures of small nanoparticles have numerous implications in nanoscale fabrication [35]. The most remarkable manifestation of this effect is significantly lower melting points of nanoclusters, which makes them ideal building units for epitaxial recrystallization on relatively cold surfaces [36, 37].

We hope that the reader already realized that ultrasmall objects have fundamentally different properties from those of bulk materials of essentially the same chemical composition. These properties can be tuned by adjusting the sizes, shapes, chemical structure, elemental composition, and surface-to-volume ratio.

Moreover, if arrays of a large number of nanostructures are concerned, there appear a few other essential issues such as *size and shape uniformity* of individual nanostructures across the pattern/array, controlled ordering of the nanostructures in the array, and interstructure spacings. These important parameters of individual nano-objects and their arrays need to be precisely controlled during the nanofabrication process. We are thus fully aware of the most important issues we will be looking at when trying to synthesize nanoscale assemblies in plasma-aided nanofabrication.

This is the stop point for now and it would be prudent to ask if there are any questions as what we normally do at the end of every lecture. And as it often happens, we sometimes forget to define some most essential things we were talking about in the lecture. Just like in this case: we will not be surprised if a reader would look again at the title of this book and finds out that we have not yet properly defined what the nanofabrication is in the first place! Well, our commonsense can help again: nanofabrication apparently means fabricating something very small, with the sizes in the nanometer range. So far, we have not discussed all the possibilities in this regard. Indeed, there might be a very large number of approaches, techniques, and processes to fabricate different sorts of nanoscale objects. Taken into account that the number of such nanoassemblies rapidly increases, so is the number of suitable fabrication and characterization tools and processes, the number of possibilities can be virtually infinite.

From the very fundamental perspective, the nanofabrication can be defined as the process of assembly of subnanometer-sized building units (e.g., atoms, ions, radicals) into larger objects (e.g., nanostructures, nanopatterns, nanostructured films) with feature sizes in the nanometer range [36]. This approach is commonly known as a bottom-up approach and heavily relies on self-organization (also commonly termed self-assembly) of building units (BUs) in a suitable nanofabrication environment or manipulation and stacking of such BUs externally (e.g., by a tip of a scanning tunneling microscope).

The other fundamental possibility is understood as a top-down approach, which relies on carving larger objects to reduce their size (this, e.g., can be achieved by using chemical etching, which removes some BUs) and also includes creation of nanosized void features such as tiny trenches in silicon wafers with sub-100 nm linewidths. Put in a simple language, the nanofabrication is a set of means to manipulate (e.g., move, stack, remove, control self-organization) the building units in a suitable environment and create nanoscale assemblies.

Wait a minute! We have just said "in a suitable environment" and started this introductory chapter with defining a plasma. Thus, we have no choice but to pose a now clearly obvious question: is the (low-temperature) plasma environment suitable for nanoscale applications or not and how to choose the right plasma for specific nanoscale applications?

1.3
Plasma-Assisted Synthesis of Nanomaterials

The answer to the first part of the question posed in the last sentence of Section 1.2 is a definite "yes"! There is a plethora of convincing evidence to this effect and it will be a futile attempt to try to produce an exhaustive review of all nanofilms, nanoparticles, nanostructures, porous nanofeatures, etc. ever fabricated by using plasma-based tools. It is amazing that the groundbreaking Iijima's discovery of carbon nanotubes was made in a carbonaceous soot synthesized by arc discharge plasmas [13]! Let us now consider a few representative examples of most common nanostructures and nanofilms synthesized by using low-temperature plasmas.

The first common example of plasma applications in nanotechnology is the synthesis of materials with nanoscaled features, commonly referred to as nanostructured materials. Such features can include some elements of surface morphology (e.g., humps, bumps and valleys of the nanometer size), internal inclusions, granular structure and some others. One such example is given in Fig.1.5 showing a high resolution transmission electron micrograph of ultrasmall (∼4–5 nm in size) ultrananocrystalline inclusions in an amorphous silicon matrix. This nanostructured, mixed-phase material has been synthesized in the Plasma Sources and Applications Center, NIE, Nanyang Technological University, Singapore by using low-temperature plasmas sustained in reactive mixtures of silane, hydrogen, and argon gases. Under certain plasma-based process conditions, it turns out possible to achieve quite high crystalline fractions and elemental purity, which is often quite challenging, if possible at all, to implement via other, nonplasma-based routes. Quite similar nanostructured materials, showing a very strong dependence of their nanocrystalline contents and degree of hydrogenation on the plasma parameters, have been synthesized by a number of research groups including but not limited to the University of Orleans and Ecole Polytechnique (France), Kuyshu University (Japan) and the University of Minnesota (USA) [38–50].

Work in this direction emerged at the cross-roads of physics of semiconductors and dusty plasmas, the two research fields that hardly can be put close to each other by our commonsense. Briefly, since the mid-1980s amorphous silicon (a-Si) has been among the most promising materials for photovoltaic applications. Among the main issues in the fabrication of amorphous silicon-based solar cells there have always been (and in fact, still remain!) a low deposition rate, relatively poor photostability, and a few others. Amorphous silicon films for microelectronic applications are commonly synthesized from reactive silane (SiH_4)-based plasmas (some other gases such as argon and hydrogen are used for passivation and activation of the growth surface). Amazingly, but in the late 1980s to early 1990s it was observed that silane-based plasmas are prone of small solid particles, which are now commonly termed dust grains or simply dusts; the plasma that contains such particles is known as the "dusty" (or "complex") plasma [11, 51–54].

The observations showed that under some conditions large clouds of ultrasmall particulates are formed in silane-based plasmas; such conditions were commonly termed as "dust/powder generation regime(s)." Some of the dusts are too small to be detected by optical means and/or seen by a naked eye. The size of such particles is very broadly distributed but in most cases remains within the nanometer range. The first thing that comes into mind is

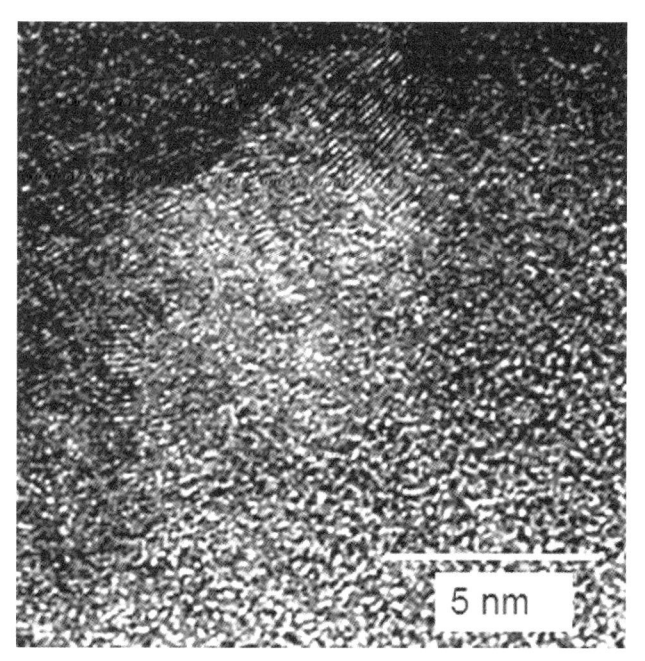

Fig. 1.5 High resolution scanning electron micrograph of Si nanocrystallites embedded in an amorphous silicon matrix and synthesized in low-temperature silane-based plasmas (photo courtesy of the Plasma Sources and Applications Center, NIE, Nanyang Technological University, Singapore, unpublished).

that such particles need to be removed, by whatever means, from the plasma reactor, to avoid their incorporation into the a-Si films. However, everything depends on the sizes and structure of such particles. Apparently, 50- to 100-nm-sized cauliflower-shaped agglomerates made of a large number of smaller particles (very frequently observed in experiments with silane plasmas) and some other large dusts would be nothing else but a nuisance and definitely need to be removed. At that time these "large" particles were deemed killers of microelectronic integrated circuitry, which had comparable feature sizes. Special means of removal of such particles have been developed, tested, and commercialized over the years.

In the mid-1990s this problem was solved. At that time no one really cared about even smaller (truly nanometer-sized) grains, also generated in silane-based plasmas. The main reason is that they were too small to be of any concern for the microelectronic manufacturing. Another reason lies in the enormous difficulty to detect such particles. Indeed, while in the plasma, they are largely invisible by even most sophisticated optical tools and can only be detected by sensitive materials analytical tools (e.g., scanning electron microscopy) after their deposition onto a substrate.

The most exciting thing was that even without being properly detected, such small particles grown in the plasma, led to a remarkable increase in the deposition rates and also to a quite substantial improvement of the performance of amorphous silicon in solar cells. Transmission electron microscopy enabled researchers to reveal the presence of crystalline inclusions in an amorphous matrix of hydrogenated silicon [46]. The appearance of such a microstructure (which is quite similar to what is shown in Fig. 1.5) has been convincingly related to the generation of ultrasmall, nanometer-sized powder particles in specific dust/powder generation discharge regimes. More importantly, the plasma-generated nanoparticles are extremely small in size (typically ~2–10 nm) and have a clear crystalline structure.

However, in most cases these nanoparticles did lack elemental purity because of a quite significant hydrogen content. As is understood nowadays, the particles nucleate and crystallize in the ionized gas phase and then are transported through the plasma sheath and eventually deposit on the surface. In the meantime, the growth of the amorphous matrix continues and the nanocrystals eventually become "buried" in amorphous silicon material. This new sort of mixed-phase materials, wherein crystalline grains are embedded into an amorphous matrix, has been termed "polymorphous" (*pm*-) materials (e.g., *pm*-Si:H in the above example). Such materials with ultrasmall but highdensity ultrananocrystalline inclusions are presently of a considerable interest for the development of new-generation quantum dot lasers and light emitting devices.

As we have noted in the above examples, the plasma-generated nanoparticles are very small (ultrananocrystalline) but are embedded into a thick layer of amorphous silicon. It is notable that plasma-based methods also allow one to synthesize freestanding silicon nanoparticles, which are not buried in an amorphous matrix [55–60]. One such example is shown in Fig. 1.6. Bapat et al. [58] have recently reported on a new and efficient plasma process that is capable of producing single crystal, highly oriented (faceted) and defectfree silicon nanoparticles with a highly monodisperse size distribution. Such nanoparticles are synthesized for applications in nanoelectronic devices such as single nanoparticle transistors. To be compatible with current lithography capabilities the nanoparticles should not exceed several tens of nanometers in size.

Other experiments have also revealed that low-pressure (thermally nonequilibrium) inductively coupled plasmas can be used to synthesize silicon nanoparticles with excellent control over the particle size [59]; however, the spread of nanoparticle size distributions is not always suitable for nanodevice applications. In these experiments an inductive GEC reference cell with a planar coil in pure SiH_4 at a pressure of 10–12 mTorr and an RF power of 200 W was used. More details of the experimental setup can be found elsewhere [59].

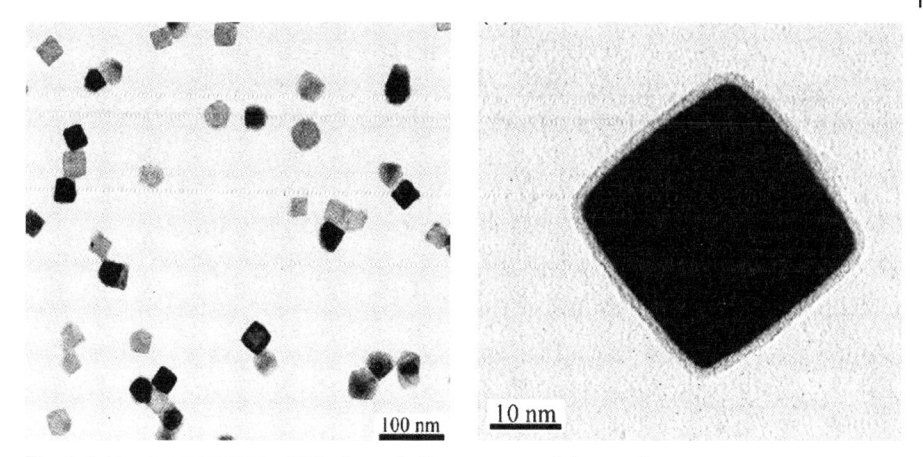

Fig. 1.6 Single-crystalline, cubic-shaped silicon nanoparticles synthesized in silane-based reactive plasmas at the University of Minnesota (USA) [58].

Interestingly, the nanoparticles fabricated in such a way appear to be highly monodisperse and nonagglomerated and their size can be effectively controlled by the duration of plasma discharges (plasma-on time). However, an extreme caution should be taken when synthesizing Si nanoparticles from reactive silane plasmas as they often appear amorphous and feature unacceptably high defect densities.

To improve the crystallinity of the nanoparticles, it is often required to provide process conditions that disfavor or effectively eliminate the possibility of formation of amorphous deposits, matrices, particles, or continuous films. Among many practical ways to improve crystallinity of thin solid films and particles, postannealing at high temperatures ($\sim 1000\,°C$) is probably one of the most commonly used techniques. However, such high temperatures can result in a substantial thermal damage of the materials, structures, and substrates involved. This problem can even be more severe in microelectronic manufacturing as such temperatures can easily exceed the melting points of metallic interconnects (e.g., $1083\,°C$ for bulk copper; copper nanolayers can melt at substantially lower temperatures).

Another effective way is to operate the plasma discharge at high power densities to ensure that neutral gas temperatures are high enough for efficient crystallization. By using an inductively coupled plasma reactor of a smaller size and a flowthrough configuration, wherein the plasma can be sustained by a helical inductive coil at high RF power densities of up to $2\,W/cm^3$, Bapat et al. [55] managed to synthesize single-crystalline, faceted, defect-free, highly oriented silicon nanocrystals in a reactive plasma-based process in a mixture of 5 % silane and 95 % helium at a total pressure of 700 mTorr. However, the particle size distribution in such silane-based experiments often appears polydisperse; moreover, a great variety of different particle morphologies can be obtained under quite similar process conditions. Although some of the particles are useable, potential problems with nonmonodisperse nanoparticle separation make this process less suited for nanodevice applications.

In a recent study Bapat et al. [58] reported on an innovative reactive plasma-based process that yields single-crystal silicon nanoparticles with a mainly well-defined cubic shape and a rather monodisperse size distribution. A representative example of the cubic and single-crystalline nanoparticles analyzed by transmission electron microscopy (TEM) is shown in Fig. 1.6. In this figure, the nanoparticles have an almost perfect cubic shape and do not show any noticeable dislocations or other planar defects. One can also notice a 1–2 nm thin amorphous layer surrounding the particle [58]. This substantial improvement of the nanoparticle quality was achieved by using a constricted, filamentary capacitively coupled low-pressure plasma, as opposed to the inductively coupled plasmas used by the University of Minnesota's group in their earlier experiments [55, 59].

While it appears perfectly clear that the plasma properties do matter in the above case, the exact plasma-related mechanism of the particle origin, growth, and shape development still need to be elucidated. It is interesting to note that the equilibrium cubic shape of the silicon nanoparticles indicates on their surface termination by hydrogen [61].

In a more recent report [60], a new and very efficient reactive plasma-based process for single-step fabrication of ultrasmall (from 2 to 8 nm in size) luminescent silicon nanoparticles was proposed. Such particles can be synthesized on time scales of a few milliseconds, thus enabling the output of the corresponding continuous process in the range 14–52 mg/h of nanoparticles.

It is worth mentioning another effective plasma-based technique to reduce the presence of the amorphous phase in the films, namely chemical etching by reactive radicals generated in the ionized gas phase. For example, in the synthesis of various carbon-based nanostructures considered in detail in Chapter 4, it appears essential to make sure that the growth of carbon nanostructures outpaces that of the unwanted amorphous deposits. We will return to this important issue in Section 1.4.

It is remarkable that low-temperature thermally nonequilibrium plasmas are suitable for synthesizing nanocrystalline films and structures made of various materials. For example, ultrananocrystalline diamond is another evidence of successful use of low-temperature weakly ionized plasmas in nanofabrication [62, 63]. It is commonly known that diamond is a less stable allotrope of carbon compared with graphite. Thus, synthesis of this material usually requires strongly nonequilibrium conditions, such as high pressures and temperatures of precursor gases and substantial additional heating of deposition substrates. Synthesis of nanocrystalline modifications of diamond is even more challenging since it is extremely difficult to control the formation of numerous sp^3 atomic networks when the number of constituent nanocrystals becomes large. This is why the results of Gruen and his colleagues [62,63] on different hydrocarbon-based plasma-related possibilities of synthesizing, in a highly controllable fashion, ultrananocrystalline diamond with crystalline grain sizes as small as ~1 nm sound very encouraging.

These processes are also related to unique properties of nonequilibrium plasmas to accumulate substantial amounts of energy with a high density, without actually pushing the limits and using unusually high gas temperatures or pressures. We emphasize that these are the only two options that can be used in a neutral gas-based process. On the other hand, low-temperature nonequilibrium plasmas can offer a better deal of synthesis of nanomaterials that require essentially nonequilibrium process conditions, such as the ultrananocrystalline diamond of our interest in this section or crystalline silicon carbide quantum dots considered in Chapter 6. This point will also be elaborated further in Section 1.4.

Plasma-based techniques have also been extensively used to fabricate nanoparticles of various composition, chemical structure, shapes, etc. For example, quasispherical carbon nanoparticles with a variable crystalline phase content and extent of hydrogenation can be synthesized and unambiguously detected in low-pressure hydrocarbon-based discharges [64–67]. In this way it turns out possible to synthesize, in a laboratory plasma environment, terrestrial analogs of "astrophysical" carbon-based nanoparticles from stellar environments and interstellar gas. Another example is the highly efficient and controllable assembly of ultrasmall titanium dioxide nanoparticles and nanoclusters in a low-temperature plasma-assisted diode sputtering process [68–70]. Interestingly, by manipulating the sputtering process conditions, such as the working pressure, plasma density and also changing the particle collection point, one can assemble nanoclusters of a relatively small number of atoms, typically ranging from a few hundred to a few thousand atoms.

Plasma-assisted synthesis of carbon-based nanostructures is probably represented by the largest number of publications from numerous research groups from virtually all over the world; with only a small fraction of these works mentioned here [71–82], with some more introduced and discussed elsewhere in this book. This is not a surprise since carbon nanotubes and related nanostructures have recently been recognized as the hottest research topic in the last 5 years, judging by the citation impact of relevant publications.

Without trying to provide an exhaustive overview of the plasma-assisted synthesis or carbon-based nanostructures or going deeply into details of such processes, here we will only highlight some of the most important issues in this research direction. As we have already mentioned above, carbon nanotubes were discovered in plasma-synthesized carbonaceous soot [13]. Since the mid-1990s, there has been an increased number of attempts to synthesize carbon nanotubes and related nanostructures with specific features required for their applications. Nonplasma grown nanotubes are in most cases quite disordered and randomly oriented with respect to solid substrates resembling a bunch of interwoven wires or threads. Such carbon nanotube networks found numerous applications as reinforcing scaffolds in polymer-nanotube nanocomposite materials.

Another popular challenge for nanotube applications is related to the development of electron field microemitters. To enable such functionality, the nanotubes need to be properly positioned on a solid substrate and also oriented with respect to it. The best practical realization of such carbon nanotube-based microemitter arrays requires vertical (normal with respect to the substrate) alignment of the nanotubes, similar to what is seen in Fig. 1.7. Extensive studies of electron field emission properties of objects of different shapes revealed that capped cylindrical nanotubes, appropriately positioned in an array on metal-catalyzed (e.g., Ni/Fe/Co) surfaces have an outstanding potential for electron microemitter device applications. The optimized positioning turns out to be very sensitive to the actual nanotube dimensions, which are characterized by their diameter and length. It is commonly accepted nowadays that the nanotube diameter is determined by the sizes of metal catalyst particles formed after fragmentation of the catalyst layer as a result of substantial (up to $\sim$700–900 °C and even higher) external heating of the substrate. On the other hand, the nanotube length should depend on the actual kinetics of the growth process, which is determined by the balance between the delivery and consumption of suitable building units from the nanofabrication environment to the growth site [36].

Thus, one would expect that the plasma can merely affect the growth rate of the nanotube-like structures. The most amazing observation made by a large number of researchers was that the nanotubes grown in low-temperature plasma environments are vertically aligned and the direction of their alignment is the same as that of the electric field in the plasma sheath (this is also the case in Fig. 1.7). The origin, strength, and polarity of such a field have been discussed in Section 1.1. Although this amazing plasma-related effect still awaits its conclusive explanation, the basic understanding of this phenomenon is related to two main issues. One of the issues is the electric field-mediated strain in the lattice of a nanotube slightly bent with respect to the normal to the growth substrate direction. This mechanism was found most relevant to the nanotubes grown with a catalyst particle either on their top or at their bases [77]. As a result, carbon nanotubes grow perfectly aligned with the electric field in the plasma sheath.

Fig. 1.7 Vertically aligned carbon nanotubes synthesized in a methane-based PECVD process (photo courtesy of the Plasma Sources and Applications Center, NIE, Nanyang Technological University, Singapore).

Another issue is to elucidate the actual role of the ion fluxes that are most responsive to the electric fields. In fact, despite relatively lower densities compared to neutral species the ions can arrive at the nanostructure growth site faster and under certain conditions their flux can exceed that of neutral radicals. An interesting thing is that positively charged ions always follow the direction of the electric field, which drives them toward the nanostructured substrate.

Moreover, since the electric field usually converges near sharp tips of high-aspect-ratio and aligned nanostructures, the plasma ions can be effectively driven straight to the upper sections of the nanotubes as has recently been confirmed by numerical simulations [21, 23, 24]. Thus, if the nanotube grows in the "catalyst particle on top" mode, large numbers of the plasma ions can be deposited onto the catalyst particle and diffuse through it, at a faster rate, to incorporate into the nanotube structure.

Therefore, vertical alignment of nanotube-like structures in plasma-based nanofabrication environments is a commonly accepted yet still not completely explained phenomenon, which is attributed to the alignment effect of the electric field in the plasma sheath. What is even more amazing is that the actual alignment of the nanostructure growth can also be controlled by external (not only those due to charge separation or external DC substrate bias) electric fields. For instance, by applying an external DC electric field parallel to the substrate surface, carbon nanotubes can be bent in sharp, predetermined angles (e.g., $90°$) to form L-shaped nanotubes [83].

It is remarkable that the electric field within the plasma sheath and the ions also significantly contribute to the growth and reshaping of other, and not only carbon-based nanostructures. A striking example is the shaping up of zinc oxide nanorods from small island-like nuclei into relatively large nanostructures with relatively large aspect ratios [85]. Interestingly, columnar nanocrystalline AlN structures considered in detail in Chapter 7 also develop under a strong influence of the electric fields and plasma ions. Indeed, as the structures grow predominantly in one direction (e.g., c-axis or (002) crystallographic direction) and elongate, ion fluxes from the plasma focus onto sharper (near the top) sections of the growing nanostructures and eventually result in further sharpening and elongation of the structures. This is also the case in the plasma-assisted fabrication of vertically aligned gallium doped zinc oxide nanorods [85].

Numerical simulations also suggest that selective manipulation of ions fluxes can be instrumental in maintaining a steady growth (with a predetermined shape) and/or reshaping of capped cylindrical nanorods into conical spike-like microemitter structures [22, 23]. This important feature can also be used for postprocessing (e.g., coating with nanofilms, functionalization, or doping) of plasma-grown nanostructures of various dimensionality [24]. The range of nano-objects that can be treated in this way also includes various voids, cavities, and pores with the dimensions in the nanometer range [84].

It is also important to note that in plasma-based processes it appears possible to control the actual number of walls in carbon nanotubes. Initially, plasma nanotools enabled one to synthesize mostly multiwalled carbon nanotubes (MWCNTs) [72] rather than their single-walled counterparts, which is commonly accepted as a much more gentle process compared to the synthesis of MWCNTs. However, with the rapid advance in the control and diagnostic of plasma processes, it became possible to routinely fabricate arrays of SWNTs and many other carbon nanotube-related nanostructures, such as nanofibers, nanotips, nanoneedles, nanowalls, nanoribbons, nano-onions, etc. [82].

In the meantime, significant advances in sample preparation and catalyst prepatterning enabled one to design intricate arrangements of carbon nanotubes in dense arrays and nanopatterns. An example of dense nanotube forests grown on selected areas on nickel-catalyzed silicon surfaces is shown in Fig. 1.8. In this micrograph one can see that the nanotube forests are arranged in a readable 30×15 μm-sized micropattern "NIE," which stands for the National Institute of Education of Nanyang Technological University of Singapore.

The nanostructures discussed above are usually made of the same element, in most cases carbon. It is remarkable that plasma-based techniques also make it possible to fabricate, with great precision, nanoassemblies made of a larger number of elements with very different chemical struc-

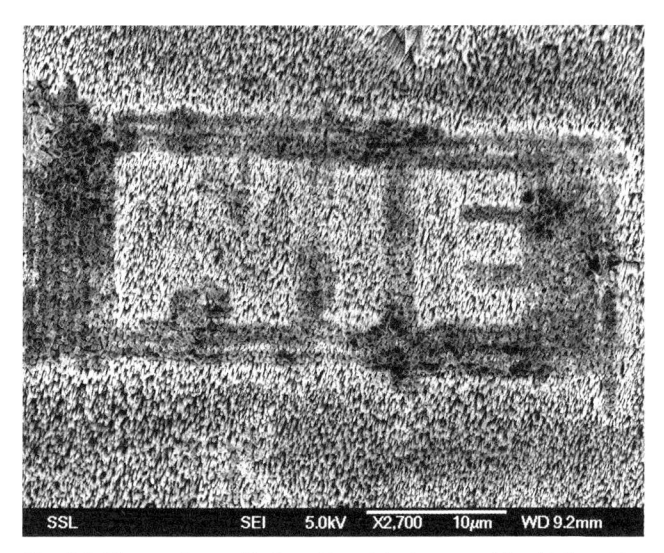

Fig. 1.8 Nanopattern with the abbreviated name of the National Institute of Education (NIE), cut from a dense forest of vertically aligned carbon nanotubes synthesized in methane-based plasmas on nickel catalyst (photo courtesy of the Plasma Sources and Applications Center, NIE, Nanyang Technological University, Singapore).

tures, shapes, alignment, arrangements in nanopatterns, etc. Complex binary/ternary/quarternary semiconductor quantum confinement structures of different dimensionality (which is 0 for ultrasmall quantum dots, 1 in the case of one-dimensional, high-aspect-ratio nanowires and nanotube-like structures, 2 for nanowall-like structures, nanowells and layered heterostructures, and 3 for differently shaped nanoparticles and also smaller-aspect-ratio nanostructures (e.g., nanopyramids, nanocones, nanorods)) is perhaps the most striking example of such successful applications. Interestingly, by using plasma-based tools, it becomes also possible to control the arrangement of ultrasmall nano- and sub-nanosized objects into ordered spatial arrays, similar to the high-resolution SEM image of SiC quantum dots fabricated by the plasma-assisted reactive RF magnetron sputtering of high-purity solid SiC targets in plasmas of $Ar+H_2$ gas mixtures (see Fig. 1.9). These nanodot structures will be considered in more detail in Chapter 5.

An alternative way to synthesize binary semiconductor quantum dots is to use plasma enhanced chemical vapor deposition from mixtures of reactive gases such as silane (discussed above in relation to nanofabrication of nanocrystalline silicon and single-crystalline silicon nanoparticles) or germane. For example, highly uniform large-area patterns of germanium quantum dots on silicon have been successfully fabricated by using reactive plasmas of GeH_4 and H_2 gas mixtures [86].

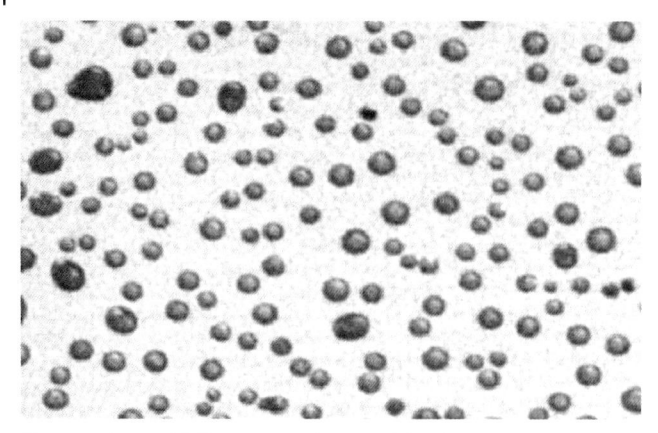

Fig. 1.9 Array of SiC quantum dots on Si(100) substrates with AlN lattice matching interlayer synthesized by reactive plasma-assisted RF magnetron sputtering deposition (photo courtesy of the Plasma Sources and Applications Center, NIE, Nanyang Technological University, Singapore, unpublished). The average size of the quantum dots within the array is approximately 23 nm.

Interestingly, the composition of the plasma species also plays a prominent role in the growth kinetics of nanostructures. One interesting example is shown in Fig. 1.10, which displays ultra-high-aspect-ratio SiCN nanowires synthesized in a quite similar, Ar+H$_2$ reactive plasma environment, with the minor differences being in the additional inlet of nitrogen gas and the use of nickel-catalyzed substrates (we recall that a similar technique is used in nanofabrication of carbon-nanotube-like structures).

At this stage we should make a stop and proceed with the introduction of the main ideas and aims of this monograph. Let us just recall that the purpose of this section has been to familiarize the reader with typical and widespread applications of plasma nanotools for materials synthesis and processing at nanoscales. The number of already existing and emerging processes, techniques, and technologies is in fact much larger and is continuously increasing at a relatively high pace. This section aimed to give a few typical examples of uses of low-temperature plasmas to fabricate nanostructures of various dimensionality; in no way it is an exhaustive review of all existing plasma-based techniques and processes. For a wider coverage of current advances and major challenges in the area the reader can be referred to topical review articles [36, 72, 82]. Nevertheless, we do believe that the reader has already received a reasonable exposure to most important uses of low-temperature plasmas in nanotechnology and an awareness of the importance and present-day status of this undoubtedly hot research area.

At this stage our logic suggests us to pose the next important question: what features should the plasmas and the plasma sources have to be not only use-

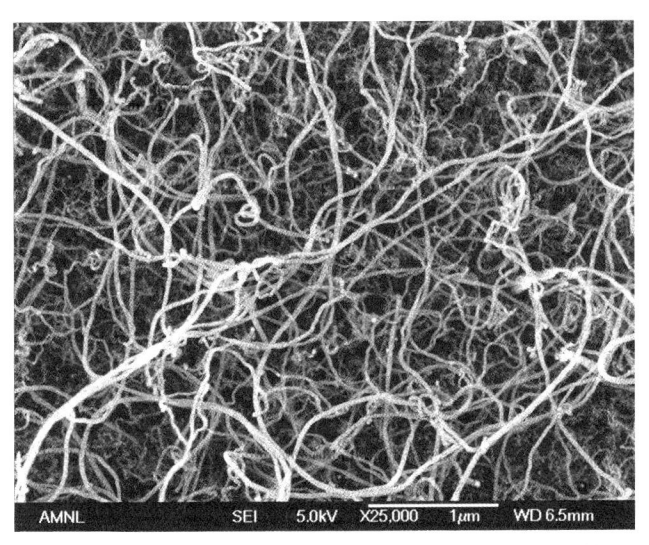

Fig. 1.10 Ultra-high-aspect-ratio SiCN nanowires synthesized by reactive plasma-assisted RF magnetron sputtering deposition on the Ni-catalyzed p-Si(100) substrate (photo courtesy of the Plasma Sources and Applications Center, NIE, Nanyang Technological University, Singapore). The aspect ratio (length to width) of the nanowires typically ranges from 250 to 1000.

ful but also efficient in specific nanoscale applications? On this note we will proceed to the next section and briefly discuss some of the issues that might facilitate the proper choice of the plasmas and plasma facilities suitable for the envisaged nanoprocesses.

1.4
How to Choose the Right Plasma for Applications in Nanotechnology?

At this stage we already have some basic knowledge about the fundamental properties of low-temperature plasmas, most important issues and building blocks of nanotechnology, and most common uses of such plasmas for the synthesis and processing of nanomaterials and nanostructures. In this brief section, we will try to put the most essential bits of information together and work out which plasmas would be right to use to fabricate a specific nanoassembly.

First of all, there is no general answer to this nontrivial question. In fact, the answer is in most cases process specific and depends on many factors, such as the nanostructure's or nanomaterial's type, internal structure, size, specific features, elemental composition, shape, targeted application, and some others. However, the knowledge we already have will enable us to narrow our choices and look at the plasmas and their sources from a more focused perspective.

From Section 1.1, we have already learned that plasmas can be hot and cold, thermal and nonequilibrium. Out of four possible choices, we should eliminate the two possibilities related to hot plasmas, which we recall are considered suitable to sustain nuclear fusion reactions at millions of degrees and are thus far too hot to be useable for creation of nanosized objects.

After this elimination, we are left with no choice but to use low-temperature plasmas; however, which kind of plasmas to use, thermal or nonequilibrium? It is now prudent to recall that low-temperature plasmas are in most cases weakly ionized and plasma discharges naturally contain two (ionized and neutral) gas components. In thermally equilibrium plasmas, all species usually have very similar temperatures ($T_e \sim T_i \sim T_n$), which is obviously not the case for nonequilibrium plasmas. If we focus on the gas-phase nucleation of atomic, molecular, or radical species into nanoclusters or small nucleates, the temperature of the environment would be the most important factor. And since the plasma is weakly ionized, we arrive at the conclusion that the neutral gas temperature is what most likely controls the process.

We now have two options: either to go ahead with the thermal or nonequilibrium plasmas. In a deterministic approach, the choice will depend on the specifics of the targeted nanoassemblies. Otherwise, one can discuss what sort of nano-objects one can expect to grow from both plasmas. First of all, gas temperatures are quite different in thermal and nonequilibrium plasmas. Indeed, in thermal plasmas with $T_e \sim T_i \sim 1$ eV, one can reasonably expect the neutral gas (which dominates in the environment) to have a temperature of a few thousand degrees and even higher. On the other hand, in a nonequilibrium plasma with the same electron temperature, neutral gas temperatures very rarely exceed 1000 K and in most cases remain close to room temperatures. This remarkable difference in gas temperatures can result in a variety of possible sub-nanosized and nanosized objects that can be synthesized under so different conditions.

The easiest thing to notice is that temperatures ($\sim$300 K) of a neutral gas in nonequilibrium plasmas are below melting points of most of bulk solid materials. On the other hand, gas temperatures in thermal plasmas are high enough to cause melting and even evaporation of solid objects. Therefore, if a small nanosized object has been nucleated in a gas phase, a chance of finding it in a solid state is much higher in nonequilibrium rather than in thermal plasmas. At lower temperatures small objects have a better chance to have an irregular (e.g., cauliflower- or fractal-like) shapes than at higher temperatures. Moreover, as the temperatures get higher, any irregular features of gas-phase-borne nanoassemblies often disappear as a result of coagulation and spheroidization effects.

In simple terms, it means that thermal plasmas are best suited for synthesizing spherical nanoparticles! Such spherical nanoparticles are very different from perfectly faceted cubic nanocrystals (Fig. 1.6) synthesized in nonequilibrium cold plasmas. Thus, if specific targeted nano-objects can benefit from high-temperature droplet-like states or are expected to be of a perfectly spherical shape, then using thermal plasmas is a definitive must. However, if nanoassemblies need to be grown at low temperatures, then nonequilibrium plasmas would certainly be a better choice. For example, it would be problematic to synthesize in thermal plasmas polymer nanoparticles, which often have melting points as low as 400–450 K.

We hope that this simple example convincingly shows that by using different kinds of plasmas one can obtain very different results in terms of nanofabrication. Also by using plasmas of different sorts (e.g., inductively or capacitively coupled plasmas) one can make a major impact on actual outcomes of the nanoassembly processes. Moreover, within the same sort of plasmas, the plasma and discharge parameters can be very different depending on the process operation conditions. And what is even more important is that the actual properties of nanoassemblies are extremely sensitive to such choices and/or variations.

The next important thing to take into account when choosing the most suitable plasma and process conditions is to note that the growth of most of the presently known nanoassemblies in low-temperature plasmas does require a suitably prepared solid substrate. More importantly, this substrate faces the plasma and is separated from it by the plasma sheath discussed in detail in Section 1.2. Therefore, the interaction of the plasma species with the growth substrate is what controls the growth of the nanoassemblies in question. In particular, surface bombardment by the plasma ions can significantly increase the temperature of the surface and/or affect the fragmentation/nanostructurization of catalyst layers, which are widely used to synthesize carbon nanotubes and a many other common nanostructures made of variety of materials.

A knowledgeable reader would agree but perhaps could question the relevance of the previous paragraph to the issue of the appropriate choice of the most suitable plasma and optimized process parameters. First of all, the impact of the ion bombardment critically depends on the plasma and substrate bias used in the nanoassembly growth experiments. Different plasmas can have very different compositions and contain a variety of atomic, molecular, and radical species that depending on their functions will often be termed "building units" or "working units" throughout this monograph. Some of the species are reactive and can chemically modify the surface, while nonreactive ones usually cannot. Some of the species are meant to stack into nanoassemblies being grown whereas some others are unwanted or even deleterious.

The number of possibilities is virtually infinite and it is not the purpose of this work to reveal them.

We emphasize that it is extremely important to properly select the right kind, sort of plasmas and most essential conditions in the plasma discharge, working gases, and deposition surfaces. These conditions should be optimized for the most effective and controlled growth of the required nanosized objects.

It is a common belief that the choice of the working gas is the easiest thing to do. In a sense, yes: if one wants to grow carbon nanotubes then there is no point to use silane precursor gas as it does not contain carbon atoms the nanotubes are expected to be made of. However, which carbon-bearing precursor gas to use? Methane? Acetylene? Higher hydrocarbons? Or use solid targets to introduce carbon material via plasma-assisted sputtering of pulsed laser ablation? Ultimately, everything depends on the most suitable building units of the targeted nanostructures [36]. For example, if a nanostructure can be assembled by carbon atoms only, it is advisable to use a nanofabrication technique and a precursor gas that most effectively generate carbon atoms in the required energetic and/or bonding state.

It should be pointed out that the number of the suitable building blocks should also be reasonable, as significant oversupply of carbon material to the substrate surface can result in undesired growth of unwanted carbon deposits rather than well-shaped and structured nano-objects. This issue has already been mentioned in this chapter. For example, the target is to grow crystalline carbon nanostructures, which are believed to grow via stacking of CH_3 radical building units. In this case methane-based plasmas would be the most obvious choice since the required BUs can be easily produced by electron impact dissociation of methane molecules

$$CH_4 + e \rightarrow CH_3 + H + e$$

which usually has quite high reaction rates and is very effective in generating methyl radicals. On the other hand, using C_2H_2-based plasmas would not be effective since a long chain of chemical transformations is necessary to create CH_3 reactive radical from C_2H_2.

On the other hand, unwanted amorphous pile-ups can be effectively removed by selective chemical etching, which affects amorphous carbon but does not affect crystalline carbon nanostructures. Therefore, balancing the rates of deposition of useful building units and etching of unwanted phase (a-C in this case) is another important issue to note when selecting the most suitable plasmas and developing relevant process specifications and plasma facilities. For the purposes of etching amorphous carbon, one can effectively use atomic hydrogen working units, which are chemically reactive and widely used for surface passivation and etching.

Therefore, the ideal plasma for nanofabrication of crystalline carbon nanostructures with reduced amorphous carbon contents should be sustained in a CH_4+H_2 gas mixture. Or is it really ideal or can still be improved? This is where the plasma–surface interactions come along. Numerous experiments have shown that the process outcomes can be much improved by adding an inert gas such as argon, and in quite substantial amounts. One of the reasons is that it is usually easier to sustain plasma discharges in argon than many other gases. In fact, argon can be effectively ionized by using "tail" electrons as discussed in Section 1.2; populations of such energetic electrons can be quite large in argon, as compared to other gases. In this case, it is often possible that the plasma in such a mixture can be dominated by argon ions as compared with hydrogen and hydrocarbon ions. Moreover, argon ions are heavy, and their large concentrations can be used to create intense ion fluxes onto the growth surfaces, which mostly increase the surface temperature without elevating the neutral gas temperature and etching the surface. From this point of view, argon ions can be also considered important working units, which serve for the purpose of plasma maintenance and controlling surface temperatures and other surface conditions such as distribution of surface stresses.

Thus, when choosing the right gas mixture, one should have a very clear idea what species and in what state (e.g., charged or neutral, radical or nonradical, energetic or nonenergetic) are actually required for each particular process stage and/or component involved in nanofabrication of the required nanoassembly.

As has probably became clear, choosing most suitable plasmas and optimizing the process conditions involves a large number of other parameters and issues. In fact, every single process parameter in most cases significantly affects the outcomes. This is why it is so important to convert these parameters into effective turning knobs of plasma-aided nanofabrication. In this monograph we will consider several examples of how a variation of the plasma parameters makes a major difference in the plasma-assisted growth of various nanoassemblies. For example, discharge input power can be used to effectively control the number densities of electrons, ions, and radicals in the discharge. This can dramatically modify the balance between the delivery and consumption of the nanoassembly building blocks and also the surface conditions (e.g., passivation and bond availability and temperature). On the other hand, a DC substrate bias can be used to control the fluxes of positive ions and radicals onto the surface, including surfaces of individual nanostructures.

Strict requirements for the plasma sorts and specific process parameters pose a great challenge for the development of plasma-aided nanofabrication tools and facilities. Generally speaking, some of the most essential requirements for the plasma source design and operation parameters are quite similar to the existing demands of the microelectronic industry. For example, to be useable for nanofabrication purposes, low-temperature plasmas should be stable and quiescent, show excellent reproducibility of its parameters (e.g., electron/ion temperature, densities and surface fluxes of electrons, ions, and other major species), be easy in handling and operation, be easily sustained in large ranges of process parameters such as the working gas pressure and input power, and several others. Moreover, the fluxes of the ion and radical species should be uniform over the areas where the nanoassembly is actually conducted. This usually implies high degrees of uniformity of densities of numerous species over large volumes and surface areas. These and other requirements for the development of suitable plasma sources are discussed in detail in Chapters 2 and 3.

However, this is not yet the end of the story. Amazingly, but plasma facilities that show excellent performance in microfabrication of integrated circuits or deposition of superhard wear-resistant coatings may not necessarily be suitable for successful and deterministic synthesis of nanoscale assemblies. This is why it is so important to properly identify the most effective controls of each specific process and tailor the plasma and process parameters and, if necessary, specifically design a new or redesign an existing plasma facility to make such a synthesis a reality.

For example, to synthesize semiconductor quantum dots, plasma facilities should be able to generate small (typically submonolayer) numbers of building units; such numbers are usually much less than those used to deposit continuous thin films in plasma-assisted processes. In many cases a simple reduction of a working pressure and input power does not help much and a new way of controlled generation of the building units in the reaction chamber is required. Such ways of creating and delivering various species make a substantial difference in terms of the performance of plasma facilities in nanofabrication.

It is worth mentioning that a large number of species, especially in reactive plasmas, adds an additional level of complexity to the application of low-temperature plasmas in nanotechnology. This is why it is so important to properly understand the ways to control those parameters that can affect the nanoassembly process and in the way we actually want. We reiterate that such an approach is commonly termed deterministic as opposed to widespread trial and error practices. Finally, we emphasize that most of the presently available plasma nanotools have not yet reached such a level and it is one of the aims of intense ongoing research efforts worldwide and of this monograph to show a viable pathway from the development of high-performance plasma sources to their uses in plasma-aided nanofabrication.

1.5
Structure of the Monograph and Advice to the Reader

We are now ready to outline what exactly can the reader (who should already be quite puzzled and intrigued) find in this monograph. Again, as the title suggests, it shows a pathway from choosing the right plasma and developing a suitable plasma facility to actual implementation of plasma-based nanoassembly processes.

In addition to this introductory chapter, this monograph also contains six main chapters, Chapters 2–7 and one concluding chapter, Chapter 8. The References section contains 496 relevant references. The first logical part of the book (Chapters 2–3) is devoted to generation of low-temperature plasmas and development of advanced plasma sources suitable for nanofabrication purposes. These chapters introduce the approach for tailoring the plasma nanofabrication environment via developing versatile plasma sources, which are capable to generate, at specified rates, the required number of the desired plasma species.

Physical properties and operation of inductively coupled plasma sources, benchmark plasma reactors of the present day microelectronic and other industries, are considered in Chapter 2. These plasma sources have found successful applications for plasma-enhanced chemical vapor deposition of various carbon-based nanostructures, including nanotubes, nanotips, needle-like structures, quasi-two-dimensional nanoflakes, and nanowalls and other important nanoassemblies.

In Chapter 3, the issues of the improvement of the plasma parameters in inductively coupled plasma sources and development of versatile nanofabrication facilities are critically examined. This chapter shows different examples of how the performance of inductively coupled plasma sources can be improved to meet the requirements of nanofabrication processes. In particular, we explain how the uniformity of fluxes of various species in inductively coupled plasmas can substantially be improved by introducing an internal radiofrequency antenna that excites flat unidirectionally oscillating RF current sheets.

In another example, we detail how an introduction of a DC/RF magnetron inside a conventional inductively coupled plasma reactor can give extra flexibility in terms of generation of additional metal species, which are quite difficult, if possible at all, to generate in a PECVD process. By combining the advantages of large-area plasma processing, stability and parameters of inductively coupled plasmas generated by either external or internal inductive coils, with the flexibility of selection of sputtering targets of DC/RF magnetron sputtering electrodes, it turns out possible to develop an advanced combinatorial plasma nanofabrication facility, which was used to fabricate a large variety of semiconductor nanostructures of various dimensionality.

The second logic part of this monograph details numerous applications of low-temperature plasmas for the synthesis of exotic nanoassemblies, nanostructured materials, functional coatings, processing submicrometer and nanometer features in porous materials and microelectronic devices, and other relevant technologies.

In Chapter 4, we address the benefits of using reactive plasmas of hydrocarbon-based gas mixtures to fabricate carbon nanostructures, including but not limited to ordered arrays of vertically aligned single crystalline nanotips, multiwalled nanotubes, nanofibers, nanopyramids, nanowalls, and nanoflakes. This chapter also includes extensive results on numerical simulations of atomic structure of single-crystalline carbon nanotip structures and experimental advances in postprocessing (e.g., coating, doping, functionalization) of carbon nanotubes and related nanostructures.

Plasma-aided nanofabrication of low-dimensional semiconductor quantum confinement structures, such as quantum dots, nanoparticles, nanowires, nanorods, nanoparticle films, superlattices, heterostructures, and several others, as well as intricate patterns of low-dimensional quantum confinement structures is the main focus of Chapter 5. Most of the nanostructures and nanopatterns considered in this chapter have been synthesized in the Integrated Plasma-Aided Nanofabrication Facility introduced and described in detail in Chapter 3. By using various operation modes of this facility, it turns out possible to assemble exotic quantum confinement structures with different dimensionality, which range from 0 for tiny quantum dots to 3 for rod- or pyramid-like structures. Many useful properties of such nanostructures such as electron confinement and photoluminescence are discussed alongside with the detailed description of the analytical characterization of structural, chemical, and other properties of such nanoassemblies.

In Chapter 6 we show how a "plasma-building unit" approach [36] can be successfully used for the plasma-assisted synthesis of nanostructured hydroxyapatite bioceramics for orthopedic and dental applications. Nanostructured hydroxyapatite fabricated in such a way has an excellent degree of crystallinity, stoichiometric elemental composition, grain structure, surface morphology, and other useful features, not achievable via other fabrication techniques.

Chapter 7 deals with a range of materials and nanoscale objects synthesized or processed by using plasma nanotools. The examples considered include doping of SiC quantum dots by atoms of rare-earth metals, plasma-assisted reactive magnetron sputtering deposition of Ti–O–Si–N nanocrystalline films, using the building unit approach to fabricate nanocrystalline AlCN films, highly oriented columnar AlN nanocrystalline films, plasma-assisted synthesis of nanocrystalline vanadium pentoxide films, and plasma-based treatment of micro- and nanoporous materials. This chapter ends with a brief overview of other examples of successful uses of low-temperature plasmas for nanofabrication

The monograph concludes with Chapter 8, which contains further examples of advanced applications of (mostly thermally nonequilibrium) low-temperature plasmas for nanoscale materials synthesis and processing (Section 8.1). Section 8.2 is devoted to summarizing, by using specific examples considered elsewhere in this monograph, benefits and challenges of using plasma-based tools in nanofabrication. The monograph concludes with a brief summary of some current and emerging issues of the plasma-aided nanofabrication and outlook for future directions in this exciting research area (Section 8.3).

From now on, the narration will become a lot more technical and will mainly contain specific scientific results, with the overwhelming majority of them published in prime international research journals. However, this does not mean that only scientists, engineers, and postgraduate students should read this monograph further on. In fact, anyone who is interested in either plasma- or nanorelated research advances should at least browse this book and pay attention to any microphotographs or other images or figures that catch their eye and then try to read the narration around. If the level of technical description turns out to be difficult, it is worthwhile to read a couple of introductory paragraphs to the relevant section; such paragraphs summarize the advances described in the subsections and are written in a less technical language than the rest of the subsections. Well, if this cannot help, there is always a way out: ask us a question and we will be most happy to explain! Finally, good luck with the reading and we hope that you will also enjoy the following highly technical part of this monograph.

2
Generation of Highly Uniform, High-Density Inductively Coupled Plasma

Recently, high-density, low-temperature RF plasma sources have been increasingly attractive for numerous industrial applications ranging from the traditional highly selective dry etching and microstructuring of silicon wafers in ultra large scale integration semiconductor manufacturing to recently reported synthesis of carbon-based nanostructures in the fabrication of electron field emitters for the development of advanced flat display panels [72, 87–89]. High number density and excellent uniformity of fluxes of reactive species over large volumes and surface areas are crucial for the improvement of the efficiency of plasma processing. Among various sources of low-temperature nonequilibrium plasmas, inductively coupled plasma (ICP) devices have attracted a great deal of attention because of their excellent properties to generate high-density, large-volume, and large-area plasmas. Presently, low-pressure, low-temperature ICP sources are used by several industries as reference plasma reactors for numerous applications in semiconductor manufacturing, optoelectronics, and synthesis and processing of advanced functional films and coatings [52, 90].

This chapter is devoted to one of the most common configurations of ICP sources with an external flat spiral RF coil that was actually used for various nanofabrication applications discussed elsewhere in this book. Because of the space limitations, we purposely focus here on a detailed description of the design and operation of the selected plasma source instead of reviewing all possible configurations of ICP plasma sources. Our choice has been justified by the following reasons:

- this book highlights the role of the plasma in process-specific nanoscale applications; hence, even minor details of the plasma generation can become crucial for tailoring the plasma environment in each specific application;

- since this type of plasma sources has been used in a large number of applications detailed in other chapters, it is instructive to provide the details of the plasma source in a separate chapter other than repeating common descriptions in each specific process;

Plasma-Aided Nanofabrication. Kostya (Ken) Ostrikov and Shuyan Xu
Copyright © 2007 WILEY-VCH Verlag GmbH & Co. KGaA, Weinheim
ISBN: 978-3-527-40633-3

- the results presented elsewhere in the book thus become more easily reproduceable by other researchers and can even be adopted for Honours and graduate research projects and undergraduate laboratory courses.

This chapter begins with the description of the design and operation of the experimental facility and diagnostics of low-frequency inductively coupled plasmas (LFICPs, Section 2.1). In Section 2.2, we discuss the unique attribute of ICP sources, namely, the ability to operate in two distinctive discharge modes that differ by the prevailing mechanism of RF power coupling. The bimodal source operation offers a better deal of flexibility in materials processing and also poses a number of challenges to the stability of discharge operation near mode transition points. Section 2.3 focuses on the electromagnetic field distribution and nonlinear phenomena in LFICP sources. The results of optical emission spectroscopy of ICPs in complex gas mixtures are presented in Section 2.4. Modeling of particle and power balance in the plasma discharge is a very useful tool to predict the spatial distributions of densities, energies and fluxes of the plasma species, as discussed in Section 2.5. This chapter concludes with a summary of the most salient features of low-frequency inductively coupled plasmas as benchmark industrial plasma reactors (Section 2.6). The material of this chapter is primarily based on original publications [91–97] of the authors and their colleagues.

2.1
Low-Frequency ICP with a Flat External Spiral Coil: Plasma Source and Diagnostic Equipment

In common sources of inductively coupled plasmas the configurations and positioning of RF current driving antennas can be quite different. For example, the inductive coil can be either placed externally or internally with respect to the discharge chamber, and also have a flat or a helical geometry [15]. In one of the most common embodiments considered in this chapter, the inductively coupled plasma is sustained by the RF power deposited by an external flat spiral inductive coil ("pancake coil") installed externally to and separated by a small air gap from a dielectric window that seals a (usually cylindrical) vacuum chamber [93, 98–100] as shown in Fig. 2.1.

Most of the commercial ICP reactors feature a fused silica or reinforced glass window sealing the chamber in its r–ϕ cross-section from the top. The RF current driven through the flat spiral inductive coil generates the electromagnetic field that features the azimuthal electric E_ϕ as well as the radial H_r and axial H_z magnetic field components [101]. We emphasize that high uniformity of densities and fluxes of ions and reactive species, high product yield, process efficiency, selectivity, and reproducibility are the common requirements for

2.1 Low-Frequency ICP with a Flat External Spiral Coil

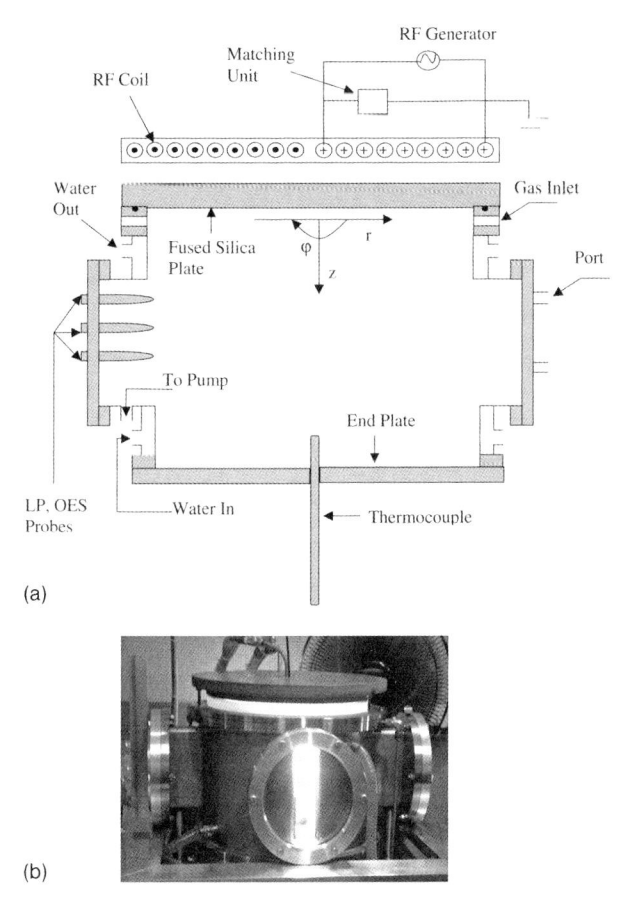

Fig. 2.1 Sketch and photograph of the low-frequency inductively coupled plasma source.

plasma processing [15]. The sources of inductively coupled plasmas with an external planar coil have proven to meet the above requirements. Indeed, the ICP sources are capable of generating large-area, large-volume, high-density plasmas with low sheath potentials near processing surfaces, which enables the possibility of independently controlling the plasma density and ion energy [102–106]. It is remarkable that the ICP sources can operate at low gas feedstock pressures and have demonstrated and outstanding potential in generating high-density plasmas with moderate input powers. Moreover, the plasma features a high level of spatial uniformity, which is crucial for large-area materials processing.

Generally, inductively coupled plasmas are produced in a low-aspect-ratio cylindrical vacuum chamber by an external flat spiral coil (Fig. 2.1). The induced RF currents sustain the discharge maintaining the ionization of neutral gas at a required level. The coil is most commonly powered by a 13.56 MHz RF generator via a matching network. However, low-frequency (typically < 1 MHz) ICPs possess several advantages that make them especially useful for the development of large-area high-density plasma sources [91, 93, 107]. In addition, upscaling of 13.56 MHz reactors may prove difficult since the RF wavelength becomes shorter than the coil length, which may result in poor plasma uniformity [98, 101, 108]. One of possible solutions of this problem is to excite traveling waves in the coil [108]. The alternative way, discussed in this chapter, is to operate at lower frequencies with a higher number of turns of the RF coil, yet without increasing the coil reactance.

2.1.1
Plasma Source

A schematic diagram of the ICP plasma source is shown in Fig. 2.1. The vacuum chamber has a cylindrical shape with the diameter and height of 32 cm and 23 cm, respectively. The plasma reactor is made of stainless steel and is of a double-walled construction to allow cold water circulation to remove the excessive heat dissipation during the plasma discharges. Four rectangular ports (two of them are shown in Fig. 2.1) are symmetrically arranged around the circumference of the plasma chamber to facilitate visual monitoring of the discharge and enable the access of various plasma diagnostic tools, such as magnetic, single RF-compensated Langmuir, and optical emission probes, considered in detail in Section 2.1.2. The diagnostic probes can be inserted radially at different vertical positions in the side ports, each of them has seven portholes separated by 2 cm in the axial direction. There are similar portholes in the aluminum bottom endplate of the plasma chamber. The endplate contains a set of 15 holes each separated by 2 cm, which allows the diagnostic probes to be inserted axially at various radial positions.

The plasma chamber is pumped though a small side port located at the lower portion of the vessel between two rectangular side ports. A KYKY (type 2XZ) turbo-molecular pump (with a pumping speed of 450 l/s) backed by a two-stage rotary pump was used to evacuate the vessel. A typical routinely achievable base pressure can be as low as 5×10^{-5} Torr. A Pirani gauge (Edward model RM 10 with the measuring pressure range from 10^3 to 10^{-3} Torr) and a Penning gauge (Edward model CP25K with the measuring pressure range from 10^{-3} to 10^{-8} Torr) were used to measure the base pressure. Both gauges were controlled by an Edward Pirani Penning readout (model 1005). For basic studies of the plasma source operation and parameters, pure argon (99.99 % purity) was used as a working gas. In order to control the flow rate of the working gas and the equilibrium pressure within the plasma chamber, MKS Flow Controllers 1100 series connected to MKS Type 247C four-channel Readout are used. A MKS Baratron capacitance manometer (model 122AA), which is connected to a MKS Type PDRC-2C Power Supply Digital Readout, is used to monitor the pressure inside the chamber. The working gas pressure p_0, which is typically in the range of 1 to 100 mTorr, is controlled by a combination of a MKS Flow Controller and a manual gate valve equipped in the pumping line.

An Advanced Energy (model PDX 8000, 460 kHz) RF generator is used to drive the RF current in the antenna. The maximum output power of the generator is 8000 W into a 50 Ω, nonreactive load. However, the power supplied to the inductive coil in most experiments varied from 100 to 2500 W. The generator is connected to the coil via a specially designed π-type matching network. Two cooling units of type EYELA Cool Ace CA-1100 are used to supply cold water to the vacuum vessel, RF generator, and turbomolecular pump.

The equivalent circuit of the plasma discharge is given in Fig. 2.2, where V_0 and R_0 are the open circuit voltage and resistance of the RF generator, V_i and I_i are the input voltage and current into the matching network, R_c and L_c are the resistance and inductance of the unloaded coil, ΔL_p is the variable part of the circuit inductance due to the plasma load, R_p is a reflected plasma resistance, I_c is the coil current, and $L = L_c - \Delta L_p$. Capacitors C_1 and C_2 are used to match the RF generator to the plasma load. In calculating the power deposited in the plasma, the power dissipated in the coil has been deducted. We note that $X = (V_i/I_c)\sin\phi = \omega L - (1/\omega C_1) = X_c - \Delta X_p$ is the total circuit reactance, X_c and ΔX_p are the unloaded circuit reactance and a reactance change due to the plasma, and ϕ is the phase shift between V_i and I_c.

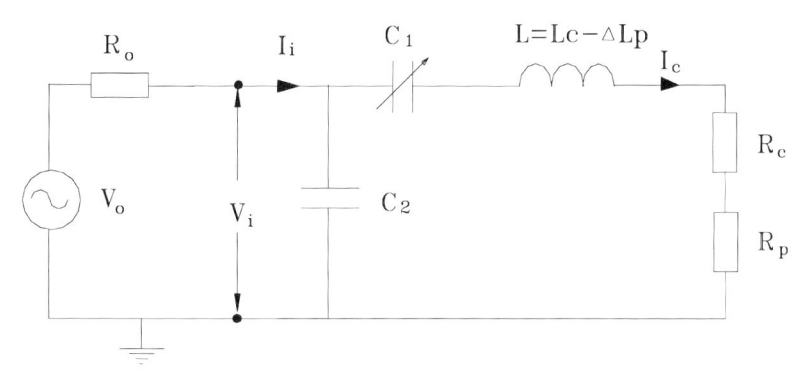

Fig. 2.2 Equivalent circuit diagram of the low-frequency inductively coupled plasma source [93].

2.1.2
Diagnostics of Inductively Coupled Plasmas

A set of RF voltage and current probes, miniature magnetic probes, RF-compensated Langmuir probes, optical emission spectroscopy (OES), and quadrupole mass spectrometry (QMS) is used to investigate the properties of the ICP plasma source.

2.1.2.1 RF circuit diagnostic

The RF voltage V_i and the circuit currents I_i, I_c are measured using Tektronix P6009 voltage probes and the two Pearson current transducers (mode 1025), respectively. The RF signal is fed to Tektronik TDS 460 digital storage oscilloscopes via a 50 Ω triaxial cable. The measured RF signals are processed by the amplitude and phase detecting circuits.

2.1.2.2 Miniature magnetic probes

The electromagnetic properties of the ICP plasma source can be studied by using two custom-designed miniature magnetic probes. The schematics of two different miniature magnetic probes is shown in Fig. 2.3. Briefly, the key element of the probes is a miniature coil (with a different number of turns) wound around a teflon frame mounted at the end of an aluminum tube with the internal diameter of 3 mm. Depending on the orientation of the probes, one of them can pick up the azimuthal or axial components of the magnetic field (Fig. 2.3(a)), while the other can sense the radial magnetic field component (Fig. 2.3(b)). A continuous flow of compressed air can be used to cool the probes. The air flows in through the aluminum tube and exits through the periphery space between the aluminum tube and the inner surface of the

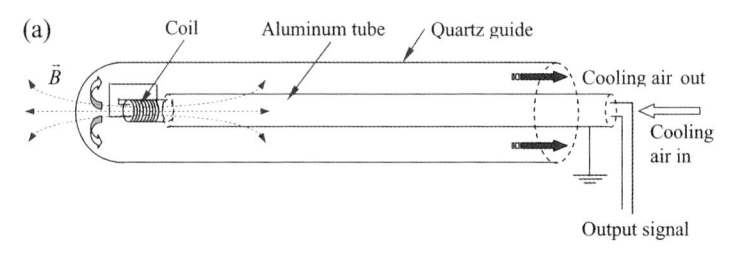

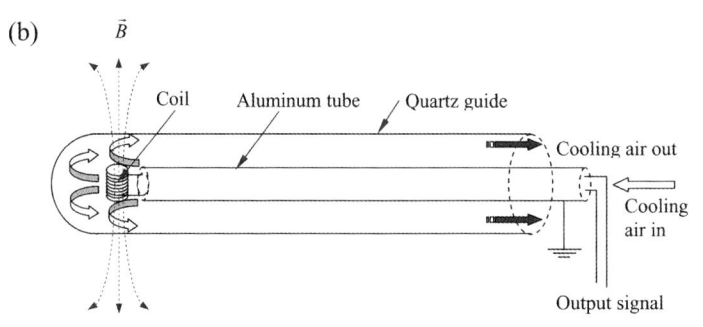

Fig. 2.3 Sketch of miniature magnetic probes for radial (a) axial and azimuthal (b) magnetic field components [97].

quartz feedthrough. To minimize the effect of the RF interference on the magnetic probe signals, the aluminum tube is properly grounded. The spatial resolution of the probes is 0.6 cm, which enables a detailed mapping of the RF magnetic field inside the plasma chamber.

2.1.2.3 Langmuir probe

Global plasma parameters can be obtained from the time-resolved measurements by a single RF-compensated cylindrical Langmuir probe (Fig. 2.4). The probe is powered by AC (50 Hz) voltage in the range from -40 to $+40$ V through a variable transformer. To isolate the electric connection between the probe and main power supply an additional isolation transformer (with 1:1 ratio) is used. The probe voltage across the plasma load resistance can be measured by a Tektronix voltage probe and monitored on a digital storage oscilloscope (Tektronix model 380) via a 1.0 MHz low-pass filter. The probe current is obtained by measuring the voltage drop across a 0.26 Ω resistor and monitored on the same oscilloscope via a 0.2 MHz low-pass filter. All the obtained signals are transmitted to a PC via the GPIB port of the oscilloscope. A PC-based data acquisition system is used to record and process the data and obtain the electron density, effective electron temperature, plasma potential, and electron energy distribution/probability functions (EEDF/EEPF).

The main plasma parameters can be determined by using the second derivative of the Langmuir probe current–voltage characteristics (Druyvestein routine) [15,107]. We note that the reproducibility of the data collected in the experiments of our interest here is excellent. All the diagnostic tools and data acquisition system are electrostatically shielded to minimize any RF interference from the generator and antenna RF fields. Since the plasma chamber provides a good reference ground for the single RF-compensated Langmuir probe measurements in the electron collection voltage range, the plasma potential and effective electron temperature can be accurately obtained from the

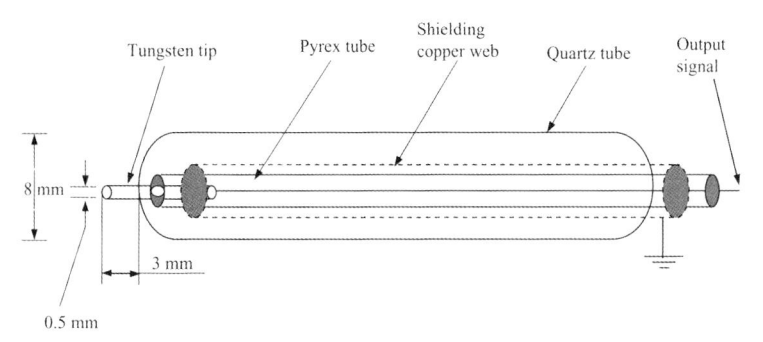

Fig. 2.4 Sketch of a cylindrical Langmuir probe used for the ICP diagnostics.

measured I–V probe characteristics. The plasma density n_e, effective electron temperature T_{eff}, and plasma potential V_p were determined by using the Druyvestein routine [15, 107]. In particular, n_e and T_{eff} can be expressed as

$$n_e = \int g_e(V)\,dV$$

$$T_{eff} = \frac{2}{3n_e} \int V g_e(V)\,dV$$

and the electron energy distribution function $g_e(V)$ can be obtained through the second derivative of the probe current over the voltage $d^2 I_e/dV^2$ as

$$g_e(V) = \frac{2m_e}{e^2 A}\left(\frac{2eV}{m_e}\right)^{1/2} \frac{d^2 I_e}{dV^2}$$

where e, m_e, and A are the elementary charge, electron mass and the probe surface area, respectively. The plasma potential V_p can be found as the maximum of the first derivative of the dependence $I(V)$ or as the zero crossing point of the second derivative of the probe current [15].

2.1.2.4 Optical emission spectroscopy

Throughout this monograph, optical characteristics of the plasma generated in various plasma sources are extensively used for process diagnostics. In most cases, the optical emission spectra are collected in the wavelength range of 300–900 nm by using the optical emission spectroscopy measurement system shown in Fig. 2.5. The variation of the optical emission intensity of different atomic, ionic, molecular, or radical lines in the plasma can be monitored in real time while the coil current or other discharge parameters are varied. The optical emission of different spectral lines of the excited/ionized species produced by the plasma discharge is collected by a collimated optical probe inserted radially or axially into the plasma chamber. An optical fiber is used to transmit the collected signal to the entrance slit of a monochromator (Acton Research SpectraPro-750i model, 0.750 Meter Focal Length Triple Grating Imaging Monochromator/Spectrograph). The emission is amplified by a photo-multiplier (PMT, THORN EMI electron tube) and then dispersed and analyzed by a monochromator (with spectral resolution of 0.023 nm) in the preset wavelength range. The amplitude of the output signal from the photo-multiplier could be changed by adjusting the voltage output from a high-voltage generator. In order to monitor the output signal on the computer in real time, the signal is continuously digitized by an A/D convertor built-in in the scan controller. The scanning and data acquisition process was controlled by the data acquisition and analysis software Spectrasense[TM] (Acton Research Corporation). Using the data acquisition system, one can record broad spectral bands or selectively monitor certain spectral lines.

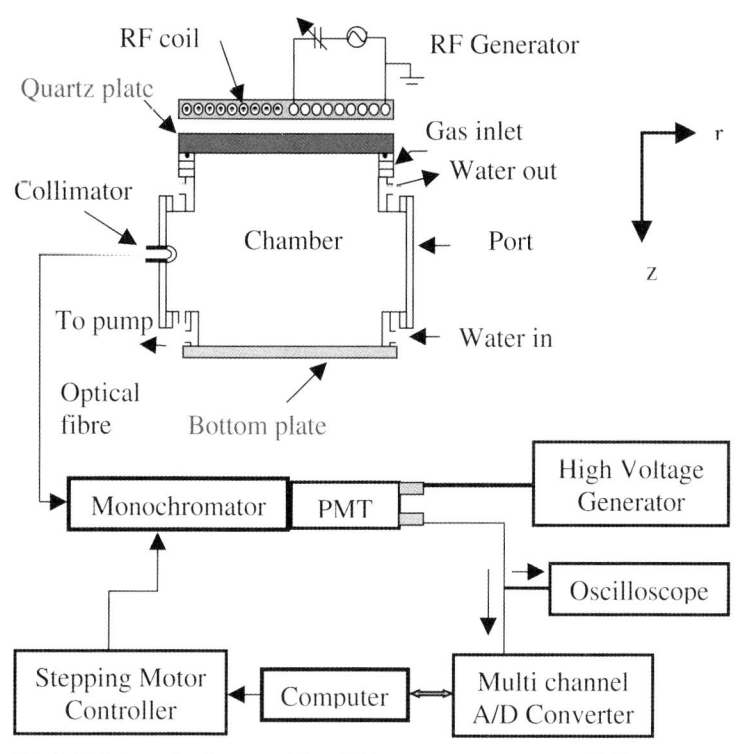

Fig. 2.5 Schematic diagram of the OES measurement system [94].

2.1.2.5 Quadrupole mass spectrometry

The quadrupole mass spectrometry (QMS) is used for real-time *in situ* diagnostics of neutral species in the gas phase of the discharge (see Fig. 2.6). The ICP facility of our interest here is equipped with a complete quadrupole mass analyzer system Microvision Plus manufactured by MKS Instrument Spectra Products.

This system enables one to measure the concentrations of neutral species with atomic masses not exceeding 200 amu. The spectrometer consists of an ionization chamber, a quadrupole analyzer, a pumping system, and a control unit. The entire measurement process is computer controlled. The quadrupole mass analyzer is connected to the ICP chamber with a 23 cm long and 24 mm in diameter metallic tube. The gas species are first ionized in the ionization chamber and then analyzed by a mass detector. The partial pressures of different species are usually recorded as the intensities of the ion fluxes in the detector. A Faraday cup with a single filter is used as the ion detector. The sensitivity of the Faraday cup detector is in the range of 10^{-4} A/Torr to 10^{-3} A/Torr, which allows one to process the output signals from the detector without any additional amplification. Data acquisition is controlled by a

Fig. 2.6 Connection of the OES and QMS measurement systems to the discharge chamber.

multipurpose Process Eye 2000 software, which enables the user to monitor, in real time, density variations of a selected species or scan any preset mass ranges below 200 amu.

2.2
Discharge Operation Regimes, Plasma Parameters, and Optical Emission Spectra

In this section, we describe the operation of the plasma source and give insight into measurements of electromagnetic properties of the LFICP discharge [93]. Radial and axial profiles of the plasma potential, electron density and temperature will be considered. It will also be demonstrated that the plasma source can efficiently operate in broad pressure ranges.

2.2.1
Electromagnetic Properties and Mode Transitions

The ICP source performance tests have been conducted in argon gas in the pressure range of 0.3–1000 mTorr [91, 93, 102]. Prior to the discharge, the chamber was filled with argon at the predetermined pressure. The matching/tuning capacitors and the generator bank voltage were set to initial levels. The RF generator was then turned on. It is noteworthy that, to avoid damage of the generator by strong RF currents reflected by poorly matched circuits, the ICP discharge is in most cases started with low input powers. The matching is then adjusted when the power is increased. Thus, the discharge always starts in a faint electrostatic (E) mode sustained by the potential drop between the inner and outer turns of the inductive coil [102, 103]. Gradual increase of the RF generator output results in discharge transition to a high-density and visually much brighter regime commonly referred to as the H-mode. Thereafter, the power input is slowly decreased back to the minimum starting level. The peak-to-peak voltage, coil current, and the phase difference between them are continuously recorded.

Figure 2.7 displays the plasma load resistance R_p and reactance change ΔX_p as a function of the power dissipated in the plasma P_p in a 12 mTorr Ar discharge. The process starts at point (1) corresponding to a faint E-mode, the discharge remains faint until reaching the state (2). A subsequent increase of the power P_p leads to a sudden transition to a bright H-mode point (3). It is clearly seen from Fig. 2.7 that the $E \rightarrow H$ transition is accompanied by a substantial increase in the plasma resistance R_p and a significant decrease of the coil current I_c and the plasma reactance ΔX_p. When the input power is gradually increased, the discharge remains in the H-mode. As the RF power

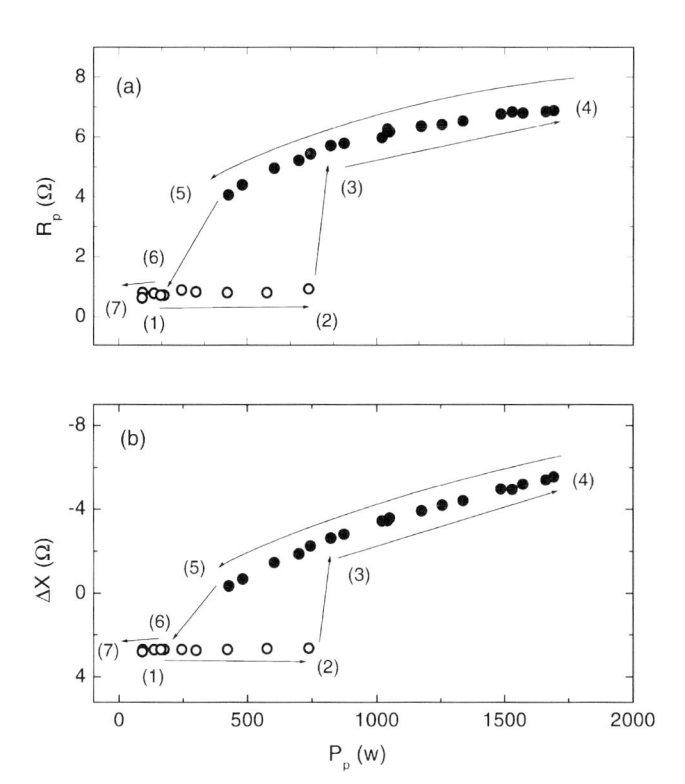

Fig. 2.7 Reflected plasma resistance (a) and change in load reactance (b) versus power absorbed by the plasma for $p_0 = 12$ mTorr [93].

is increased further, the value of ΔX_p further decreases, while the plasma resistance R_p steadily increases (Fig. 2.7).

After reaching the state (4), where P_p is maximal, the coil current is lowered and the discharge still remains in the H-mode even after passing point (3) of the original $E \rightarrow H$ transition. The inverse $H \rightarrow E$ transition occurs at the coil current I_c (point (5)) less than that of $E \rightarrow H$ transition. Point (6) corresponds to the E-mode discharge phase. The discharge is extinguished at point (7). We note that the discharge is bistable if 220 W $< P_p <$ 840 W. In this case the two different values of circuit parameters correspond to the same RF power. The observed process is cyclic and highly reproducible.

The global electric quantities R_p and ΔX_p can be used to obtain the spatially averaged plasma parameters. Using a simple transformer model, El-Fayoumi and Jones [102] interpreted their electrical measurements by considering the RF coil to form the primary, and the plasma to act as a single-turn secondary coil, of an air-core transformer. The measured circuit quantities are then linked to the electromagnetic fields through the power balance equation. This model provides an efficient and simple way in design of the RF coupling circuit and in understanding of the general properties of LF ICP discharges. For further details, we refer the reader to original works [102, 103].

2.2.2
Plasma Parameters

Here we show the results of detailed Langmuir probe scans revealing spatial profiles of the electron density n_e, temperature T_e, and plasma potential V_p for varying total input power and filling pressure. The axial profiles of the plasma parameters are measured at seven available axial positions in the side ports. The radial distributions are obtained by moving the probe along the diameter of the chamber at the central port ($z = 10$ cm) position. Figure 2.8 shows the radial profiles of n_e, T_e, and V_p for the total input power $P_{tot} \sim 1.7$ kW in 22 mTorr, H-mode argon discharge. These profiles reveal that a high-density, highly uniform plasma is generated over a large area in the discharge cross-section. The radial nonuniformity of the electron density is estimated to be less than 8 % over the radii of the chamber, $R < 15$ cm. The value of the electron number density is very high ($n_e \sim 9 \times 10^{12}$ cm^{-3}). The plasma potential (13–17 V) is quite low, so is the electron temperature ($\sim$2.5 eV).

The axial profiles of n_e, T_e, and V_p measured at $r = 0$ for the same discharges are displayed in Fig. 2.9. It is seen that the electron density, temperature, and plasma potential gradually decrease along the axial direction. This result agrees with Chakrabarty's heterodyne microwave interferometer measurements [109].

Another important feature of the LF ICP discharge is that it can be operated in a broad range of the filling gas pressures. Figure 2.10 depicts how

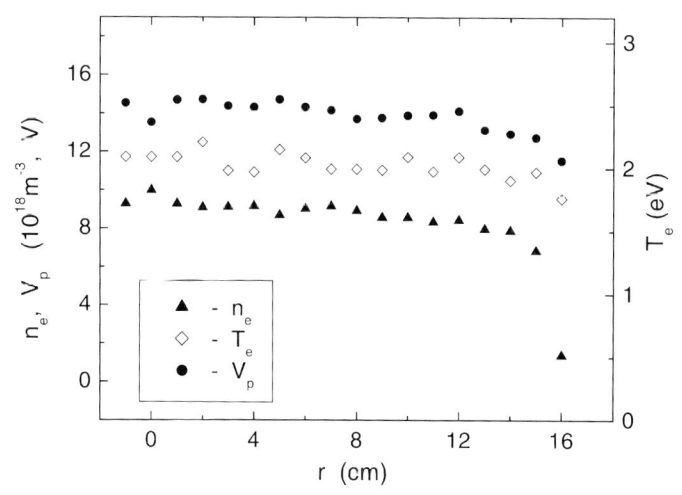

Fig. 2.8 Radial profiles of the electron density n_e, temperature T_e, and plasma potential V_p in a 22 mTorr, H-mode Ar discharge. The RF input power is 1.7 kW [93].

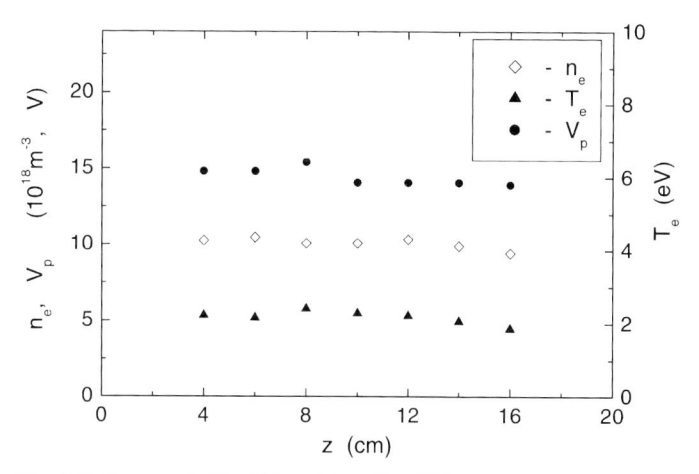

Fig. 2.9 Same as in Fig. 2.8, axial profiles [93].

the electron density, temperature, and plasma potential vary with the argon gas pressure. It is seen that n_e increases with pressure if $p_0 < 200$ mTorr and declines in the succeeding range. The electron temperature and plasma potential also diminish with increasing filling gas pressure. It is worth emphasizing that the source operation is highly reproducible in a broad gas pressure range of approximately 0.3–900 mTorr and beyond [93].

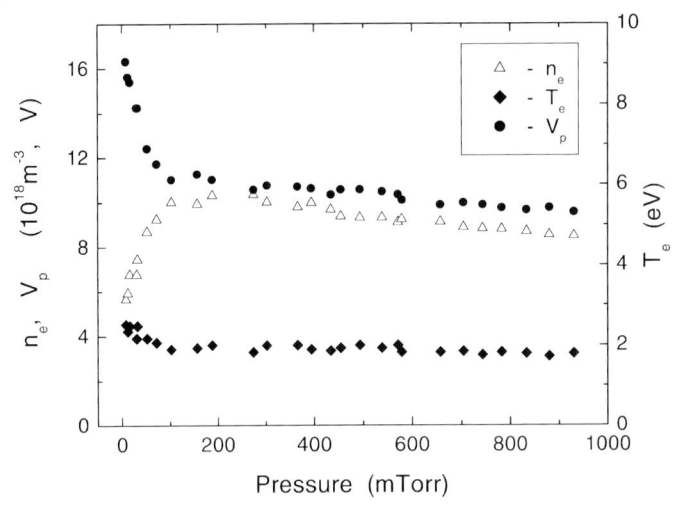

Fig. 2.10 Electron density, electron temperature, and plasma potential versus argon pressure for $P_{\text{tot}} = 1700$ W, measured at the chamber center [93].

2.2.3
Discharge Hysteresis

As was mentioned above, the cyclic variation of the input power results in strongly nonlinear hysteresis of the circuit parameters. The plasma parameters and the optical emission intensities also feature a hysteretic behavior. Likewise, the $E \rightarrow H$ transition is accompanied by generation of strongly nonlinear electromagnetic fields, a process that will be considered in detail in Section 2.3.

2.2.3.1 Variation of plasma parameters

An important feature of the LF ICP is a discontinuous transition of the plasma parameters during the $E \leftrightarrow H$ mode transitions. Figure 2.11 reveals that in the 22 mTorr discharge, the electron density instantaneously increases by about two orders of magnitude from $n_e \sim 6 \times 10^{10}$ cm^{-3} to $n_e \sim 8 \times 10^{12}$ cm^{-3} in the vicinity of the $E \rightarrow H$ transition. The inverse $H \rightarrow E$ transition, which occurs if the input power diminishes to ~ 400 W, is accompanied by a sharp down-jump of n_e. Within the H-mode discharge, the electron temperature remains almost constant ($T_e \sim 2.5$ eV) and is not affected by the variation of the input power in the range 400–1700 W. In the E-mode, the average value of T_e (~ 8.5 eV) turns out to be higher than that in the electromagnetic mode.

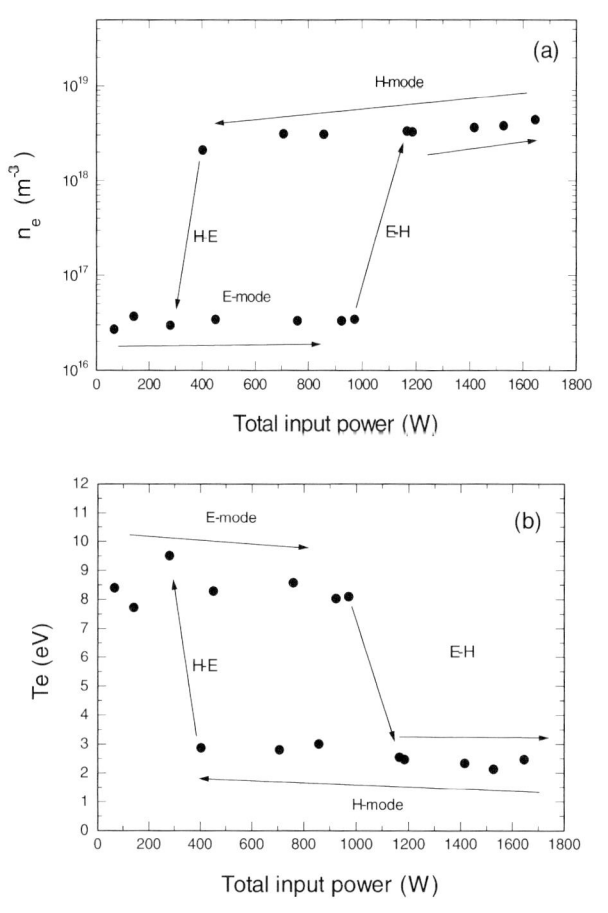

Fig. 2.11 Changes of n_e (a) and T_e (b) in the course of the $E \leftrightarrow H$ transitions. Parameters are the same as in Fig. 2.8 [93].

2.2.3.2 Optical emission spectra

The intensity of optical emission from neutral and ionic plasma species also undergoes considerable changes during the $E \leftrightarrow H$ transitions. The cyclic variation of the OEI corresponding to an 833.22 nm line of the neutral Ar (denoted Ar I throughout) atom with the coil current is depicted in Fig. 2.12 for different gas pressures. It is clearly seen that the $E \rightarrow H$ transition is accompanied by an instantaneous rise of the OEI. In the H-mode regime, the OEI further increases with the coil current.

Similar tendency has been reported [110] for mode transitions in a helical inductively coupled plasma source. However, the intensity diminishes linearly with reducing coil current, and the discharge is still in the bright H-mode even for coil currents smaller than the $E \rightarrow H$ transition current. Near the point of

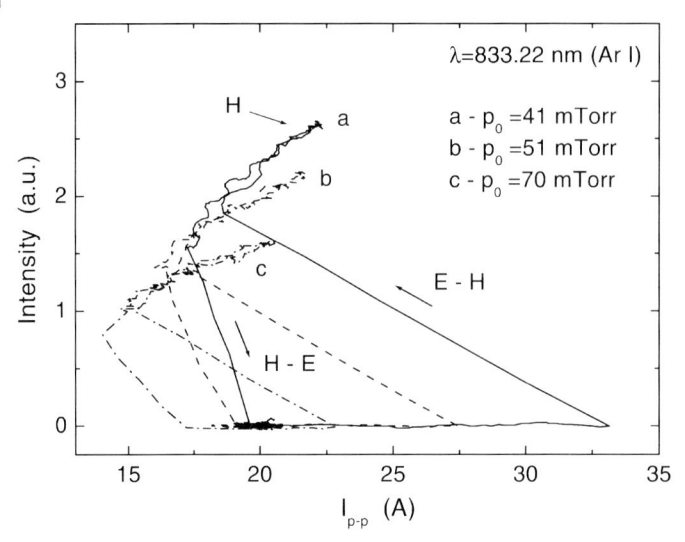

Fig. 2.12 Variation of the optical emission intensity of a neutral argon (ArI) line (833.22 nm) during $E \leftrightarrow H$ mode transitions. [93].

the $H \to E$ transition the OEI decreases to the level corresponding to the dim E-mode. In the electrostatic regime, the OEI does not change much and remains low.

Using the OEI data, one can obtain the dependence of the minimal starting current that initiates the $E \to H$ transition, and the minimal H-mode maintenance current (threshold current for the $H \to E$ transition) on the operating gas pressure [92]. The $E \to H$ and $H \to E$ transitions are initiated at different values of the gas pressure for the same coil current I_c, which confirms that hysteresis can also be observed with variation of the filling gas pressure.

2.3
Electromagnetic Field Distribution and Nonlinear Effects

From previous sections, we recall that in a common cylindrical embodiment with a dielectric window atop (in r–ϕ cross-section) (Fig. 2.1), an RF current spirally-driven in the coil predominantly excites the azimuthal RF electric field, as well as the poloidal (with B_z and B_r components) magnetic field [101–103]. Generation of the above currents and electromagnetic fields has been a subject of extensive modeling and simulation research. However, most of the available results are valid within the framework of the linear, with respect to the RF field amplitude, approximation. This approximation is usually valid at low RF input power levels.

However, the linear approximation can often be inconsistent even at low powers, e.g., in the low-frequency discharge operation regime considered in this chapter. Indeed, the nonlinearity parameter, which is the ratio of the electron quiver velocity

$$V_E = eE/m_e\omega$$

in RF electric field E of frequency ω to the electron thermal velocity V_{Te}, increases in approximately 27 times when the operating frequency is decreased from the conventional one 13.56 MHz to 500 kHz.

Hence, at low frequencies, one should expect significant distortions of the electromagnetic field/current patterns, which can be due to generation of nonlinear signals at higher Fourier harmonics or static nonlinear (ponderomotive) responses. Hysteretic effects in mode transition phenomena considered in Section 2.2 is yet another intrinsically nonlinear feature of inductively coupled plasmas [93, 100].

Here, following the original work [97], we examine the distribution of nonlinear electromagnetic fields in a low-frequency (500 kHz) source of inductively coupled plasmas. It will be shown that the nonlinear plasma response results in generation of the azimuthal magnetic field component, which does not appear in the linear regime at lower RF powers. A simple theoretical explanation of the persistent nonlinear second-harmonic signal is also provided.

2.3.0.3 Experiment

The radial distribution of the magnetic fields in the chamber is studied separately in the electrostatic (E) and electromagnetic (H) discharge regimes by sweeping the magnetic probes radially inside a quartz tube inserted through the closest to the quartz window porthole ($z = 4$ cm). The details of the ICP plasma source and magnetic probe diagnostics are given in Section 2.2. In the electrostatic discharge mode sustained with low (~ 170 W) RF powers, the electromagnetic field pattern includes all three components of the RF magnetic field, namely B_r, B_ϕ, and B_z (Figs. 2.13(a)–(c)).

Apparently, the field pattern features only the fundamental frequency signals. Specifically, the second-harmonic signals appear to be typically 10–100 times weaker than the fundamental frequency ones. Furthermore, in the areas close to the chamber center, the B_z component is a dominant one, whereas at distances ~ 8 cm apart from the walls, B_r becomes more pronounced. The azimuthal magnetic field component, although non-negligible, appears to be much smaller than the other two components. It is remarkable that in the low-power case the electromagnetic field pattern is consistent with the theoretical results of El-Fayoumi and Jones [102, 103]. However, in the linear (with respect to the field amplitudes) theory that also approximates the real flat spiral inductive coil by the set of concentrical azimuthal RF currents, the B_ϕ component does not persist. Nevertheless, since no other Fourier harmonics were detected in the E-mode discharge, the measured azimuthal magnetic field sig-

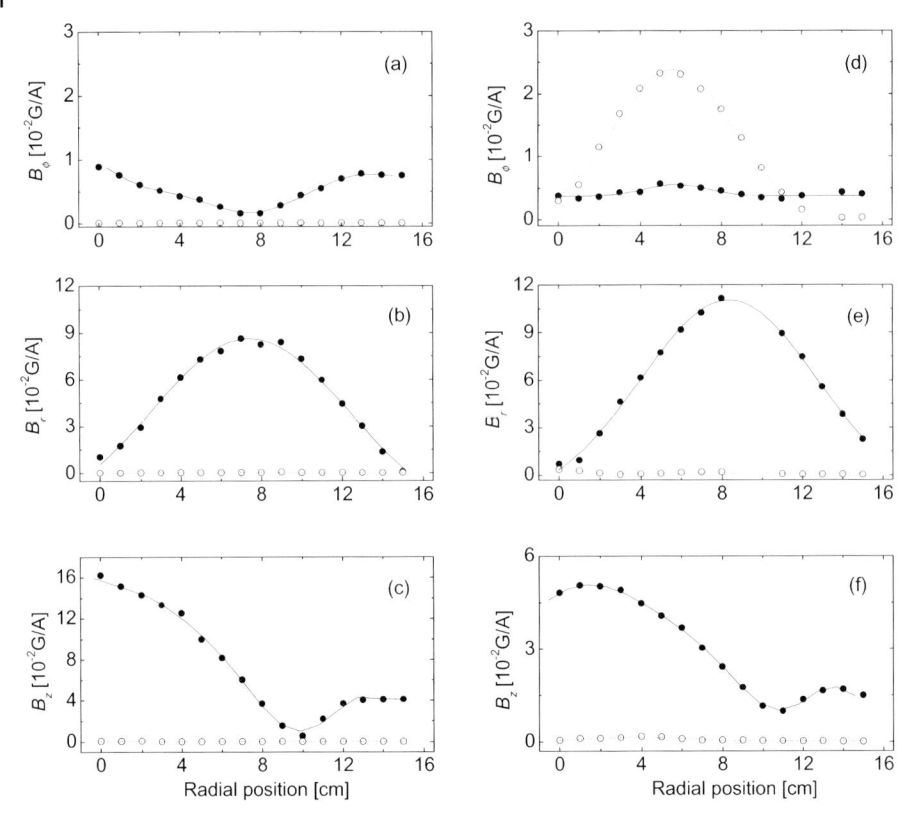

Fig. 2.13 Radial profiles of B_ϕ (a), B_r (b), and B_z (c) components in the E-mode discharge sustained with 170 W RF powers at 50.8 mTorr. Solid and open circles correspond to the fundamental frequency and second-harmonic signals, respectively. (d)–(f) same as for (a)–(c) but for the H-mode discharge sustained with $P_\text{in} = 1130$ W [97].

nal can be attributed to the nonvanishing radial RF electric field generated by the real planar spiral coil. Thus, we note that the nonlinear effects are negligible in the low-power electrostatic regime.

After transition to the electromagnetic mode, the field pattern changes dramatically as depicted in Figs. 2.13(d) and (e). In particular, the fundamental component of B_z becomes even smaller than in the E-mode discharge. More importantly, a pronounced generation of the second Fourier harmonics of B_ϕ was clearly observed (Fig. 2.13(d)). It is remarkable that the amplitude of the second-harmonic nonlinear signal $|B_\phi^{2\omega}|$ appears to be 4–6 times higher than that of the fundamental harmonics. The maximum of the amplitude of the second Fourier harmonic is located at approximately 5–6 cm from the chamber axis. One can also observe that the magnitude of the nonlinear magnetic field increases with RF power. It is thus reasonable to presume that the observed effect is due to the nonlinear plasma response, which normally increases with the amplitude of the fundamental harmonics.

Non-negligible second-harmonic signals also appear for the B_r and B_z components. However, they are much smaller than the second Fourier component of B_ϕ. Likewise, the harmonics higher than the second Fourier harmonics do not affect the electromagnetic field profiles. It is worthwhile to note that the resulting magnetic field in the electromagnetic mode has a linear (fundamental frequency) poloidal (r–z) component and a nonlinear (second-harmonic) azimuthal component. Thus, the dominant linear RF current features a predominant azimuthal component j_ϕ, whereas the nonlinear RF current mostly flows in the poloidal cross-section. This conclusion will further be elucidated theoretically below.

However, linear magnetic field components feature hysteretic effects (different behavior for increasing and decreasing RF power). In particular, when the power increases, the linear B_ϕ component increases slowly. However, when the input power goes down, the latter decreases faster to smaller, than original values [97]. The second harmonics of B_ϕ also features a hysteretic behavior. However, in contrast to the linear component, the nonlinear signal exceeds the original values in the E-mode while the power input diminishes. One can also note that the hysteretic effects for the $B_\phi^{2\omega}$ signal are less resolved in the H-mode than in the E-mode. The third Fourier harmonics appears too weak to be of any concern and is not depicted.

Hysteretic effects are also pertinent to the purely linear B_r and B_z components. In the H-mode, the corresponding values are quite reproducible in the H-mode at increasing and decreasing powers. On the other hand, the field values in the E-mode appear to be noticeably lower when the input power is decreasing [97]. Thus, both the linear and nonlinear components of the magnetic field vary in a hysteretic (essentially nonlinear) manner during the process of the $E \to H \to E$ mode transition.

2.3.0.4 Fluid model and discussion

Possible reasons for the nonlinear plasma response can be elucidated by using a simple fluid model discussed below [97]. This model accounts only for nonlinear current effects and sidesteps any nonlinearities associated with pressure gradient forces. Thus, assuming $T_e, n_e = \text{const}\,(r, \phi, z)$ (which is fairly correct at $z = 4$ cm for $r < 10$ cm in the H-mode discharge [93]), from the electron momentum equation

$$m_e \partial_t \mathbf{v}_e + m_e \nu_e \mathbf{v}_e + e\mathbf{E} = \mathbf{F}_{\text{NL}} \tag{2.1}$$

one can estimate the direction of the nonlinear force

$$\mathbf{F}_{\text{NL}} = -m_e[(\mathbf{v}_e \nabla)\mathbf{v}_e + (e/m_e c)\mathbf{v}_e \times \mathbf{B}]$$

and relevant nonlinear currents $\mathbf{j}_{NL}$ resulting in the observed nonlinear second-harmonic response, where $\mathbf{E}$, $\mathbf{B}$ are the RF electric and magnetic fields. Here, $\mathbf{v}_e$, ν_e, e, and m_e are the electron fluid velocity, effective collision frequency, charge, and mass, respectively.

Expanding all the time-varying quantities in series

$$G = G_0 + (1/2)[G_1 \exp(-i\omega t) + G_2 \exp(-2i\omega t) + \cdots + \text{c.c.}]$$

where G is either of $\mathbf{v}_e$, $\mathbf{E}$, or $\mathbf{B}$, and c.c. stands for complex conjugate, one obtains

$$\mathbf{F}_{NL}^{(2\omega)} = -\frac{1}{4}\left[m_e(\mathbf{v}_{e1}\nabla)\mathbf{v}_{e1} + \frac{e}{c}\mathbf{v}_{e1}\times\mathbf{B}_1\right]\exp(-2i\omega t) \tag{2.2}$$

for the nonlinear force at the second harmonics and

$$\mathbf{F}_{NL}^{(0)} = -\frac{1}{4}\left\{m_e[(\mathbf{v}_{e1}^*\nabla)\mathbf{v}_{e1} + (\mathbf{v}_{e1}\nabla)\mathbf{v}_{e1}^*] + \frac{e}{c}[\mathbf{v}_{e1}^*\times\mathbf{B}_1 + \mathbf{v}_{e1}\times\mathbf{B}_1^*]\right\} \tag{2.3}$$

for the static nonlinear response (ponderomotive force [111]), where the asterisk denotes complex conjugate quantities. Using the set of conventional cylindrical coordinates (r,ϕ,z) and noting that in a linear approximation $\mathbf{B}_1 = (B_{r1}, 0, B_{z1})$, $\mathbf{E}_1 = (0, E_{\phi 1}, 0)$, $\mathbf{v}_{e1} = (0, v_{e\phi 1}, 0)$ [95], we obtain

$$F_{NLz}^{2\omega} = \frac{e}{4c}v_{e\phi 1}B_{r1}$$

and

$$F_{NLr}^{2\omega} = \frac{m_e}{4}\left[\frac{v_{e\phi 1}^2}{r} + \frac{e}{m_e c}v_{e\phi 1}B_{z1}\right]$$

for the axial and radial components of the second-harmonic nonlinear force. It is notable that there is no nonlinear force acting on the plasma electrons in the azimuthal direction ($F_{NL\phi}^{2\omega} = 0$). Thus, the nonlinear second-harmonic currents are indeed driven in the poloidal (r–z) cross-section.

The components of the nonlinear second-harmonic force can be derived by using the expressions for the linear electromagnetic fields in the cylindrical metal chamber of radius R and length L sealed at its top by a dielectric window with dielectric constant ε_d derived elsewhere [95]. The solution of the set of Maxwellian equations with relevant nonlinear terms yields the following equation for the nonlinear azimuthal magnetic field at the second Fourier harmonic

$$\frac{\partial}{\partial r}\frac{1}{r}\frac{\partial}{\partial r}(rB_{\phi 2}) + \frac{\partial^2 B_{\phi 2}}{\partial z^2} + \frac{4\omega^2}{c^2}\varepsilon_{2\omega}B_{\phi 2} = i\frac{2\omega\chi_{2\omega}}{ce}\left[\frac{\partial F_{NLr}^{2\omega}}{\partial z} - \frac{\partial F_{NLz}^{2\omega}}{\partial r}\right] \tag{2.4}$$

where $\varepsilon_{2\omega} = 1 - \chi_{2\omega}$ and $\chi_{2\omega} = \omega_{pe}^2/2\omega(2\omega + i\nu_e)$. The nonlinear signal at the second harmonic also includes the radial

$$E_{r2} = -\frac{ic}{2\omega\varepsilon_{2\omega}}\frac{\partial B_{\phi 2}}{\partial z} - \frac{\chi_{2\omega}}{e\varepsilon_{2\omega}} F_{\text{NLr}}^{2\omega}$$

and axial

$$E_{z2} = \frac{ic}{2\omega\varepsilon_{2\omega}}\frac{1}{r}\frac{\partial}{\partial r}(rB_{\phi 2}) - \frac{\chi_{2\omega}}{e\varepsilon_{2\omega}} F_{\text{NLz}}^{2\omega}$$

components of the electric field. It is important to note that other components of the electromagnetic field do not feature the second-harmonic terms. In the approximation adopted here, the nonlinear poloidal (r–z) current at the second harmonic can be calculated as

$$\mathbf{j}_{\text{pol}}^{2\omega} = \mathbf{r}j_r^{2\omega} + \mathbf{z}j_z^{2\omega}$$

where $j_{(r,z)}^{2\omega} = \sigma E_{(r,z)2}$, $\sigma = \omega_{pe}^2/4\pi(\nu_e - i\omega)$ is the conductivity of the uniform collisional plasma. Here, $\mathbf{r}$ and $\mathbf{z}$ are the unit vectors in radial and axial directions, respectively.

Therefore, the second-harmonic nonlinearities appear as a result of nonlinear interactions between the linear azimuthal current $j_{\phi 1}^{\omega}$ with the linear magnetic fields B_{r1} and B_{z1}. This interaction leads to pronounced generation of the poloidal second-harmonic RF currents that in turn self-consistently generate the nonlinear magnetic field at the second Fourier harmonics $B_{\phi 2}^{2\omega}$.

The fluid model fairly accurately describes the general tendencies in the radial profile of the azimuthal magnetic field at the second harmonic, obtained by solving Eq. (2.4). However, the discrepancies between the theoretical and experimental results increase with distance from the center of the vacuum chamber.

One of the possible reasons for the apparent discrepancy is the limitation of the weak nonlinearity approach. Indeed, since the nonlinear second-harmonic signal $B_{\phi}^{2\omega}$ at higher powers can be several times larger than the corresponding linear component B_{ϕ}^{ω}, the weak nonlinearity approximation used here can eventually become invalid. The nonlinear-to-total signal ratio $\vartheta = B_{\phi}^{2\omega}/[(B_z^{\omega})^2 + (B_r^{\omega})^2 + (B_{\phi}^{2\omega})^2]^{1/2}$ reflects the strength of the plasma nonlinearities in the electromagnetic mode of the low-frequency inductively coupled discharges. The analysis of the radial dependence of ϑ suggests that the nonlinear magnetic field at the second harmonic is significant at radial distances $2 < r < 7$ cm. Accordingly, the series expansion of the total nonlinear signal into higher harmonics with the amplitudes $A_{j+1} \ll A_j$, which was used to derive Eq. (2.4), remains fairly accurate except for the radial area $2 < r < 7$ cm, where the nonlinear parameter ϑ exceeds 0.2, a qualitative marginal parameter for the weak nonlinearity approximation to remain valid [112].

Moreover, the simple fluid model adopted in this study assumes the axial and radial uniformity of the plasma density and electron temperature. In reality, assumption $T_e = \text{const}(r,z)$ is usually justified better than $n_e = \text{const}(r,z)$. For moderate-power H-mode discharges, the radial profiles of the electron number density which are flattened in the central areas ($r < 8$–10 cm in the chamber with $R = 16$ cm) and decline towards the walls are more realistic [95]. Radially, the plasma density is typically distributed cosine-wise with the minima near the chamber bottom endplate and the dielectric window [15]. Thus, the values of the above nonlinear terms can be considered fairly accurate in the central areas of the plasma glow (approximately $r < 8$–10 cm and $5 < z < 15$ cm). Nevertheless, the model in question does correctly predict generation of poloidal RF currents and azimuthal magnetic field at the second Fourier harmonics and can be used for qualitative explanation of the observed effect.

It is worthwhile to mention that the direction of the resulting nonlinear force $\mathbf{F}_{\text{NL}}^{2\omega} = \mathbf{r} F_{\text{NLr}}^{2\omega} + \mathbf{z} F_{\text{NLz}}^{2\omega}$ also changes along the radial direction. It is interesting that for the experimental parameters of Fig. 2.13 the azimuthal component of the nonlinear Lorentz force is always larger than the radial one. Furthermore, for radial positions $10 < r < 13$ cm the resulting nonlinear force $\mathbf{F}_{\text{NL}}^{2\omega}$ is directed along the chamber axis. In this case the radial component turns out to be 6–9 times less than the axial one.

It is remarkable that the nonlinear radial second-harmonic force $\mathbf{F}_{\text{NLr}}^{2\omega}$ contains contributions from the nonlinear Lorentz force ($\sim v_{e\phi 1} B_{z1}$) and nonlinear (also nonuniform) electric field $\sim v_{e\phi 1}^2 / r$. However, the estimates show that for the typical parameters of the experiments concerned (in the H-mode discharge), the contribution of the latter term is 1–2 orders of magnitude smaller and can be neglected.

Due to apparent similarity in the procedure of derivation of the ponderomotive force-caused nonlinear terms, one can also expect pronounced generation of the nonlinear static azimuthal magnetic field $B_{\phi 2}^{(0)}$. Theoretical quantification and experimental verification of the nonlinear static responses will shed more light on the nature of strong nonlinearities in low-frequency inductively coupled plasmas.

2.4
Optical Emission Spectroscopy of Complex Gas Mixtures

Recent progress in the plasma-aided materials nanofabrication and, more generally, in materials synthesis and processing, stimulates an increasing demand for a thorough study of RF discharges in molecular and reactive gases

[113–115], and reactive gas mixtures [116]. In fact, various surface processing techniques require atomic and molecular neutrals in excited states [117,118]. It is thus important to characterize the reacting species and understand their roles in certain processes without noticeably affecting the discharge performance. The optical emission spectroscopy appears to be an efficient tool for the above purpose, and can yield the kind, temperature, and concentration of the excited species [119].

The purpose of this section is to investigate the optical emission spectra of atomic, molecular, and ionic species in low-frequency, high-density ICP discharges in pure nitrogen, argon gases, and gas mixtures Ar+H_2, N_2+Ar, and N_2+H_2 [94]. We will also discuss how the optical emission spectroscopy (OES) technique can be used to investigate the mode transition and hysteresis phenomena that follow cyclic variations of the RF input power. The results of this section also suggest that the capacity of different gases to sustain nonlinear hysteresis can be quite different. Meanwhile, a combination of the OES and RF current measurements can be a convenient technique to record the turning points for the mode transitions. Measurements of the radial and axial profiles of the OEI of neutral and ionized species can also be indicative to the corresponding spatial distributions of the species' densities in the plasma reactor.

It is interesting that by adding argon or hydrogen to nitrogen plasmas, one can effectively control the optical emission intensities and thus the energy stored by certain states of neutral and ionic nitrogen species. Furthermore, the effect of the working gas composition on the $E \rightarrow H$ transition threshold is detailed and the underlying physics associated with heavy particle kinetics, rearrangement of the electron energy distribution function, and variation of the net RF power dissipated in the plasma, is also discussed.

2.4.1
Optical Emission Spectra and Hysteresis

Optical emission spectra of pure argon, nitrogen, and gas mixtures Ar+H_2, N_2+Ar, and N_2+H_2 have been recorded in the electromagnetic (H) and electrostatic (E) discharge operating regimes. Our focus here is on H-mode discharges, which are most important for applications. However, the optical emission characteristics of E-mode discharges have been measured as well, to illustrate the mode transitions and discharge hysteresis.

Representative OES for the H-mode discharges in pure argon and nitrogen, and gas mixtures are displayed in Fig. 2.14. Reference data [119–121] have been used to identify the species. In the argon discharge, the highest OEIs belong to the wavelength range $\lambda = 415.0$–435.0 nm. The highest intensity corresponds to the 420.07 nm line of the neutral argon (Fig. 2.14(a)). The max-

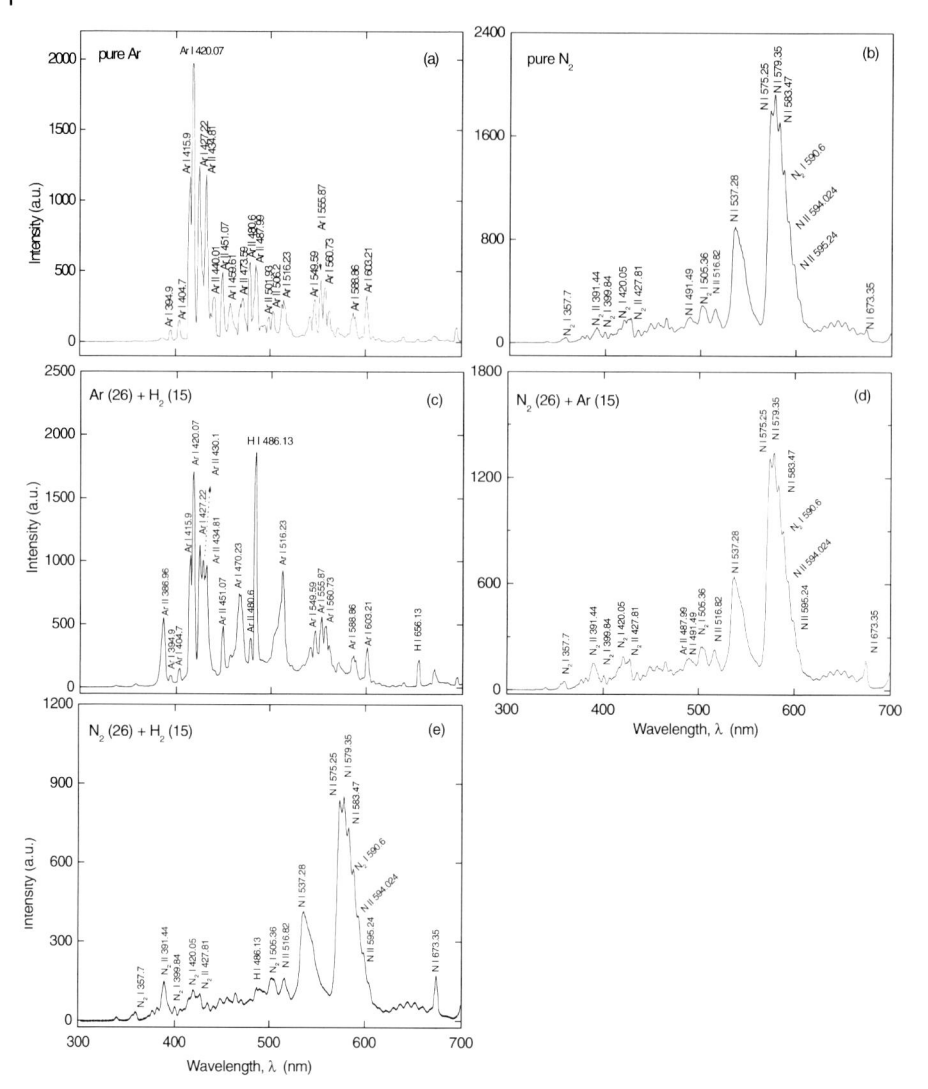

Fig. 2.14 Optical emission spectra of the inductive discharges in pure
(a) Ar ($P_p = 1.17$ kW, $p_0 \sim 28$ mTorr), (b) N$_2$ ($P_p = 1.5$ kW, $p_0 \sim 30$ mTorr), and gas mixtures (c) Ar+H$_2$ ($P_p = 1.5$ kW, gas flow rates: Ar – 26 sccm, and H$_2$ – 15 sccm), (d) N$_2$+Ar ($P_{in} = 1.4$ kW, N$_2$ – 26 sccm, and Ar – 15 sccm), and (e) N$_2$+H$_2$ ($P_{in} = 1.4$ kW, N$_2$ – 26 sccm, and H$_2$ – 15 sccm), respectively [94].

imum intensity for the nitrogen spectrum is in the range $\lambda = 550.0$–600.0 nm, with the highest intensity for the NI$\sim$579.35 nm line of a neutral nitrogen atom (Figs. 2.14(d) and (e)). Comparing the OEIs in Figs. 2.14(d) and (e), one can notice that nitrogen plays a dominant role in plasmas of N$_2$+Ar and N$_2$+H$_2$ discharges.

In Fig. 2.14, the optical emission spectra have been recorded for fixed values of the input power P_{in}. However, variation of the input power can lead to mode jumps and corresponding changes in the OEI. Moreover, under certain conditions the dependence of the OEI on the input power (or coil current) can exhibit hysteresis. It is remarkable that the features of hysteresis are different in pure gases and gas mixtures. To observe mode transitions between the electrostatic and electromagnetic modes, the input power has been smoothly raised from 50 W to 2.5 kW, and then further decreased [94]. In this experiment, the optimal power transfer from the RF generator to the plasma source was controlled by the automatic matching unit, which prevents the adverse discharge-coil mismatch in the process of the mode transitions [122–124].

However, it appears quite effortful to record the exact moment for the mode transition, as the coil/plasma currents change instantaneously during

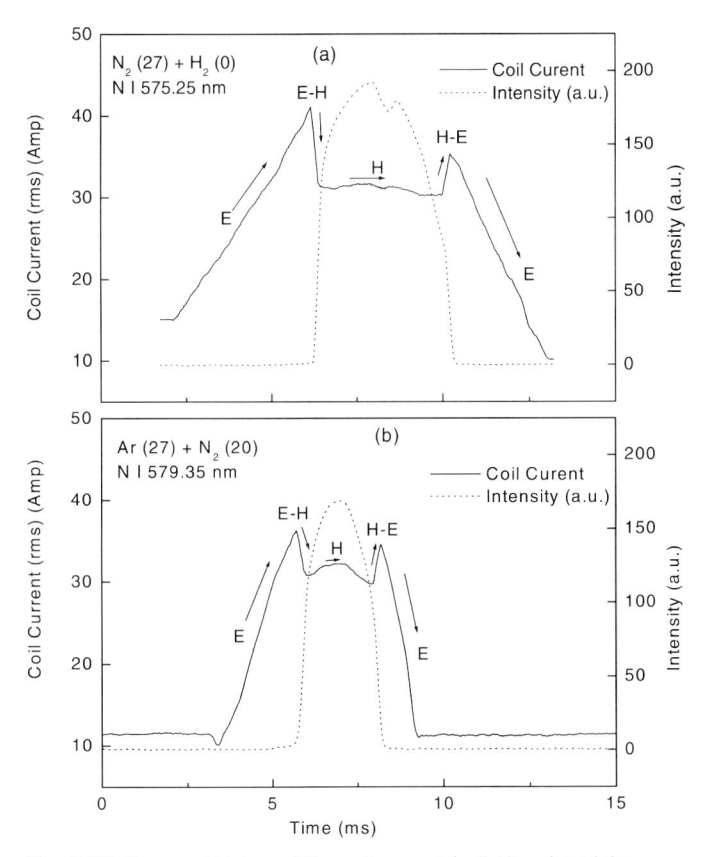

Fig. 2.15 Temporal history of the coil current (solid lines) and the OEI (dashed lines) during the $E \leftrightarrow H$ mode transitions in (a) N_2 (27 sccm, $p_0 = 30.55$ mTorr), and (b) Ar (27 sccm)+N_2 (20 sccm) ($p_0 = 59.21$ mTorr) discharges. The OEI has been recorded for the nitrogen lines (a) NI 575.25 nm and (b) NI 579.35 nm, respectively [94].

the E→H transition. This point has been resolved by recording separately the time histories of the optical emission intensity $\mathcal{J}(t)$ and the coil current $I_c(t)$ [94]. Two examples corresponding to the 575.25 nm line of the atomic nitrogen in pure nitrogen and the 579.35 nm line in Ar+N$_2$ discharges are shown in Fig. 2.15, which elucidates the way of data collection. One can clearly see that in the process of the $E \to H$ transition the coil current diminishes while the OEI raises. The coil current is fairly constant in the H-mode regime, and the optical emission intensity is maximal. The inverse $H \to E$ transition is accompanied by the OEI down-jump and increase of the coil current. This time-resolved technique enables one to resolve the turning points for the mode transitions, obtain the dependence $\mathcal{J}(I_c)$, and investigate the mode transitions and hysteresis in pure argon and nitrogen, as well as in Ar+H$_2$, N$_2$+Ar, and N$_2$+H$_2$ gas mixtures.

2.4.2
$E \to H$ Transition Thresholds

We now turn our attention to the dependence of the $E \to H$ transition thresholds on gas composition. In our analysis of particular reasons leading to certain tendencies in mode transition thresholds, several factors will be considered. Physically, the mode transition threshold is obtained from the balance of powers absorbed by the plasma from the external RF field P_{abs} and lost for maintenance of major elementary processes in a discharge P_{loss}, including electron impact ionization, associative ionization, excitation/de-excitation of higher atomic/molecular levels through electron impact and heavy particle collision processes, escaping of positive ions to the discharge walls through the sheaths to the walls, thermal motion, etc. [15, 125].

In the following, we will comment on relative impacts of heavy particle inelastic collisions, changes in the electron energy distribution functions, and RF power deposition (the latter mostly changing because of variations of the electron number density and skin length), which inevitably accompany variations in the gas feedstock.

Variation of the $E \leftrightarrow H$ transition thresholds $I_{\text{th}}^{E \to H}$ and $I_{\text{th}}^{H \to E}$ with working pressure in the ICPs has been reported previously for 13.56 MHz [110] and 460–500 kHz [91, 93] argon discharges. Similar results recorded in the experiments of Ostrikov et al. [94] are presented in Table 2.1. One can clearly see that $I_{\text{th}}^{E \to H}$ and $I_{\text{th}}^{H \to E}$ diminish with increasing pressure. It has been understood that the ion mobility decreases with p_0, which results in depleted loss of ions to the chamber walls and in excessive RF power which is usually proportional to the plasma density. Thus, the discharge in argon can be sustained at elevated pressures with lower input powers, and, generally, lower coil currents. Note that for the discharge pressure of 33.8 mTorr, which is the argon pressure before addition of any other gases, the $E \to H$ transition can be initiated by an RF current of 27.2 A.

Tab. 2.1 Thresholds of mode transitions in argon [94].

p_0 (mTorr)	$I_{th}^{E \to H}$ (A)	$I_{th}^{H \to E}$ (A)
12	43.9	36.8
16	40.1	34.2
22	33.3	28.1
30	27.9	23.8
33.8	27.2	23.1
56	25.9	21.9
75	24.1	21.2
100	22.9	20.0

Table 2.2 shows the mode transition thresholds in nitrogen discharges. It is seen that in the pressure range below 100 mTorr, the thresholds $I_{th}^{E \to H}$ and $I_{th}^{H \to E}$ increase with working pressure. Our understanding is that the rates of inelastic collisions in nitrogen increase with pressure, with relative effects of quenching of excited molecular states (naturally leading to an increase in the discharge maintenance fields and thus mode transition thresholds) prevailing over the effects of associative and electron impact ionization. The latter two processes usually contribute to lowering thresholds of transitions to discharge modes with higher electron/ion number densities.

Langmuir probe measurements in similar experimental conditions [126] suggest that the effective electron temperature (by definition being $T_{eff} = 2/3 \langle \mathcal{E} \rangle$, where $\langle \mathcal{E} \rangle$ is an average electron energy [15], see also Section 1.1) in nitrogen diminishes with working pressure [126]. This is consistent with the reported "cooling of the EEDF" with increasing nitrogen partial pressures [127]. Thus, the rates of the electron impact processes are brought down, also depleting the electron number density. In this case, it is natural to expect that in the H-mode discharge, P_{abs}, which is a decaying function of n_e [125], will change to enhance RF power deposition to the plasma.

Tab. 2.2 Thresholds of mode transitions in nitrogen [94]

p_0 (mTorr)	$I_{th}^{E \to H}$ (A)	$I_{th}^{H \to E}$ (A)
10	37.9	33.8
15	39.1	35.2
20	41.3	37.3
30.55	42.8	38.3
38	45.0	40.6
57	47.1	42.5
64	48.4	44.3
80	50.2	47.5

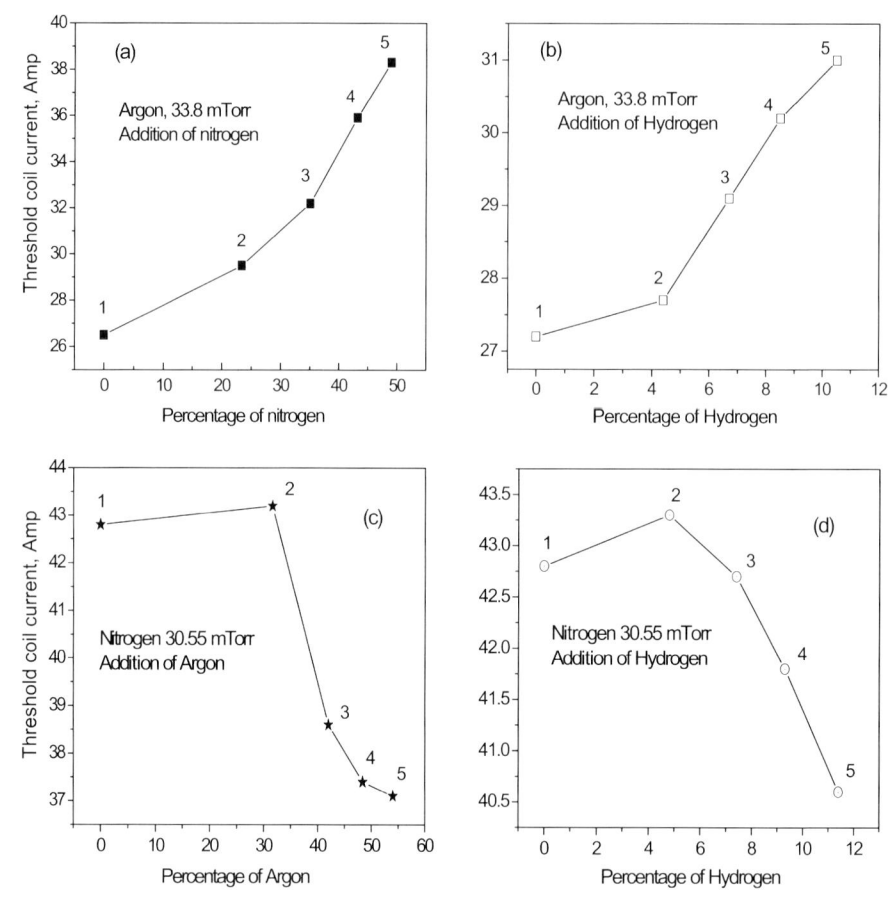

Fig. 2.16 Threshold coil current for the $E \rightarrow H$ transition, in the 33.8 mTorr argon discharge (point 1) with nitrogen addition ((a), points 2–5). Points 2–5 correspond to N_2 partial pressures 10.3, 18.26, 25.7, and 32.42 mTorr, respectively. In (b), points 2–5 stand for additional inlet of hydrogen with partial pressures 1.55, 2.43, 3.15, and 3.95 mTorr, respectively. (c) and (d) show similar dependences for a nitrogen discharge at 30.55 mTorr with additions of argon (c) and hydrogen (d). Partial pressures of hydrogen are the same as in (b). In (c), points 2–5 correspond to partial pressures of argon 14.1, 22.1, 28.66, and 35.9 mTorr, respectively [94].

Figure 2.16(a) shows that addition of nitrogen to the 33.8 mTorr discharge in argon results in the rise of the $E \rightarrow H$ transition threshold from 27.2 A in pure Ar to $\sim$ 38 A in 51 % Ar + 49 % N_2 mixture. We note that the electron collisional inelastic loss rates in nitrogen are higher than those in argon and hydrogen [127–131]. As can be seen from Fig. 2.16(a), an increase in pressure results in a dramatic increase in the threshold coil current. Hence, the H-mode threshold follows the "nitrogen" scenario, implying higher discharge maintenance fields and importance of heavy particle inelastic collisions.

It is remarkable that the rate of the threshold elevation also increases (the slope of the curve in Fig. 2.16(a)) with relative percentage of nitrogen, which means that the role of "argon" effects diminishes. Furthermore, as a result of addition of nitrogen to argon, the electron number density and temperature diminish ("cooling of the EEDF") [126]. Thus, the rates of the electron impact processes also diminish, which leads to depleted electron/ion pair production, and hence, elevated mode transition thresholds.

On the other hand, the depleted electron number density improves the RF field penetration into the chamber and the net value of the power transferred to the plasma electrons, which would be favorable for lowering the minimum current for transition to the H-mode discharge. However, the latter tendency has not been observed in original experiments [94], whence one can infer that in Ar+N_2 mixtures (with gradually increasing proportion of nitrogen) relative impact of heavy particle collisions and rearrangement of the EEDF prevail over the electromagnetic power absorption effects.

In Fig. 2.16(b), as percentage of hydrogen increases, $I_{th}^{E \to H}$ also increases. It is common that addition of small ($< 10\%$) number of hydrogen to nitrogen reduces population densities of vibrationally excited states and electronic metastable states [129]. Similarly, one can expect that argon metastables are also quenched by H_2, and the mode transition threshold increases, in particular, due to weakening of the stepwise ionization processes that can under certain conditions contribute up to 40% of the electron/ion number densities [128]. Meanwhile, addition of hydrogen elevates the total gas pressure, which is expected, according to the "argon" scenario (see Table 2.2), to diminish. However, this does not happen in the case reported, which means that variation of the working pressure is clearly not enough to sustain any remarkable mode transition threshold variations.

One can thus presume that heavy particle quenching effects dominate over total pressure variation effects. It is notable that increase in the threshold is not as large as it used to be in the Ar+N_2 case (Fig. 2.16(a)) since inelastic loss rates in hydrogen are much smaller than those in nitrogen [127]. However, a clear tendency for saturation (points 3–5 in Fig. 2.16(b)) indicates that additional inlet of hydrogen can lead to the subsequent change in the character of variation of thresholds with H_2 percentage. Unfortunately, due to relatively small hydrogen inlets, this possible effect is yet to be verified. Likewise, additions of hydrogen normally "warm up the EEDF" [127, 130], with consecutive enhancement of the electron impact ionization, which should be regarded as an additional argument that $I_{th}^{E \to H}$ can be diminished with larger additions of hydrogen to argon.

Figure 2.16(c) displays variation of the $E \to H$ thresholds in the nitrogen-based discharge with addition of argon. One can see that with argon proportions up to 32%, threshold coil current in the mixture slightly increases followed by a noticeable drop afterwards. Apparently, there is a competition between the "nitrogen" and "argon" scenarios of development of the mode transition thresholds with addition of argon. At the beginning, the "nitrogen" tendency controlled by a dominance of heavy particle inelastic collisions prevails, i.e., increasing threshold with pressure, which changes noticeably in Fig. 2.16(c). Afterward, changes in pressure become sufficient to affect ion mobility, and hence P_{loss} and transition thresholds.

Meanwhile, as Langmuir probe measurements in Ar+N$_2$ mixtures in a similar setup [126] suggest that additional inlet of argon is accompanied by an increase in the electron number density and diminishing of the electron temperature. Therefore, the rates of the electron impact ionization and associative ionization in nitrogen decrease so that electron/ion creation processes slow down, which is reflected by a tendency of the threshold to saturate. Meanwhile, an increase in n_e leads to shrinking of the skin layer and depletion of the actual power absorption in the plasma. This naturally should result in somewhat elevated mode transition thresholds. Thus, rearrangement of the EEDF and changes in power deposition act as factors limiting subsequent depletion of the $E \rightarrow H$ thresholds.

Variation of $I_{\text{th}}^{E \rightarrow H}$ with hydrogen addition is shown in Fig. 2.16(d). At small inlets of hydrogen, the threshold increases mostly because of the dominant heavy particle quenching effects [129]. Meanwhile, assuming that addition of hydrogen to nitrogen usually elevates the average electron energy (and hence T_{eff}), we arrive at a conclusion that the rates of the electron impact and associative ionization, as well as dissociation, increase, and the total electron number density should also become higher. An explanation of diminishing the maintenance electric field in a DC discharge as the percentage of H$_2$ grows ($> 10\%$), includes the EEDF rearrangement effects, as well as the effects of temperature and density of neutrals [129,130]. An increase of the electron number density, in turn, diminishes P_{abs} and RF field localization. However, this factor is believed to be of little importance since the actual addition of hydrogen was quite small.

We now turn our attention to radial profiles of the OEI in the low-frequency inductively coupled plasmas [94]. The radial profiles are monitored by inserting a collimator in eight radial positions through the bottom portholes in the chamber. Radial distribution of the OEI in pure argon discharge is given in Fig. 2.17. One can observe that the intensity maximum of most of argon lines is located few centimeters apart from the chamber axis. This tendency is most apparent for neutral argon lines, especially for the Ar I 420.07 nm line (Fig. 2.17). The radial OEI profiles of the argon ions appear to be more flat than those of the neutral species. We also remark that similar tendencies were observed in nitrogen plasmas [94]. The axial profiles of the optical emission (studied for seven available positions in the side flange) also show an interest-

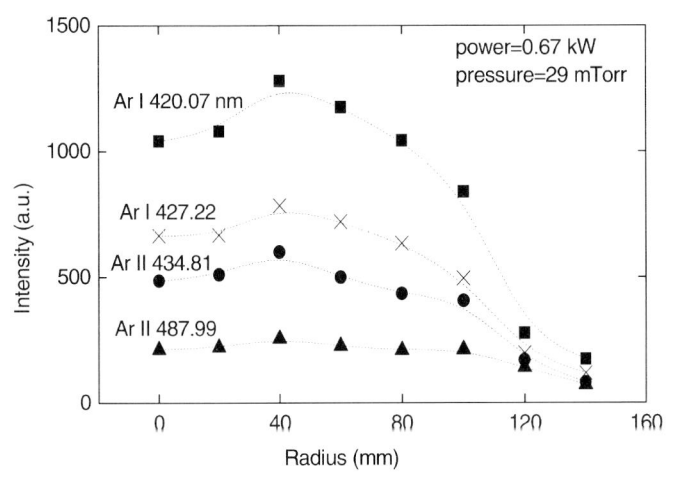

Fig. 2.17 Optical emission intensities of neutral and ionized argon atoms versus radial distance for Ar discharge with $P_0 = 29.3$ mTorr and $P_p = 670$ W [94].

ing dependence on the discharge parameters. Interested reader can be referred to the original work [94].

Therefore, the capacity of several gases/gas mixtures for hysteresis appears to be different. In fact, in pure nitrogen discharge the $H \rightarrow E$ transition is smooth [94]. Indeed, the inverse transition is not necessarily discontinuous [124], although in certain cases it is accompanied by abrupt jumps in circuit and plasma parameters [110, 122]. In particular, the discontinuous $H \rightarrow E$ transition is characteristic to discharges in pure argon [91].

It should also be noted that the pressure dependence of the OEI loop width can provide valuable information about the mechanisms of hysteresis [124]. It is interesting that in argon discharge the hysteresis loop width diminishes with pressure ($p_0 < 100$ mTorr) [94]. However, in nitrogen plasmas, pressure variations result in much weaker changes in the OEI loop width. Thus, in nitrogen inductively coupled plasmas the pressure does not seem to be a dominant factor for the hysteresis. It is also instructive to mention that the electron density n_e is an increasing smooth function of p_0 in argon ($p_0 < 200$ mTorr). In nitrogen discharge, the plasma density rises in the range $p_0 < 30$–40 mTorr, and then decreases. Note that the OES of nitrogen discharge have been recorded in the pressure range 45–55 mTorr, where slopes of $n_e(p_0)$ appear to be different in Ar and N_2 plasmas. Furthermore, a tendency of the hysteresis loop to shrink with pressure is different from that in the internal-coil 13.56 MHz pulsed-regime inductive discharge [124]. This discrepancy can be attributed to a difference in discharge operating conditions and coupling regimes. Thus, a detailed investigation of the effect of frequency on multistability and hysteresis in Ar, N_2, and their mixtures seems to be a worthwhile exercise.

The results discussed in this section suggest that the intensity of optical emission in the nitrogen discharge can be controlled by adding certain admixtures of Ar and H_2. Hence, power accumulated in certain nitrogen excited states can be varied. Indeed, this can be important in processes where composition of excited molecules/atoms plays the role. Furthermore, plasma applications for nanoscale processing require precise control of a number of excited species in certain states. As was mentioned in Chapter 1, the energy of the plasma species in the vicinity of nanostructured surfaces is critical for the adequate insertion of the plasma-generated building units into nanoassemblies being synthesized [36].

An interesting conclusion is that one can reduce the intensity of the optical emission from molecular nitrogen species by adding hydrogen to the discharge. In this case the contribution of electronically excited and vibrational states of N_2 rapidly decreases when either H_2 or O_2 is added to the discharge, and the involved excited states are then destroyed by H_2, H, O_2, O, and NO species [132,133]. Alternatively, addition of easily ionized argon enhances the excitation of nitrogen via a number of electron-impact effects.

We emphasize that some of the nanofabrication processes discussed in this monograph use nitrogen plasmas, e.g., plasma-aided synthesis of $Al_xIn_{1-x}N$ quantum dot arrays by co-sputtering of Al and In targets in $Ar+N_2$ plasmas (see Chapter 5 for details). Knowledge of the effect of different process parameters on the energy of the main building units will ultimately enable one to synthesize various nanostructures in a controlled fashion. From this viewpoint, low-temperature plasmas can be regarded as a useful nanofabrication environment, where the energetic states and reactivity of the species can be managed and precisely controlled, which is difficult, if possible at all, in conventional chemical vapor deposition and some other techniques.

2.5
Modeling of Low-Frequency Inductively Coupled Plasmas

In this section, we focus on numerical modeling of low-frequency (~ 460 kHz) inductively coupled plasmas generated in a cylindrical metal chamber by an external flat spiral coil [95]. A two-dimensional fluid approach is used to compute the spatial profiles of the plasma density, electron temperature, and excited argon species for different values of the RF input power and working gas pressures.

This model allows one to achieve a reasonable agreement between the computed and experimental data, the latter including the electron number densities and temperatures, electron energy distribution functions, and optical emission intensities of the abundant plasma species in low-to-intermediate pressure argon discharges. The neutral gas temperature, one of the most important factors in the plasma-assisted synthesis of various carbon-based nanostructures, also exerts a noticeable effect on the plasma parameters, especially at elevated levels of RF power input.

2.5.1
Basic Assumptions

The real vacuum chamber of the ICP plasma reactor (Fig. 2.1) is modeled by considering a metal cylinder of the inner radius R and length L, with a dielectric disk of width d and permittivity ε_d atop. The components of the electromagnetic field are calculated assuming that the chamber is uniformly filled by the plasma with the electron/ion number density equal to the spatially averaged plasma density $\bar{n}$.

The main results of this modeling are relevant to the pressure range 20–100 mTorr, where the contribution of nonlocal collisionless heating is usually smaller than that of collisional heating [134]. The model of the RF power deposition is thus solely based on the assumption of the prevailing Ohmic (collisional) heating.

To simulate the real environments of some of the plasma processing experiments [93] the study has been carried out for elevated RF powers absorbed in a plasma column $P_{\text{in}} = 0.6$–1.4 kW. Consequently, the density of electrons is large, and the electron–electron collisions are expected to strongly affect the EEDFs.

2.5.2
Electromagnetic Fields

The components of the transverse-electric (TE) electromagnetic field in the chamber fully filled by the uniform plasma with the plasma density $\bar{n}$ are

$$B_z^p = A \sum_{n=1}^{\infty} \frac{\alpha_{1n} \kappa_n^H}{D_n(TE)} \zeta_n^H(z) J_0(\kappa_n^H r) \tag{2.5}$$

$$B_r^p = A \sum_{n=1}^{\infty} \frac{\alpha_{1n} \gamma_n^H}{D_n(TE)} \eta_n^H(z) J_1(\kappa_n^H r) \tag{2.6}$$

and

$$E_\phi^p = i\frac{\omega}{c} A \sum_{n=1}^{\infty} \frac{\alpha_{1n}}{D_n(TE)} \zeta_n^H(z) J_1(\kappa_n^H r) \tag{2.7}$$

where

$$\zeta_n^H(z) = \sinh[\gamma_n^H(L-z)]/\cosh(\Gamma_n^H d)\cosh(\gamma_n^H L)$$
$$\eta_n^H(z) = \cosh[\gamma_n^H(L-z)]/\cosh(\Gamma_n^H d)\cosh(\gamma_n^H L)$$
$$D_n(TE) = \Gamma_n^H \coth(\Gamma_n^H d) + \gamma_n^H \coth(\gamma_n^H L)$$
$$\alpha_{1n} = [8\pi/cR^2 J_2^2(\rho_{1n})] \int_0^R r J_1(\kappa_n r) dr$$

ϕ is the azimuthal angle, $\Gamma_n^H = [(\kappa_n^H)^2 - (\omega/c)^2 \varepsilon_d]^{1/2}$ and $\gamma_n^H = [(\kappa_n^H)^2 - (\omega/c)^2 \varepsilon_p]^{1/2}$ are the inverse RF field penetration lengths into dielectric and plasma, respectively [91]. Furthermore, $J_j(x)$ is a Bessel function of the jth order, $J_1(\rho_{1n}) = 0$, $\kappa_n^H = \rho_{1n}/R$, $\varepsilon_p = 1 - \omega_{pe}^2/[\omega(\omega + i\nu_{en})]$ is the dielectric constant of the uniform plasma and $\omega_{pe} = \sqrt{4\pi \bar{n} e^2/m_e}$ is the electron Langmuir frequency. Here, ν_{en} is the electron-neutral collision frequency for momentum transfer, A is a constant, and e and m_e are electron charge and mass, respectively.

The averaged plasma density is computed from the particle and power balance equations (see the following subsection) and substituted into the electromagnetic fields (2.5)–(2.7), which are continuously updated when $\bar{n}$ varies.

2.5.3
Particle and Power Balance

Since the plasma column and inductive coils are fairly uniform in the azimuthal direction, the problem is two-dimensional and the plasma parameters depend on r and z only. It is assumed that the ion temperature T_i is equal to the temperature of the working gas T_g. The latter was a variable parameter in the computations. The plasma is treated within the ambipolar model that assumes the plasma quasineutrality and the equality of electron and ion fluxes. The effect of negative ions is neglected so that the overall charge neutrality condition can be written as $n_e = n_i \equiv n$, where n_e is the electron number density.

Accordingly, the particle balance equation for the electrons or ions is

$$\partial n/\partial t + \vec{\nabla} \cdot (n\vec{v}) = n\nu^i \qquad (2.8)$$

where $\vec{v}$ is the electron or ion fluid velocity, and ν^i is the ionization rate. We recall that in the ambipolar diffusion-controlled regime [15]

$$\vec{v} \approx -\nabla(nT_e)/nm_i \nu_{in}$$

where $\nu_{in} \approx \sqrt{v^2 + v_{Ti}^2}/\lambda$ is the ion-neutral collision frequency, m_i is the ion mass, and $T_i \ll T_e$. Here, $\lambda = 1/(n_N \sigma_{in})$ is the ion mean free path, where

$\sigma_{in} = 8 \times 10^{-15}$ cm^2, $v_{Ti} = \sqrt{8T_i/\pi m_i}$ is the average ion thermal velocity, and $n_N = p_0/T_g$ is the number density of neutrals, and T_e is the electron temperature. We note that in computation T_e is self-consistently derived from the set of electrodynamic, plasma particle and RF power balance equations. On the other hand, in the experiment, T_e is routinely measured through the averaged electron energy $\langle \mathcal{E} \rangle$ as $T_e = 2/3 \langle \mathcal{E} \rangle$ following Druyvesteyn's technique [15].

The RF power balance in the discharge is described by [135]

$$\frac{3}{2} n_e \frac{\partial T_e}{\partial t} + \vec{\nabla} \cdot \vec{q}_e \approx -n_e I_e + S_{\text{ext}} \tag{2.9}$$

where I_e is the collision integral for the electrons, and $\vec{q}_e \approx -(5/2 - g_u) n_e T_e / (m_e v_{en}) \vec{\nabla} T_e$ is the heat flux density with $g_u = (T_e/v_{en}) \partial v_{en}/\partial T_e$. The term S_{ext} denotes the Joule heating of the electrons by the RF field,

$$S_{\text{ext}} \approx n v_e m_e u_{\text{osc}}^2$$

where $u_{\text{osc}} \approx |eE_\phi^p|/[2m_e^2(\omega^2 + v_{en}^2)]^{1/2}$ is the time-averaged oscillation velocity of the electrons in the RF field. The equilibrium state corresponds to setting $\partial_t = 0$ in (2.8) and (2.9). Our approach is valid for fixed RF power absorption in the plasma column, with

$$P_{\text{in}} = \int_0^L \int_0^R S_{\text{ext}} 2\pi r \, dr \, dz$$

which also yields the constant A in (2.5)–(2.7). The rate of the electron-neutral collisions v_{en} that depends on T_e can be determined by using the elastic scattering rate coefficients [95, 136].

Assuming that the excitation and ionization of the neutral gas proceed mainly via the electron impact processes, one infers that the collision integral I_e is equal to the average power lost by an electron colliding with a neutral. Accordingly,

$$I_e \approx (3m_e/m_n) T_e v_{en} + \sum_j v_j \mathcal{E}_j + v^i \mathcal{E}^i$$

where m_n is the mass of the neutral, v_j is the excitation rate from the ground state to level j with a threshold energy $\mathcal{E}_j$, and $\mathcal{E}^i$ is the ionization threshold energy. Stepwise ionization and excitation processes are not accounted in the present simulation.

For argon gas, the rates for ionization and excitation to the states 4s and 4p are [128]

$$v^i = 2.3 n_N \times 10^{-8} T_e^{0.68} \exp(-\mathcal{E}_i/T_e) \quad \text{s}^{-1}$$

$$\nu_{4s} = 5.0 n_N \times 10^{-9} T_e^{0.74} \exp(-\mathcal{E}_{4s}/T_e) \quad \text{s}^{-1}$$

and

$$\nu_{4p} = 1.4 n_N \times 10^{-8} T_e^{0.71} \exp(-\mathcal{E}_{4p}/T_e) \quad \text{s}^{-1}$$

where n_N is in cm^{-3}, T_e in eV, $\mathcal{E}_i = 15.76$ eV, $\mathcal{E}_{4s} = 11.5$ eV, and $\mathcal{E}_{4p} = 13.2$ eV. The above rate coefficients have been calculated by assuming the EEDFs Maxwellian [128].

We now consider the boundary conditions for integrating Eqs. (2.8) and (2.9) [95]. Because of symmetry, the radial gradients of the electron temperature and density are equal to zero at the chamber axis ($r = 0$). At the column edge ($r = R$) the radial component of the fluid velocity satisfies the well-known Bohm sheath criterion $v_r(R, z) = \sqrt{T_e(R,z)/m_i}$ [15]. Similarly, at the side walls $z = z_s$, where $z_s = 0$ or L, one has $v_z(r, z_s) = \sqrt{T_e(r, z_s)/m_i}$. Likewise, the boundary conditions for heat flow are [135]

$$q_{er}(R, z) = T_e(R, z)(2 + \ln \sqrt{m_i/m_e}) \times$$
$$n(R, z)\sqrt{T_e(R,z)/m_i} \qquad (2.10)$$

and

$$q_{ez}(r, z_s) = T_e(r, z_s)(2 + \ln \sqrt{m_i/m_e}) \times$$
$$n(r, z_s)\sqrt{T_e(r, z_s)/m_i} \qquad (2.11)$$

where q_{er} and q_{ez} are the radial and axial components of the heat flux density, respectively.

2.5.4
Numerical Results

The set of equations (2.8)–(2.9) has been solved numerically. The details of the codes and numerical procedures can be found elsewhere [95, 137, 138]. The profiles of the electron density, temperature, and ion velocity are computed from (2.8)–(2.9). The resulting electron density distribution is used to compute the spatially averaged plasma density $\bar{n}$. Thereafter, the latter is substituted into (2.7) to obtain the azimuthal electric field $E_\phi^p(r, z)$. The computation is initialized by using profiles of n, $\vec{v}$, and T_e estimated from less accurate analytical or computational results. The computation proceeds via a number of temporal steps and is terminated when a steady state is reached.

We now comment on the spatial profiles of the electron number density and temperature in the low-frequency ICP. The computed 2D profiles of the electron number density and temperature at $p_0 = 28.5$ mTorr and $P_{\text{in}} = 612.4$ W are shown in Fig. 2.18. One can see that the electron temperature is maximal

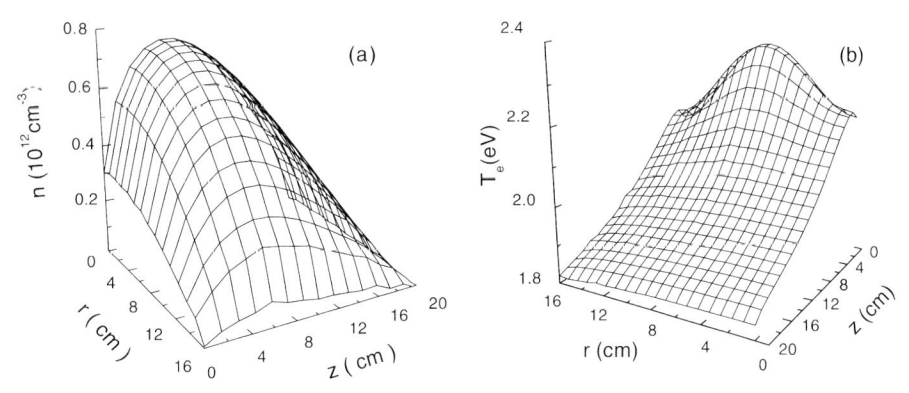

Fig. 2.18 Numerical profiles of the plasma density (a) and electron temperature (b) at $p_0 = 28.5$ mTorr, $P_{in} = 612.4$ W, and $T_g = 543$ K [95].

near the chamber top at $r \approx 8$ cm. The profile of T_e is linked to the spatial distribution of E_ϕ^p (2.7). Since the electron temperature is somehow elevated near the fused silica window, the maximum of the plasma density is shifted approximately 3 cm upward ($z \approx 7$ cm) from the central cross-section of the chamber center ($z_0 = 10$ cm). We note that in the classical case of uniform distribution of T_e, over the entire chamber, one would expect a cosine-like solutions for the plasma density, with the maximum at $z = z_0$ [15]. The computed electron number densities and temperatures have been compared with the ones measured by the Langmuir probe, with the tip positioned at $z = 5.6$ cm and $r = 4.0$ cm and found in a fair quantitative agreement.

At this stage, it is imperative to mention that the computation has been carried out under two different boundary conditions for the electron heat flux:

(i) the electron heat flux toward the boundary is governed by (2.10) and (2.11);

(ii) the electron temperature gradient vanishes at the boundary ($\nabla T_e = 0$). This boundary condition has been commonly used in simulations alongside with (2.10) and (2.11) [135, 139].

Interestingly, the best agreement with the experiment is achieved by carefully accounting for nonvanishing electron heat fluxes, thus making the boundary conditions (i) more appropriate for modeling inductively coupled plasmas at elevated powers. Specifically, application of (ii) instead of (i) would lead to higher values of the electron number density. Physically, the electron heat fluxes are capable of removing a part of the input power, which could have otherwise been gainfully used for additional electron–ion pair creation in the plasma bulk.

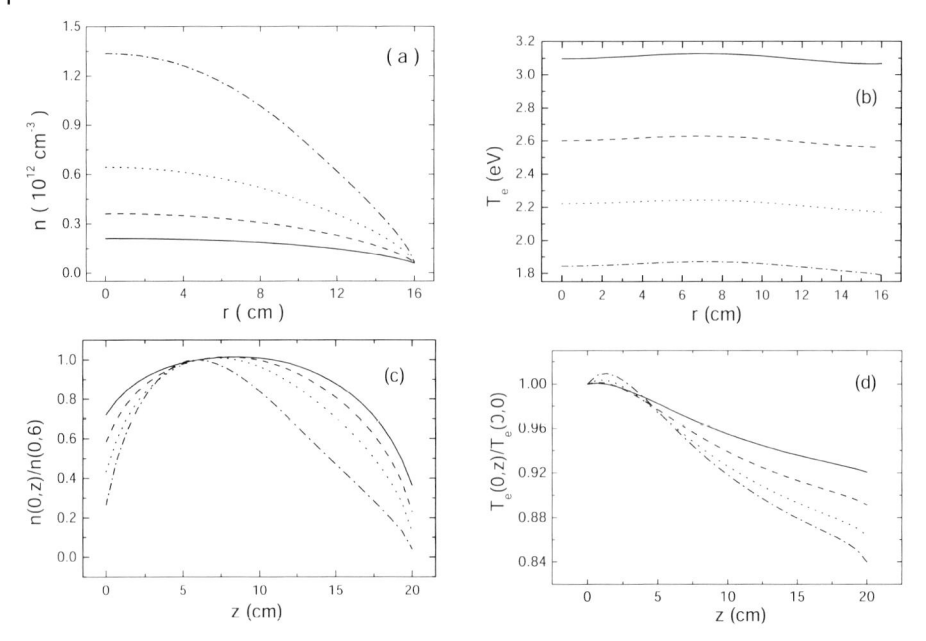

Fig. 2.19 Radial electron density (a) and temperature (b) profiles at $z = 6$ cm; the normalized axial density (c) and temperature (d) profiles at $r = 0$. Solid, dashed, dotted, and dash-dotted curves correspond to $p_0 = 5$, 10, 20, and 50 mTorr, respectively. Other parameters are the same as in Fig. 2.18 [95].

We now turn our attention to the effect of the working gas pressure on the plasma parameters and power loss in the discharge. The radial profiles (at $z = 6$ cm) of the electron number density and temperature computed for four different values of p_0 are presented in Figs. 2.19(a) and (b), respectively. The normalized axial profiles of the plasma density and electron temperature at $r = 0$ are shown in Figs. 2.19(c) and (d), respectively. The plasma density in Fig. 2.19(c) is normalized to its value at $z = 6$ cm, whereas the electron temperature is normalized to T_e calculated at $z = 0$.

It is clear from Fig. 2.19 that the working gas pressure strongly affects the plasma density and electron temperature. We note that the Langmuir probe data suggest that within the pressure range considered, the electron density n_e is proportional to the gas pressure for a given electron temperature. Hence, an increase of the gas pressure results in elevation of the average plasma density in the plasma column and in diminishing of the electron temperature. It is also seen that the axial uniformity of the electron temperature is better at lower pressures. Furthermore, the nonuniform profiles of T_e affect the electron density distribution in the chamber (Figs. 2.19(a) and (c)). Indeed, the density peak shifts toward the chamber top as the gas pressure increases. At

$p_0 = 5$ mTorr, the peak is close to the discharge center ($z \approx 10$ cm), which is a clear indication of the ambipolar-diffusion controlled regime at low discharge pressures.

However, at $p_0 = 50$ mTorr the maximum of the plasma density is shifted approximately 4 cm toward the chamber top. This can indicate a possible gradual onset of a different particle loss mode, similar to what has been reported elsewhere [95] for the elevated RF powers. The gas pressure also controls the RF power deposition process by affecting the average power loss per electron–ion pair.

An increase of the neutral gas temperature exerts an almost similar effect on the plasma parameters as a decrease of the working gas density. This can best be understood by noting the apparent link $p_0 = n_N T_g$, which means that the fixed pressure conditions require that the neutral gas density, which enters the expressions for the most of the reaction rate coefficients, has to diminish when T_g rises [95].

We now examine the spatial distribution of the excited argon atoms in the $3p^5 5p$ configuration. The profile of the excited atom density $n^*(r,z)$ can be calculated from [140]

$$n^*(r,z) \sim n(r,z) \nu^*(r,z) \tag{2.12}$$

where $\nu^*(r,z) = \nu_0^*(r,z) \exp(-U^*/T_e)$ is the excitation rate with $U^* \approx 14.5$ eV being the threshold energy for the excited level. Generally, ν_0^* is a slowly varying function of T_e [141]. Thus, evaluating $n^*(r,z)$, one can assume that ν_0^* does not depend on r and z.

The resulting spatial distributions of the excited argon species are governed by the electron density and temperature profiles. Remarkably, the radial profiles of the excited atoms are hollow near the chamber top. Meanwhile, the maximum of the OEI shifts toward the chamber axis as the axial position z increases. Likewise, when the gas pressure increases, the ratio of $n^*(r=0)/n^*_{\text{rmax}}$ decreases, where n^*_{rmax} is the maximal density of the excited species along the radius.

We note that the radial OEI profiles can be related to the integral (along the chamber axis) of the optical emission collected by the optical probe via the collimator positioned in 8 portholes in the chamber bottom plate. For this purpose, the resulting local density of the excited species (2.12) has been integrated along the z direction. Further details of the OES setup and collection of the emission can be found elsewhere [91, 94]. The emission of the 420.07 nm argon line is due to the electron transitions from 5p onto 4s levels. The radial distribution of the emission intensity is shown in Fig. 2.20(a), which reveals a good consistency of the computation and experimental results. Similarly, integrating $n^*(r,z)$ in the radial direction, the axial distribution of the intensity can be computed. Experimentally, the optical probe in this case needs to be

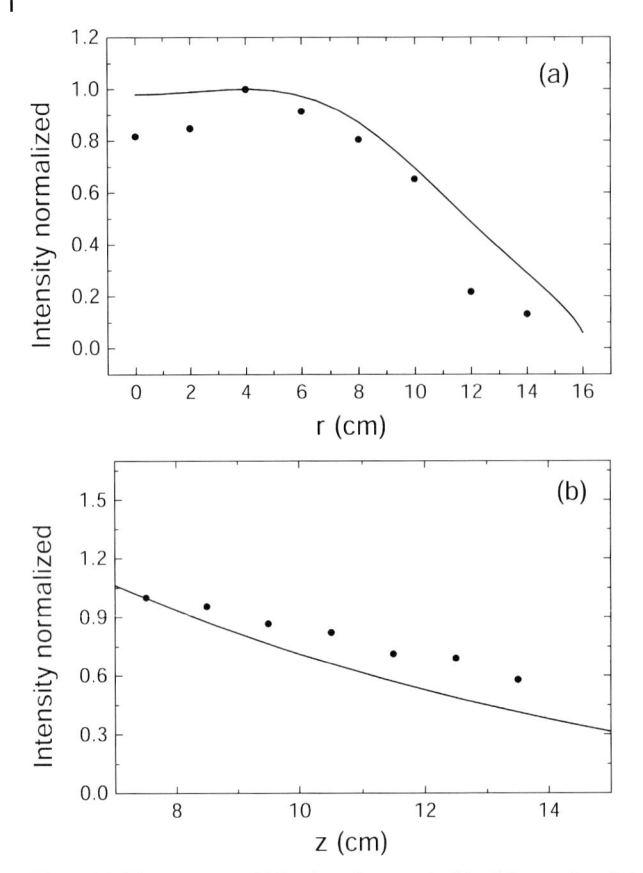

Fig. 2.20 The measured (dots) and computed (solid curve) radial ((a), $p_0 = 29.3$ mTorr, $P_{in} = 536$ W) and axial ((b), $p_0 = 40$ mTorr, $P_{in} = 960$ W) profiles of the optical emission intensity of the 420.07 nm atomic argon line [95]. In both cases $T_g = 543$ K.

positioned in the seven available portholes in the side observation port of the plasma reactor in Fig. 2.1.

Figure 2.20(b) presents the comparison of the axial OEI profiles obtained experimentally and numerically. One can notice a remarkable agreement of the calculated emission intensities with the experimental data. However, a minor discrepancy can be seen in the radial profile in the vicinity of the discharge center. We should also note that the experiment reveals that the OEI dips ($\approx 20\,\%$ less than the maximal value) near the chamber axis. The computation results, correctly following the trend, suggest a less remarkable diminishing of the OEI near the chamber axis. The deviation of the computed OEIs from the experimental data is certainly within the accuracy of the model that assumes independence of v_0^* on r and z.

Reliable knowledge of the neutral gas temperature and heat fluxes in plasma processing discharges is becoming a matter of outmost importance for a number of applications. In particular, recent results on the growth of nanostructured silicon-based films (see Chapter 1 for details) have convincingly demonstrated the neutral gas temperature as a critical factor in management of hydrogenated silicon nanoparticles in γ (powder-generating) regime or in the plasma-assisted fabrication of various carbon nanostructures. Relevant information can be found elsewhere in this book. For discussion of the relevance of this 2D fluid simulation to the experimental results and limitations of the model, refer to the original article [95].

The fairly accurate agreement between the numerical and experimental results confirms the viability of the 2D fluid discharge model in simulating the major parameters of low-frequency inductively coupled plasmas. The model can further be improved by involving both local and nonlocal kinetic approaches, stepwise excitation and ionization processes, complex gas chemistries including radicals, molecular complexes, and negative ions, details of the process (e.g., substrate bias, power, and mass transport in near-substrate areas), as well as the effects of the near-wall sheath/presheath areas. All these effects are to a certain extent critical to achieve the best results of the plasma-aided fabrication at nanoscales.

2.6
Concluding Remarks

We now discuss the main features of low-frequency inductively coupled plasmas and their applicability for industrial plasma processing. Firstly, operating in the two (E and H) modes, the LFICP plasma source (Fig. 2.1) simultaneously embodies the asymmetric capacitive and flat spiral coil inductive plasma sources. Depending on a specific problem, the operation regime can be selected accordingly. For instance, the deposition rates, species composition and reactivity, neutral gas temperature, electron/ion temperatures, etc., are quite different in the electrostatic and electromagnetic discharge regimes.

In the electromagnetic mode, the LF ICPs feature uniform, large-area, high-density plasmas with low sheath potentials near the substrate surface, independent control of the plasma density and the ion energy, high power transfer efficiency, low circuit loss and easy handling, and stable operation in a wide range of filling gas pressures. We need to stress that high density, uniform plasmas can be generated *without* any external magnetic confinement. This feature is highly desirable for low-damage, materials processing in microelectronics [142, 143]. In the electrostatic discharge mode, lower density plasmas, with higher electron temperatures and plasma potentials are produced. We note that the ICP features excellent uniformity of the electron/ion number density through the entire discharge cross-section and volume in the H-regime, and high cross-sectional uniformity in the E-mode.

It is remarkable that highly uniform, high-density (up to $n_e \sim 9 \times 10^{12}$ cm^{-3} in argon) plasmas can be produced in low-pressure discharges with moderate (in most cases below 1 kW) RF powers. It is worthwhile to mention that the LFICP plasmas can be sustained in a wide pressure range without any Faraday shield or an external multipolar magnetic confinement, and exhibit very high RF power transfer efficiency.

In the electromagnetic (H) mode, the ICP source of Section 2.1 features a high level of uniformity over large processing areas and volumes, low electron temperatures, and plasma potentials. The low-density, highly uniform over the ($z = $ const) cross-section, plasmas with high electron temperatures and plasma and sheath potentials are characteristic to the electrostatic regime. Both discharge operation modes offer great potential for various plasma processing applications, including plasma-assisted synthesis of nanostructures and nanostructured materials.

We recall that high uniformity of ions and active species, high product yield with low damage, process selectivity, and reproducibility are the common requirements for plasma processing [15]. Nanoscale fabrication puts forward a few more essential requirements on dosing the release of the necessary species from the plasma into the nanostructured surface being grown or processed. Moreover, as has been discussed in previous sections, such species should find themselves in certain energetic and chemical states that determine their reactivity. Moreover, the building units should deposit onto the surface with the energy just sufficient for their structural incorporation into the nanoassemblies being synthesized and do not cause any structural damage [36].

From the above consideration, we can argue that low-frequency inductively coupled plasmas with an external planar coil have proven to meet the most essential requirements specified above [93]. This is why the low-frequency plasma devices have recently become very attractive as efficient sources of industrial plasmas. As has been evidenced by numerous works [91–94, 98, 102, 103, 144], the LF ICPs possess a number of indisputable advantages that make them especially useful as prototypes of commercial large-area plasma reactors.

In particular, the low-frequency operation offers several practical advantages including low skin-effect-related circuit loss, easy control of the matching unit, high power transfer efficiency, easy diagnostics, and low voltage across the coil. The equivalent circuit analysis [102] shows that capacitive RF field is too weak to cause any significant sputtering of the quartz window, which can be potentially damaging in many plasma processing applications that require high purity of source material. Furthermore, lowering of the operating frequency enables one to make the RF wavelength much longer than the coil length and eliminate the standing-wave effects peculiar to 13.56 MHz ICPs. This provides a viable practical solution for upscaling of the inductively coupled plasma reactors without affecting the uniformity of electron and ion number densities [98, 108].

Another important feature of the LF ICP discharge is the wide range of operating pressures (from fractions of mTorr to a few Torr), which is very promising for multifunctional semiconductor processing. Indeed, the low pressure regime is favorable for deep-micron etching whereas higher pressures are ideal for photo-resist removal [142]. The possibility of multistage processing in broad pressure ranges is an attractive factor for the development of new recipes and specifications for integrated nanostructure synthesis and device assembly processes that include substrate pretreatment, nanopatterning, nanostructure deposition and postprocessing, removal of unwanted material, and other high-precision stages.

At the end of this chapter, it is instructive to highlight, in a bullet point format, the main indisputable advantages of low-frequency inductively coupled plasmas [93], such as

- very high plasma density;
- excellent uniformity over large cross-sectional areas;
- excellent uniformity over large processing volumes;
- low electron temperatures;
- moderate plasma potentials;
- absence of the Faraday shield;
- no need for multipolar magnetic fields to improve the uniformity;
- very low circuit loss;
- great potential for device upscaling;
- easy discharge operation and maintenance.

This list is not exhaustive and other interesting features and advantages of the LF ICP can appear in specific applications. For example, the LF ICP source is very efficient for plasma-enhanced nitriding of solid materials and PECVD of various thin films, as well as plasma-aided synthesis of various nanoassemblies discussed in more details in the following chapters. These results suggest that low-frequency inductively coupled plasmas have very attractive prospects for industrial applications, including those in micro- and nanoelectronics.

3
Plasma Sources: Meeting the Demands of Nanotechnology

Presently, a variety of inductively coupled plasma sources are used by several industries as reference plasma reactors for numerous applications in semiconductor manufacturing, optoelectronics, and synthesis and processing of advanced functional films and coatings. In the previous chapter, our focus was on the sources of low-frequency, low-temperature inductively coupled plasmas. More specifically, we have introduced the base model of the LF ICP featuring a flat spiral (also commonly termed "pancake") inductive coil (antenna) positioned externally to the main section of the vacuum chamber.

To meet the demands of the present-day plasma-aided materials synthesis and processing (including nanofabrication), the base model should be flexible enough to accommodate variously configured substrate stages, *in situ* diagnostic tools, additional electrodes, source materials in solid or sometimes liquid state, etc. Moreover, the usefulness of any particular plasma source in industrial applications is dictated by a number of requirements that specify the acceptable operation regimes, optimum ranges of the plasma parameters, process reproducibility, and several other factors.

Among them, flux uniformity of reactive species has been a major issue in the last couple of decades owing to strict quality standards of semiconductor industry demanding extra-low variation of the film thickness over the entire wafer. With the wafer sizes continuously increasing (from commonly used 4-inch wafers several years ago to currently used 12-inch wafers and projected increase to 15-inch wafers by 2010), the requirement to maintain the uniformity of the ion and radical fluxes over large surface areas becomes more and more strict. Moreover, there is a vital demand for ultrafine deep-vacuum processing of very large-area flat display panels with the surface areas exceeding 1 m^2. This means that the plasma reactors need to be upscaled to accommodate a continuous increase of the processed surfaces in size without compromising the species flux uniformity.

In this chapter, without trying to cover all the possibilities, we focus on some very important upgrades of the LFICP device of the previous chapter. In the first two subsections (Sections 2.1 and 2.2), we will discuss the new plasma source configuration with an internal inductive coil specially de-

Plasma-Aided Nanofabrication. Kostya (Ken) Ostrikov and Shuyan Xu
Copyright © 2007 WILEY-VCH Verlag GmbH & Co. KGaA, Weinheim
ISBN: 978-3-527-40633-3

signed to improve the uniformity of the RF power deposition in the plasma reactor. Furthermore, the base model of the LFICP can be upgraded by introducing a DC/RF magnetron sputtering electrode, which can serve as an additional source of solid material and ultimately improves the flexibility in terms of the range of possible coatings and thin films synthesized by using this plasma device (Section 2.3). The final section of this chapter introduces an integrated plasma-aided nanofabrication facility (IPANF), which simultaneously incorporates the features of inductively coupled plasma sources, DC/RF magnetron plasma sources and is able to accommodate both (external and internal) configurations of the RF inductive coil.

3.1
Inductively Coupled Plasma Source with Internal Oscillating Currents: Concept and Experimental Verification

In different versions of inductively coupled plasma sources the configurations and positioning of RF current driving antennas can be quite different. For example, the inductive coil can be either placed externally or internally with respect to the discharge chamber. In one of the most common embodiments discussed in the previous chapter (hereinafter referred to as the conventional ICP source or simply the ICP source), the plasma is sustained by the RF power deposited by an external flat spiral inductive coil ("pancake coil") installed externally to and separated by a small air gap from a dielectric window that seals a (usually cylindrical) vacuum chamber [98–100]. Most of the commercial ICP reactors make use of a fused silica or reinforced glass window sealing the chamber in its r–ϕ cross-section from the top. The RF current driven through the flat spiral inductive coil generates the electromagnetic field with the azimuthal electric E_ϕ as well as the radial H_r and axial H_z magnetic field components [101].

Due to the obvious symmetry of the problem, the azimuthal electric field shows a pronounced dip near the chamber axis $r = 0$. In the idealized case of concentrical purely azimuthal RF currents $E_\phi(r = 0) = 0$ [102, 103]. This feature has been confirmed by the extensive magnetic probe measurements and numerical modeling results [91, 93, 95, 97, 102, 103]. Using the linear, with respect to the RF field amplitude, approximation, which is normally valid at low input powers, it was shown that the RF power deposition to the plasma is strongly nonuniform and the actual RF power density has a well-resolved minimum near the chamber axis [145].

A representative contour plot of the RF power density in a conventional (depicted in Fig. 2.1) source of inductively coupled plasmas is shown in Fig. 3.1. This is consistent with the power density profiles derived through the map-

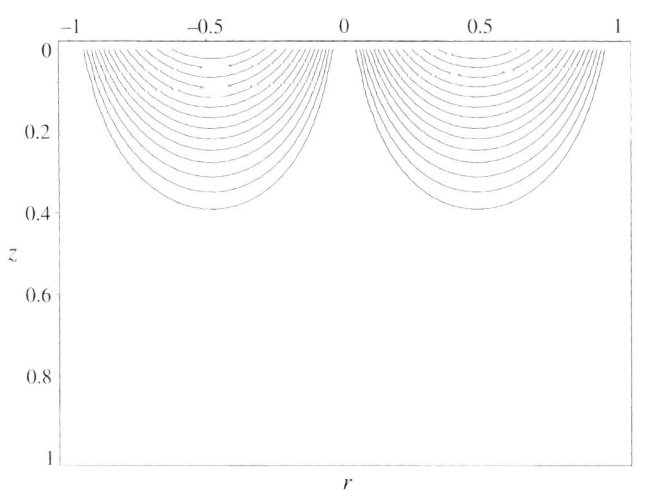

Fig. 3.1 Representative contour plot of the RF power density in a source of inductively coupled plasmas with a flat spiral ("pancake") external coil [145].

ping of the electromagnetic fields in the chamber and the results of optical emission spectroscopy [94,102,103,148]. Thus, the improvement of the uniformity of the power deposition is one of the key concerns of modern plasma processing applications and several attempts to modify the inductive coil configuration and/or adjust the RF power coupling have been reported [108,149,150].

To improve the uniformity of the RF power deposition and plasma parameters, a new source of low-frequency (~460 kHz) inductively coupled plasmas employing two orthogonal sets of eight RF currents reconnected alternately and driven inside the modified vacuum chamber has been developed and tested [146, 147, 151]. In accordance to the theoretical predictions [145], the new antenna configuration can generate more uniformly distributed electromagnetic field patterns and thus improve the uniformity of the RF power deposition. Below, our focus is on the experimental verification of this power deposition concept by the measurements of the distributions of the magnetic field and various plasma parameters in the plasma source.

3.1.1
Configuration of the IOCPS

A schematic diagram of the plasma source with the IOC antenna configuration (hereinafter referred to as the IOCPS) is shown in Fig. 3.2 [146,147]. To produce the unidirectional internal oscillating RF current inside the vacuum chamber, a new coil configuration has been designed. Contrary to conventional ICP sources with the external flat coil configuration, in the device discussed here

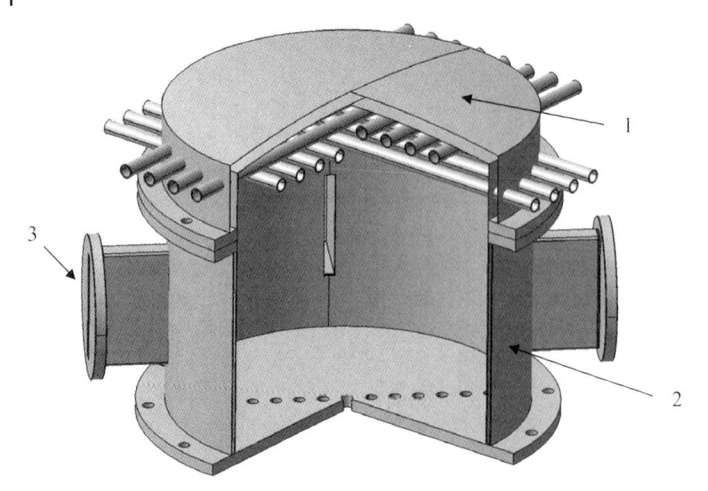

Fig. 3.2 Three-dimensional graphical representation with the 1/4 isometric cut of the plasma source. 1—top section of the vacuum chamber, 2—main section of the vacuum chamber, and 3—diagnostic/observation porthole [147].

the internal RF antenna consists of the two orthogonal sets of copper litz wires placed inside a vacuum chamber.

The wires are enclosed in fused silica tubes as shown in Fig. 3.3. The resulting RF current sheet thus oscillates in the r–φ plane of a standard cylindrical coordinate system [145], with the direction shifted $\sim 45°$ with respect to either set of coils. The total number of quartz tubes in each direction is eight and all copper wires are connected in series.

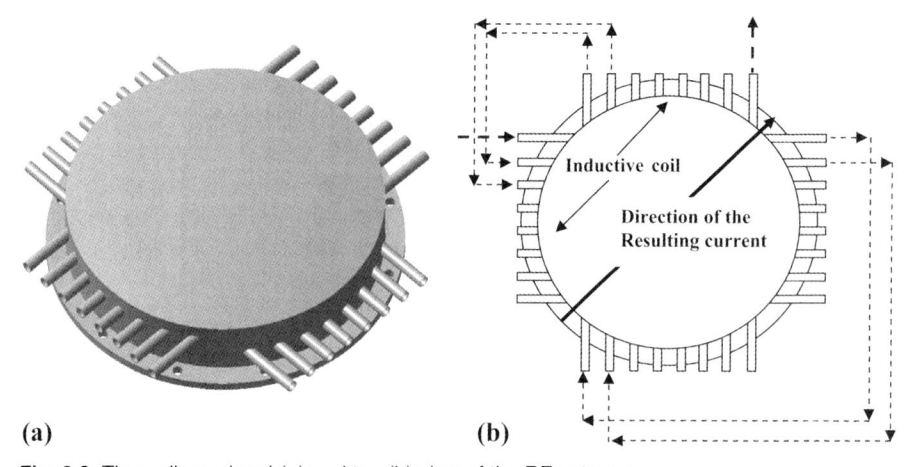

Fig. 3.3 Three-dimensional (a) and top (b) view of the RF antenna-carrying section of the plasma source. Reconnection of only two (in each direction) coil segments is shown (b). A solid arrow shows the direction of the resulting current oscillation in the r–φ plane [147].

Note that the use of the two crossed unidirectional current sheets and the way how the orthogonally directed copper wires are reconnected reduces power losses outside the plasma chamber. Due to the series connection and very low resistivity of the copper wires, the RF current is synphased in any part of the antenna. The inductive coil is made of a copper litz wire 6 mm in diameter and is enclosed in quartz tubes with the inner and outer diameters of 10 and 12 mm, respectively. Figure 3.2 shows that one of the sets of eight quartz tubes is placed 3 cm above the top flange of the main section of the vacuum chamber, while the other set of eight quartz tubes, which are perpendicular to the first one, is lifted up by 2 cm from the first set of wires.

For convenience, the horizontal plane between the lower and upper sets of the coils was chosen as the origin for the axial axis ($z=0$), i.e., 4 cm above the top flange of the main section of the vacuum chamber. Figure 3.3(b) also illustrates the connections of the first two turns of the inductive coil and the resulting direction of the RF current oscillations.

It should be noted that, contrary to the conventional case of inductively coupled plasmas, where the electric field has only a sole azimuthal component, the internal oscillating current generates an additional radial electric field and, therefore, the azimuthal magnetic field. One can thus expect that the presence of the additional electric field component can modify the power absorbed by the plasma electrons.

3.1.2
RF Power Deposition

From the arrangement of the internal coil in the IOCPS one can easily conclude that the electric field generated by the antenna oscillates in the r–ϕ plane in the direction of the bold solid arrowed line in Fig. 3.3. Thus, the electromagnetic field does not feature an axial electric field (E_z) component, which is consistent with the coil arrangement and the symmetry of the problem. Apparently, the RF current inductively driven inside the chamber also oscillates in the same (r–ϕ) direction.

However, since the currents have to form closed loops, it is reasonable to presume that the return path for this current goes through the r–z ("poloidal") cross-section. Therefore, this could be one of the major factors that cause deeper penetration of the electromagnetic field inside the plasma as compared to the conventional ICP sources with flat spiral coils that generate (at low and moderate RF powers) purely azimuthal RF electric fields and currents, the latter having the return loops in the r–ϕ cross-section [146].

In the latter case, one can attribute the field penetration mostly to the "skin effect" that controls the electromagnetic field penetration into dense plasma media with the plasma frequency exceeding the frequency of the incident wave [14]. In the IOCPS case, a combination of the two major factors, namely, the excitation of poloidal RF current loops and "skin effect" results in a pronounced penetration of the electromagnetic field inside the vacuum chamber. A deep penetration of magnetic fields into the chamber has convincingly been confirmed by the results of magnetic field measurements [146]. It is also noteworthy that this phenomenon is also affected by nonzero values of the E_z component that might appear in a real discharge due to a finite gap (neglected in numerical calculations in [145]) and hence, a potential difference between the two parallel current sheets in the coil.

The two-dimensional profiles of the RF power density in the plasma reactor chamber can be mapped via the measurements of the magnetic field topography. In the IOCPS, these have been conducted in three different regimes, namely, in a fully evacuated chamber, in low-, and high-density plasmas [146]. The low-density (rarefied) plasma, with the spatially averaged electron/ion number densities $n_{e,i} \sim 3.3 \times 10^9$ cm^{-3}, is generated in the low-input-power (~ 300 W) electrostatic (E) discharge mode at 30 mTorr.

On the other hand, the high-density ($n_{e,i} \sim 6 \times 10^{11}$ cm^{-3}) plasma is sustained with 770 W RF powers in the electromagnetic (H) mode of the discharge at the same gas feedstock pressure [146]. The magnetic field measurements complemented by the relations between the field components obtained from Maxwellian equations confirm that the actual electromagnetic field excited by the antenna configuration of our interest here indeed features two components of the electric field, E_ϕ and E_r, and three components of the magnetic field H_z, H_r, and H_ϕ (TE electromagnetic field [152]), in a remarkable agreement with the numerical results [95].

The results of measurements of radial profiles of all the three components of the RF magnetic field are shown in Fig. 3.4. The corresponding numerical radial profiles are calculated by using the model of a spatially uniform plasma [145] with the plasma density that equals the spatially averaged electron/ion number density obtained from the Langmuir probe measurements ($n_{e,i} \approx 6 \times 10^{11}$ cm^{-3}, $T_e^{\text{eff}} \approx 3.2$ eV) [146]. It is remarkable that the numerical results plotted as dashed curves in Fig. 3.4 generally reproduce the measured radial variations of the magnetic field and are also in a fairly good quantitative agreement with the experimental data. The apparent source of the remaining discrepancy between the theoretical and experimental results is the actual nonuniformity of the plasma in the chamber. The new models properly accounting for the nonuniform distributions of the plasma density in the IOCPS, are eagerly anticipated in the near future. Meanwhile, various nonlinear effects, such as generation of the second (and higher) harmonic current, ponderomotive and Lorenz forces, become more important at higher power levels and can affect the field distribution in the chamber [97, 153].

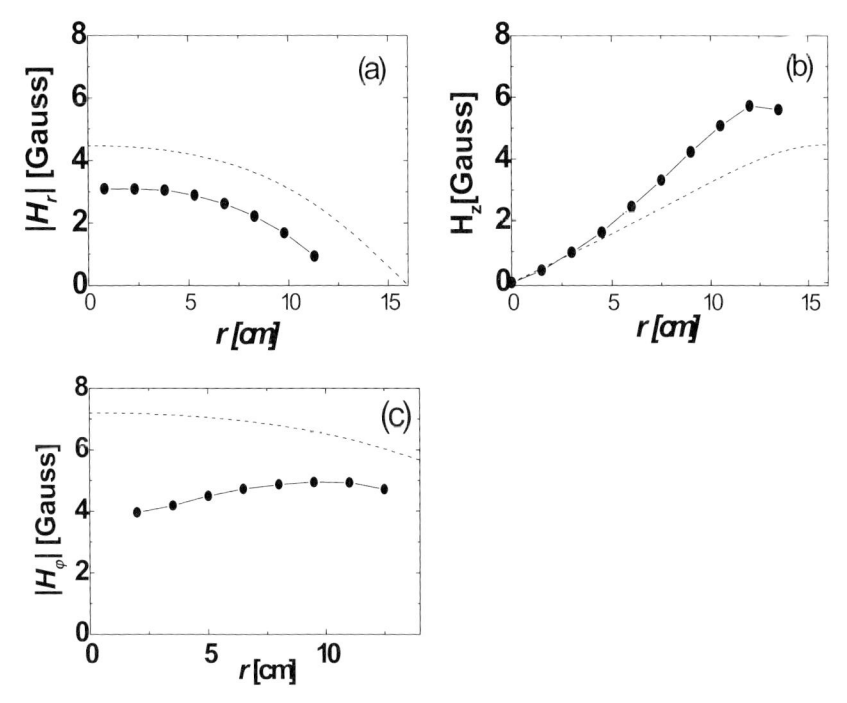

Fig. 3.4 Comparison of the measured (solid circles) and computed (dashed line) radial profiles of the magnetic field components in the 30 mTorr H-mode discharge sustained with $\sim$ 770 W RF powers. Measurements are made at the axial position $z = 8$ cm [146].

Furthermore, using the experimental map of the magnetic fields and the relations between the electric and magnetic fields following from the set of Maxwell's equations (see, e.g., equations (2)–(8) of [145]) one can obtain the distribution of the RF power deposited in the discharge chamber

$$P_p = \frac{1}{2} \int_0^{2\pi} \int_0^R \int_0^L r\mathrm{Re}(\sigma_p)|\mathbf{E}|^2 d\phi dr dz \tag{3.1}$$

where

$$|\mathbf{E}|^2 = [\mathrm{Re}(E_r)]^2 + [\mathrm{Im}(E_r)]^2 + [\mathrm{Re}(E_\phi)]^2 + [\mathrm{Im}(E_\phi)]^2$$

$\sigma_p = \omega_{pe}^2 / 4\pi(\nu_e - i\omega)$ is the conductivity of the uniform collisional plasma, and ω_{pe} is the electron Langmuir frequency. Here, ν_e is the effective rate of electron collisions and ω is the frequency of the RF generator.

A representative semiquantitative contour plot in Fig. 3.5 unambiguously confirms that the fairly uniform profiles of the RF power deposition have indeed been achieved experimentally. It is notable that the power density profiles shown in Fig. 3.5 favorably differ (especially in the areas adjacent to the

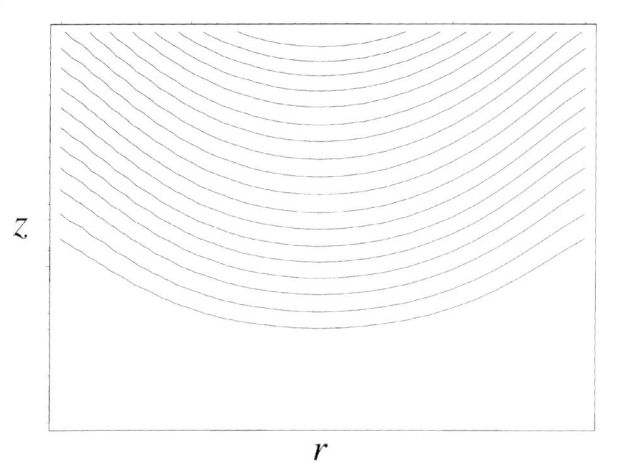

Fig. 3.5 Semiquantitative contour plot of the RF power density in the plasma source obtained from the experimental mapping of magnetic field distribution for the same parameters as in Fig. 3.4 [146].

chamber axis) from the two-dimensional contour plot in Fig. 3.1 and three-dimensional profiles shown in Fig. 9(b) of [145].

It is worthwhile to compare the spatial profiles of the RF magnetic fields in the IOCPS and conventional source of inductively coupled plasmas with the external flat spiral coil. Figure 3.6 shows the radial profiles of the nondimensional H_z, H_z, and H_ϕ magnetic field components in both plasma sources in the electrostatic (E) and electromagnetic (H) modes. In all cases the ICP discharges are sustained in the same vacuum chamber and under fairly similar conditions. The main difference is in the actual antenna configuration used.

To plot the nondimensional profiles of the magnetic field components in the conventional ICP source with the 17-turn flat spiral coil, the data of Refs. [93, 97] are used. Each of the components is normalized on its maximum value over the radial span of the chamber, i.e., $\bar{H}_j(r) = H_j(r)/H_{j,\max}$, where $j = r, z, \phi$. Despite some difference in the discharge parameters, from Fig. 3.6 one can figure out the major differences in the RF field distributions in both plasma sources.

From Fig. 3.6, one can conclude that the magnetic field profiles are generally smoother in the IOCPS. It is also seen that different field components feature quite different behavior near the chamber axis. For instance, in the vicinity of $r = 0$ the radial magnetic field component in the IOCPS is much larger than that in the ICP. On the other hand, near the chamber axis, the H_z component is close to its peak value in the ICP and is very small in the IOCPS.

From Figs. 3.6(c) and (f) one can conclude that the radial uniformity of the H_ϕ component is remarkably better in the IOCPS. Furthermore, the absolute value of the azimuthal magnetic field component (at the RF generator fre-

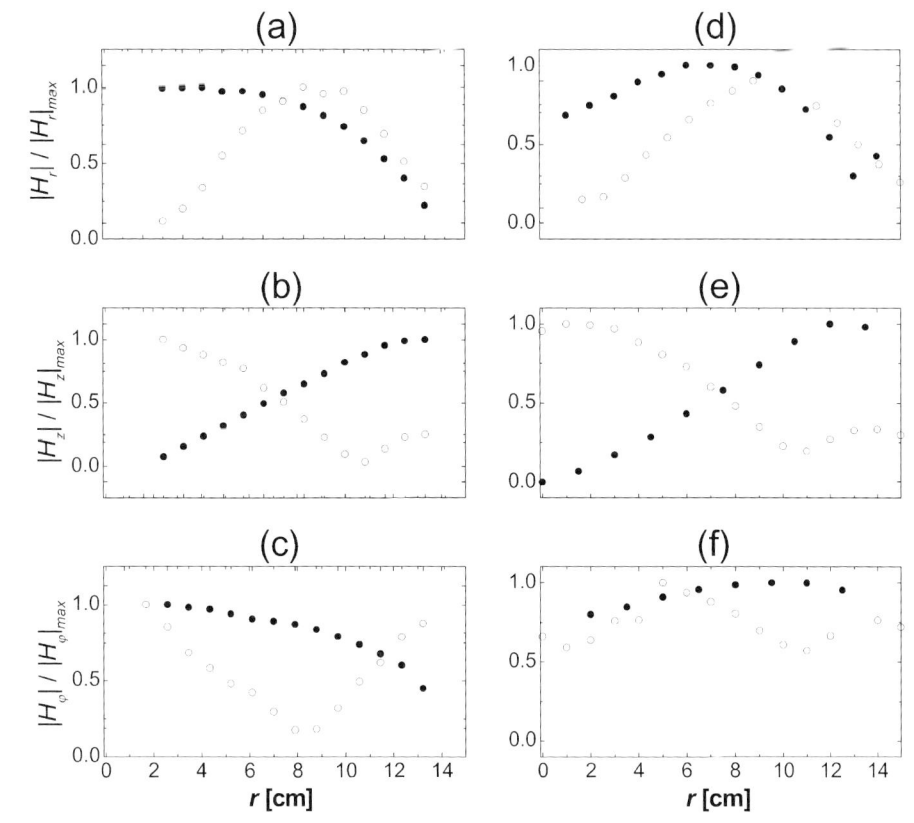

Fig. 3.6 Comparison of radial profiles of nondimensional magnetic field components in the electrostatic (E, (a)–(c)) and electromagnetic (H, (d)–(f)) operation mode of the plasma sources with the internal (solid circles, axial position $z = 8$ cm, $p_0 = 30$ mTorr; E-mode: $P_{in} = 300$ W; H-mode: $P_{in} = 770$ W) and external "pancake" (empty circles, almost the same axial position, $p_0 = 50.8$ mTorr; E-mode: $P_{in} = 300$ W; H-mode: $P_{in} = 1130$ W [146].

quency) [97] in the ICP is much smaller compared to all other magnetic field components in Fig. 3.6.

The peak locations are also quite different in the two plasma devices. In the low-power E-mode discharge in the IOCPS, the H_r and H_ϕ components peak near the chamber axis, whereas the H_z component reaches its maximum at the chamber periphery. Meanwhile, in the H-mode discharge in the IOCPS, the maximum values of the H_r and H_z components shift to the plasma bulk ($r \sim 7$–10 cm). On the other hand, the H_r and H_z components show a similar peak behavior in both modes of the ICP discharge. The radial magnetic field component peaks at $r \sim 9$–10 cm, whereas the H_z component features a well-resolved minimum at $r \sim 10.5$–11.5 cm [146].

Following the original work [146], we will now qualitatively relate the radial profiles of the magnetic field components to the uniformity of the RF power deposition in both plasma sources. First, from Fig. 2.13 of the previous chapter one can observe that $|H_r| \ll |H_z|$ near the chamber axis of the ICP source.

However, at larger radii (e.g., at the mid-radius distance), the H_r component becomes much larger than H_z. Furthermore, $|H_{r,\max}| \sim 3|H_{z,\max}|$ and $|H_\phi| \ll |H_z|, |H_r|$ [97]. Hence, the quantity

$$\eta = \sqrt{|H_r|^2 + |H_z|^2 + |H_\phi|^2}$$

that is proportional to the RF power density in the chamber does feature a dip within a few centimeters from the axis, which is consistent with the data shown in Fig. 3.1 and Fig. 9(b) of [145].

Contrary to the ICP case, the amplitudes of all magnetic field components in the IOCPS remain within the same range at different radial positions. It is also remarkable that the variations of the azimuthal component with radius are fairly slow (Figs. 4 and 6 of [146]). Furthermore, a decrease of H_r with r is "balanced" by a similar increase of H_z. Therefore, the radial dependence of η should indeed follow the pattern shown in Fig. 3.5 and feature a somewhat better uniformity compared to the ICP case.

Unfortunately, due to the difference in the operation parameters in the ICP and IOCPS devices, the actual improvement of the uniformity of the RF power deposition cannot be unambiguously quantified at this stage and a comparison under *identical* conditions is required in the future. Likewise, the absence of the detailed axial scans of the magnetic fields in the ICP source [93, 97] disables a comprehensive comparison of the dependence $H_j(z)$ at different radii. However, the maximum amplitudes of the H_r component measured through the two adjacent upper portholes (separated by 4 cm) in the H-mode ICP discharge differ at least 3–4 times, whereas a similar difference in the IOCPS does not exceed a few tens of percents [146]. This certainly evidences a deeper penetration of the electromagnetic field in the IOCPS as compared to the ICP.

3.1.3
Plasma Parameters

The plasma parameters in the IOCPS can be obtained from the time-resolved measurements by a single RF-compensated cylindrical Langmuir probe [147]. The plasma density n_e, effective electron temperature T_{eff}, and plasma potential V_p are determined by using the Druyvestein routine discussed in detail in Chapter 2. Here we recall that n_e and T_{eff} can be expressed as

$$n_e = \int g_e(V)\, dV$$

and

$$T_{\text{eff}} = \frac{2}{3n_e} \int V g_e(V) \, dV$$

whereas the electron energy distribution function $g_e(V)$ can be obtained through the second derivative of the probe current over the voltage $d^2 I_e/dV^2$ as

$$g_e(V) = \frac{2m_e}{e^2 A} \left(\frac{2eV}{m_e}\right)^{1/2} \frac{d^2 I_e}{dV^2}$$

where e, m_e, and A are the elementary charge, electron mass, and the probe surface area, respectively [147]. Furthermore, the plasma potential V_p can be found as the maximum of the first derivative of the dependence $I(V)$ or as the zero crossing point of the second derivative of the probe current [15].

Figure 3.7 shows the radial (a) and axial (b) distributions of the electron density, effective electron temperature, and plasma potential in the electromagnetic (H) mode of a pure argon discharge and for the azimuthal angle $\varphi = 0°$, which is the direction of the resulting oscillating RF current in Fig. 3.3. One can see that the value of n_e remains almost the same for radial positions r from 0.5 to 7.5 cm (Fig. 3.7(a)) and then starts to decrease. The typical values of the plasma density for the discharge are 1.1×10^{12} and 0.7×10^{12} cm^{-3} for the gas feedstock pressures 51 and 30 mTorr, respectively. Meanwhile, the plasma potential and effective electron temperature are almost constant along the chamber radius.

The measurements of the axial profiles of the plasma parameters (Fig. 3.7(b)) are performed at seven different axial positions through the available portholes in the diagnostic side port [147]. The measurements suggest that n_e, T_{eff}, and V_p remain fairly uniform along the z axis from 8 to 20 cm. Thus, one can clearly see that the introduction of the unidirectional internal RF current into the plasma indeed improves the spatial uniformity of the plasma generated.

It is interesting that the IOCPS also offers a great deal of control of the plasma parameters along the azimuthal direction [147]. In order to investigate the azimuthal dependence of the plasma density, plasma potential and effective electron temperature, the Langmuir probe measurements are carried out for the azimuthal angles $\varphi = 0$ and $90°$. The RF power density features a weak azimuthal dependence at low RF powers as shown in Fig. 3.8.

Note that this gives the possibility of controlling the RF power deposition in the processes that require the azimuthal profiling of the film thickness or etch rate. Moreover, the azimuthal dependence of the global plasma parameters usually disappears with an increase of the RF power [147]. Physically, this can be attributed to higher rates of the ambipolar diffusion, which is responsible for the establishment of the equilibrium profiles of the electron/ion number densities.

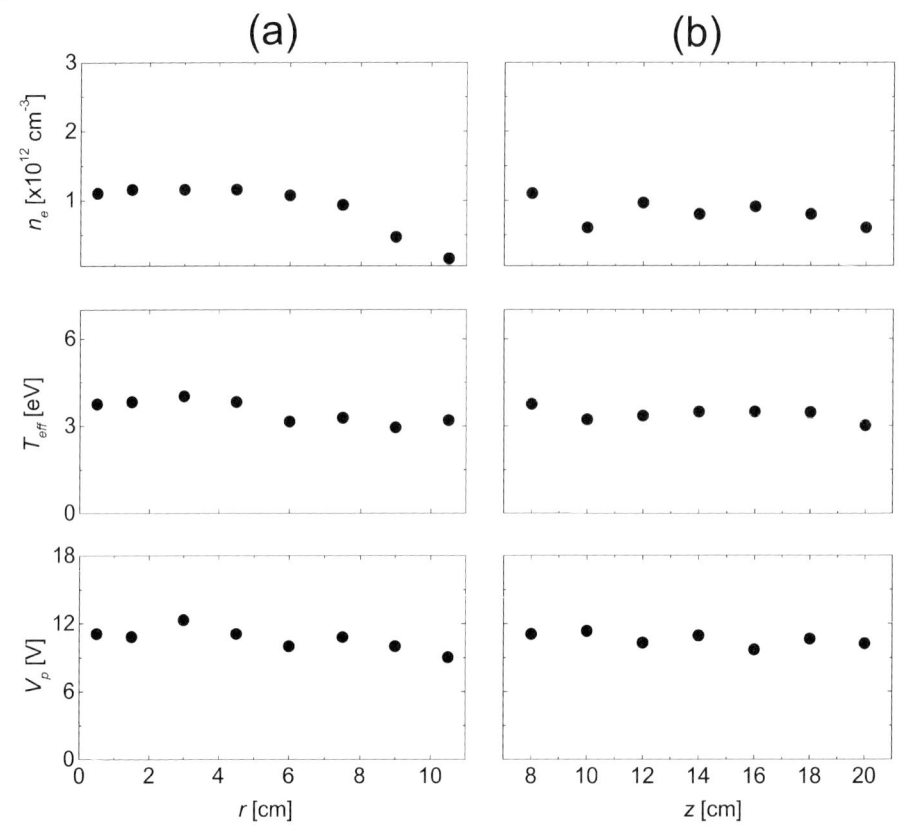

Fig. 3.7 Radial (a) and axial (b) profiles of the electron density n_e, the effective electron temperature T_{eff}, and the plasma potential V_p, for the H-mode argon discharge with RF power input $P_p \sim 1.6$ kW and gas pressure, $p_0 = 51$ mTorr. All profiles are plotted for axial (a) and radial (b) positions $z = 8$ cm and $r = 0.5$ cm, and azimuthal angle $\varphi = 0°$ [147].

The dependence of the plasma parameters on the RF input power at different gas pressures is an important performance characteristic of plasma sources. The experimental results suggest that the electron number density is linearly proportional to the net RF power input in the established electromagnetic mode within the input power range of 0.5 to 1.0 kW [147]. This is consistent with the numerical model of the plasma source with the internal oscillating currents [145] and extensive experimental and theoretical studies of conventional inductively coupled plasma sources with flat external coil configurations (see, e.g., [93] and references therein). Meanwhile, in the same RF power range the effective electron temperature appears to be a slowly increasing function of the RF power. However, the plasma potential remains almost invariable when the input power is increased.

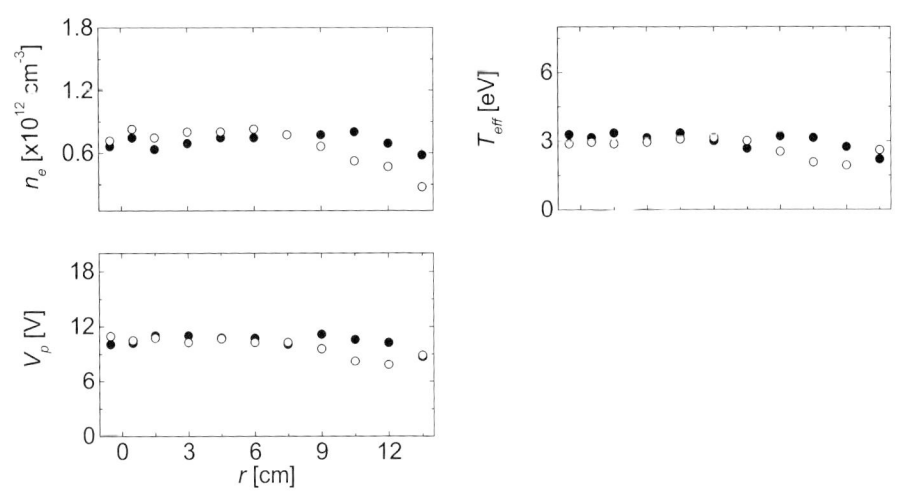

Fig. 3.8 Radial profiles of the electron density, the effective electron temperature, and plasma potential for the H-mode argon discharge with RF power input P_p = 0.62 kW and gas filling pressure p_0 = 51 mTorr. All profiles are plotted for the axial position z = 12 cm and azimuthal angle $\varphi = 0°$ (hollow circles) and $90°$ (solid circles) [147].

The effect of the gas feedstock pressure on n_e, T_{eff}, and V_p is another important factor in industrial applications of plasma sources. Figure 3.9 displays the plasma parameters as a function of p_0 for the H-mode argon discharge at a constant RF power input P_p = 0.68 kW. Note that the Langmuir probe position during the measurements is $r = 0.5$ cm and $z = 12$ cm [147]. As one can see from Fig. 3.9, in the pressure range below 51 mTorr, the electron density increases with pressure from 3.7×10^{11} cm^{-3} at 20 mTorr to 8.0×10^{11} cm^{-3} at 51 mTorr. In the pressure range exceeding 51 mTorr, the electron density starts to decline with the pressure and its value decreases from 8.0×10^{11} cm^{-3} at 51 mTorr to 5.2×10^{11} cm^{-3} at $\sim$440 mTorr. Note that at the RF input power of 0.68 kW the flex point corresponds to $p_0 = 51$ mTorr.

It should also be remarked that the rate of change of the plasma density with p_0 in the range above 51 mTorr is much slower than that in the lower pressure range ($p_0 < 51$ mTorr). As the pressure increases from 20 to 51 mTorr, the plasma potential drops rapidly from 16.1 to 10.8 V. However, V_p decreases slowly with p_0 in the range above 51 mTorr. The electron temperature follows a similar tendency. Initially, T_{eff} drops from 4.8 eV at $p_0 = 20$ mTorr to 3.7 eV at $p_0 = 51$ mTorr. Thereafter, the electron temperature continues to decrease and also experiences small fluctuations.

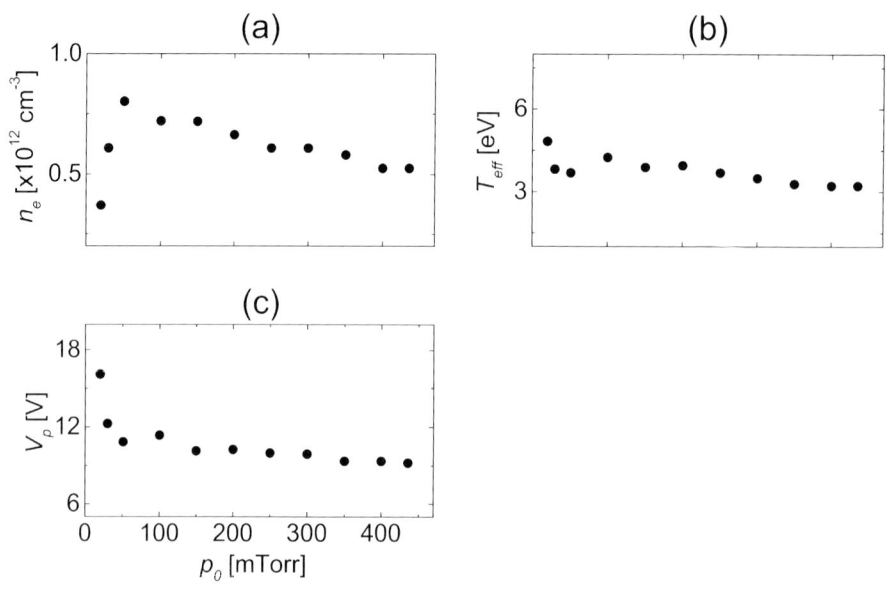

Fig. 3.9 The effect of the gas pressure on the electron density, effective electron temperature, and plasma potential for the H-mode argon discharge and for the RF input power $P_p \sim 0.68$ kW. The Langmuir probe position is $r = 0.5$ cm and $z = 12$ cm [147].

3.2
IOCPS: Stability and Mode Transitions

We now consider the issues of discharge mode stability and compare relevant phenomena to those observable in conventional sources of inductively coupled plasmas. To investigate the transitions between the two operating modes of the IOCPS, one can use the following original procedure [154]. Initially, the plasma chamber is evacuated to a base pressure of approximately 2×10^{-4} Torr. Then the argon feedstock is introduced into the reactor chamber. After applying the RF power as low as 40 W, a dim plasma glow (electrostatic discharge mode) can be clearly observed in the space between the parallel current sheets and slightly expanding through the "mesh" formed by the two layers of the current-carrying wires. The brightness of this mode progressively increased as the RF power rises up to 500–600 W, when quite abrupt and discontinuous transitions to the inductive mode take place.

We recall that in conventional ICP sources with external "pancake" antennas, the E-mode plasma glows are sustained due to the existence of the potential difference between the (usually grounded) center and outer turns of the inductive coil [93, 102, 103]. The fact that the glow of our interest here originates in the space between the two antenna layers indicates that most probably the origin of the E-mode in the IOCPS is in the potential difference between the two orthogonal RF current sheets and neighboring conducting wires. A comparison of the representative optical emission intensities of the same spectral lines under the same discharge conditions and voltages applied to the photomultiplier tube clearly suggests that the brightness of the plasma glow in the electrostatic mode is much higher in the IOCPS than in conventional inductively coupled plasmas.

After a gradual increase of the RF power, a threshold for the $E \rightarrow H$ mode transition can be eventually reached. After the transition, the glow is stable in the inductive (H) discharge mode. Thereafter, when the input power is gradually increased and later decreased back to the minimum starting level, one can observe the $H \rightarrow E$ transition and eventually the extinguishing of the discharge. Below, we will consider in more detail the variations of the optical emission intensities and the spontaneous discharge mode transitions.

3.2.1
Optical Emission

We now consider the results of the real-time studies of the variation of the optical emission intensity (OEI) in the process of a cyclic variation of the input power to illustrate the dynamics of the $E \leftrightarrow H$ mode transitions in argon plasmas. In particular, Fig. 3.10 illustrates a real-time dynamics of the OEI of the 840.82 nm line of the neutral Ar atom in 22 and 31 mTorr plasma discharges when the RF coil current changes [154]. One can easily notice that the $E \rightarrow H$ transition is accompanied by an instantaneous increase of the emission intensity. As can be seen from Fig. 3.10, the OEI further increases with the coil current in the established inductive mode of the discharge. The similar tendencies are quite common for low-frequency inductively coupled plasmas [155].

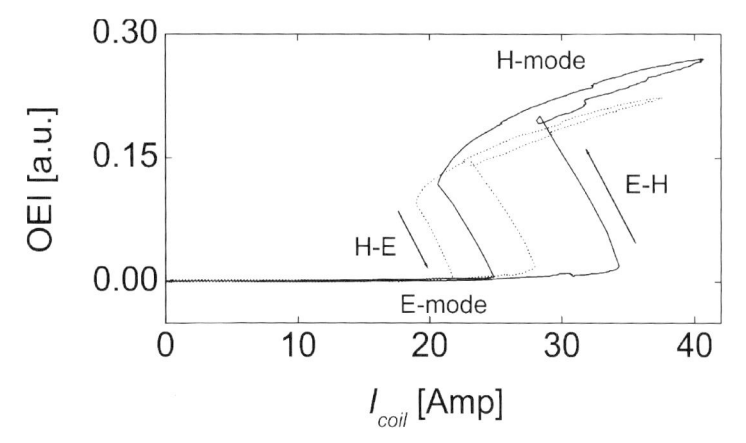

Fig. 3.10 Dynamic variation of the OEI during the $E \leftrightarrow H$ mode transitions at 22 (solid line) and 31 (dotted line) mTorr [154].

Thereafter, when the coil current is reduced, the emission intensity decreases almost linearly. It is seen from Fig. 3.10 that the discharge can still be maintained in the luminous electromagnetic mode even when the coil current is below the threshold of the original $E \rightarrow H$ transition, which indicates that the hysteresis phenomena in the mode transitions can be pronounced in the ICPs with the internal coils. Near the point of the inverse $H \rightarrow E$ transition the intensity of the optical emission falls steeply to the level corresponding to the E-mode. Thereafter, when the stable electrostatic mode is established, the OEI does not change much and remains low.

Using the OEI data, one can obtain the minimal value of RF current that ignites the inductive mode (starting current), and the minimal H-mode maintenance current (threshold current for the $H \rightarrow E$ mode transition) as a function of the operating gas pressure, in a manner similar to the LF ICP source with an external "pancake" RF coil [91, 93]. It is interesting that at different working pressures the $E \leftrightarrow H$ transitions are initiated at different values of the coil current I_{coil}, which indicates on the possibility that variations of the gas feedstock pressures can also result in hysteresis effects [154].

It is worth emphasizing that the width of the hysteresis loop in the dynamic curve for the optical emission decreases from approximately 15 A at 22 mTorr to 5.5 A at 50 mTorr. And since the actual difference between the starting and minimal maintenance currents reflects the strength of hysteresis effects, one can conclude that the nonlinear effects are stronger at lower gas feedstock pressures. The corresponding width of the hysteresis loops in the conventional ICP device with an external "pancake" coil appears to be remarkably smaller. For instance, it is ∼5 A at 22 mTorr [91]. One can thus expect a variety of nonlinear effects in the IOC plasma source. It is worth emphasizing that pronounced hysteresis phenomena have been observed in the dependence of the main IOCPS RF circuit parameters on the antenna current (and hence the RF input power).

In terms of the threshold RF coil currents, the $E \rightarrow H$ transition in a 22 mTorr argon discharge happens when the coil current reaches ∼33–34 A, when the power deposited to the plasma is ∼ 600 W. Under the same working gas pressure a similar mode transition in the conventional ICP device requires approximately 700–750 W of the power deposition [91, 93]. This is consistent with the estimates of the discharge working points in the two configurations [145] and can certainly be regarded as one of the advantages of the IOC plasma source. In the 22 mTorr H-mode discharge, the reflected plasma resistance $R_p \sim 6$ and ∼ 2.3 Ω in the IOC and ICP [91] devices (at $P_p \sim 800$ W), respectively. Under the same conditions the absolute values of the total reactance change ΔX_p in the IOCPS (∼ 2 Ω) are lower than in the ICP (∼ 7 Ω). Moreover, the total IOCPS load becomes more resistive (R_p increases) rather that reactive (ΔX_p decreases) in the established H-mode.

Hence, qualitatively comparing the ratios $R_p/\Delta X_p$ under the same gas pressure and input power, one can arrive at the conclusion that the plasma load in the IOCPS appears to be less reactive, and one can thus expect higher RF power transfer efficiency in the IOCPS compared to its conventional counterpart. It is worth noting that the $E \to H$ transition in the IOCPS appears to be discontinuous rather than smooth as has been previously reported for some other internal coil configurations [99, 100].

Nevertheless, differently to the mode transitions in conventional ICP sources with external flat spiral "pancake" coils, the $E \leftrightarrow H$ transitions in the IOCPS visually appear to be smoother. This can be attributed to presumably high plasma densities n_e in the electrostatic mode of IOCPS near the mode transition threshold. Visually, the glow in the E-mode discharge in this case is much brighter than that in the LF ICP with an external coil [91]. Thus, relative populations of argon atoms in the excited states appear to be higher in the IOCPS.

From this point of view, E-mode plasmas generated in the inductively coupled plasma source with internal oscillating current sheets can be quite attractive from the plasma processing application point of view. Furthermore, relatively high electron temperatures ($T_{\text{eff}} \sim$ 10–15 eV) of the E-mode discharges in the IOCPS [147] are favorable for the efficient dissociation of the reactive gas feedstock in various applications.

3.2.2
Self-Transitions of the IOCPS Discharge Modes

A striking observation made in the original experiments [154] is that under certain conditions the discharge originally operated in the electrostatic mode in a reasonable proximity (a few tens of watts of input power below a normal transition threshold) of the mode transition threshold, spontaneously transited to the upper stable state after a certain delay time δt_{E-H} into the discharge operation. The value of δt_{E-H} is typically in the range between 20 sec and 3 min and depended on the process parameters.

This highly unusual behavior of the 30 mTorr argon discharge is reflected in Fig. 3.11 showing the real-time dynamics of the optical emission intensity of the 840.82 nm spectral line of neutral argon, synchronized with the real-time variations of the RF coil current. It is clearly seen that the up-jump in the optical emission intensity and the abrupt fall in the coil current happen simultaneously. As Fig. 3.11 suggests, after a sharp rise immediately following the mode transition, the emission intensity steeply decreases and levels off afterward, which indicates an abrupt excitation and subsequent de-excitation ("discharge") of certain atomic levels of argon. More importantly, it is clearly seen that the OEI of the 840.82 nm line levels off in the H-mode at the level higher than in the E-mode.

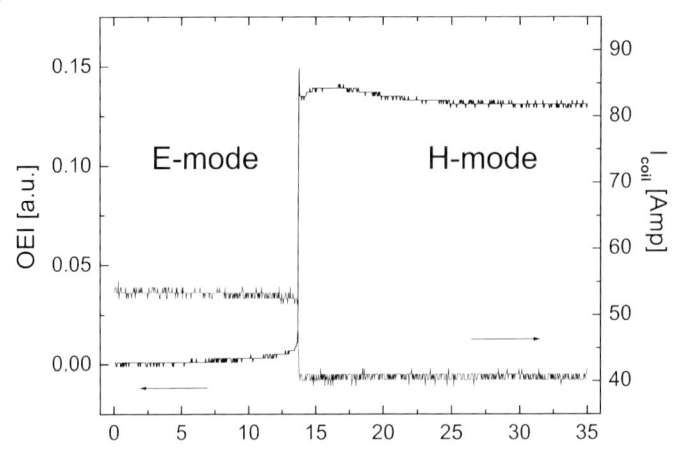

Fig. 3.11 Evolution of the OEI of Ar I 840.82 nm emission line and RF coil current during the spontaneous $E \rightarrow H$ mode transition in a 30 mTorr argon discharge sustained with 500 W input powers [154].

It is remarkable that the discharge self-transition phenomenon can be observed in a wide range of the argon gas pressure. The emission intensity and the RF coil current change by $\Delta \mathcal{J}$ and ΔI_{coil} during the discharge "self-transition", where $\mathcal{J}$ is the optical emission intensity. More importantly, these changes appear to be quite different in a range of gas feedstock pressures of 10–90 mTorr. Specifically, the actual changes in the optical emission intensity and the RF coil current become smaller when the working gas pressure increases [154]. Another interesting observation is that despite smaller changes in the OEI and RF coil current, the discharge can be operated in the E-mode further from the $E \rightarrow H$ threshold at higher gas feedstock pressures and still be able to spontaneously transit to the higher, electromagnetic regime [154].

We now comment on the possible reasons for the observed discharge mode "self-transition" phenomenon. We start the discussion by noting that spontaneous mode transitions to the electromagnetic mode still remain an issue of stability of the discharge operation in the E-mode in the vicinity of the $E \rightarrow H$ mode transition threshold.

Apparently, the initially stable E-mode discharge gradually evolves to approach the mode transition point, and once the threshold value of the RF power has been reached, a discontinuous transition to the higher density electromagnetic (H) mode takes place. A transiently unstable discharge further relaxes to a stationary state in the inductive mode. The instantaneous mode jump is accompanied by the outbursts of the optical emission.

This can be attributed to the drift of the plasma parameters caused by changes in the discharge operation conditions. One of the most likely possibilities is a slow excessive heating of the argon gas feedstock during the discharge run. Generally, the neutral gas in the chamber can be heated through the excessive heat transfer from the chamber walls, internal wires of the RF coil. In this case the slow drift of the gas temperature can follow the changes of the temperature of the internal wires and/or electrodes. It is imperative to note that the gas pressure is externally controlled and maintained constant during the experiment [154].

Under such conditions, a local increase in the gas temperature T_g (caused by any of the heating factors mentioned above) can result in a thermal expansion of the working gas, the latter is usually accompanied with a drop in the density of the neutrals n_n. Since the RF generator or circuit settings remain invariable during the experiment, this minute change in the neutral gas density would not affect the actual power deposited to the plasma electrons, and hence, the plasma sustaining electric field E_p. However, the fundamental parameter of the discharge maintenance E_p/n_n (and also usually the averaged electron energy/effective temperature) increases.

The result is an enhanced ionization and locally increased plasma conductivity, RF current density, and Joule heat release. In this case, the already hot gas is heated even more. As a result of the enhanced ionization the electron number density (and hence the RF power deposited to the plasma) can reach the threshold value corresponding to the H-mode discharge at the given conditions and a spontaneous gas breakdown in the inductive mode can be triggered. Certainly, the above-mentioned drift of the plasma and RF power deposition parameters does require the delivery of the minimum excessive heat to the working gas, which explains the recorded delay times δt_{E-H} for the mode "self-transition" to happen.

It is quite likely that the above gas temperature-induced drift of the plasma parameters can destabilize the E-mode discharge and be regarded as an initial stage of the thermal instability [20]. It is crucial to discuss the issue of stability and the most likely cause of the observed drift of the plasma parameters leading to the "self-transition" phenomenon. Notably, a quite similar drift of the plasma parameters affects the plasma confinement and impurity sources in some plasma fusion devices.

First, the excessive heat transfer from the chamber walls is quite unlikely to be an important factor that controls the observed drift of the plasma parameters [154]. The reason is that the chamber wall temperature is maintained fairly constant due to continuous heat removal from the walls by the chilled water flows driven (by a water cooling system) in a gap between the two walls of the chamber. Moreover, the water temperature is externally controlled and does not noticeably change during the experiment. It is thus quite likely that the drift of the plasma parameters can be caused by the overheating of the internal RF coil.

Another interesting observation is that intentional preheating of the gas feedstock (e.g. by running the discharge in the *H*-mode and then switching it to the *E*-mode or increasing the cooling water temperature) one can facilitate the spontaneous mode transitions. Meanwhile, if the discharge starts in a relatively "cold" *E*-mode far enough from the mode transition point, the spontaneous mode transition does not happen. Thus, the drift of the plasma parameters appears to be more pronounced at higher gas temperatures. On the other hand, this indicates that a slow drift of the plasma parameters rather than a thermal instability is the cause of the observed spontaneous mode transitions [154].

As we have discussed above, during the transition time δt_{E-H} the plasma density grows to reach the threshold value for the $E \rightarrow H$ transition. From the symbolic electron balance equation [20]

$$dn_e/dt = \Xi_{sources} - \Xi_{sinks} \qquad (3.2)$$

one can conclude that when the electron density increases ($dn_e/dt > 0$), the dynamic source terms exceed the sink terms, that is $\Xi_{sources} > \Xi_{sinks}$, which is a generalized plasma instability criterion [20]. Here, $\Xi_{sources}$ and Ξ_{sinks} denote all the combined origins and losses of the plasma electrons, including ionization, dissociation, diffusion losses to the walls, recombination in the reactor volume, etc.

In a sense, the observed mode "self-transition" phenomenon can also be regarded as an instability [20]. It is most likely that in the case considered the sources and sinks quickly reach the equilibrium due to intense collisional processes. However, one cannot rule out the possibility that if the working point is preset close enough to the $E \rightarrow H$ transition point, the *E*-mode discharge can become unstable and the discharge stabilizes only after the transition to the stable electromagnetic mode. Strong and quickly relaxing outbursts of the optical emission intensity in Fig. 3.11 are the qualitative indicators of this possibility.

It is thus likely that the stepwise ionization and accumulation of larger amounts of metastable atoms can also destabilize the *E*-mode discharge operation and trigger the $E \rightarrow H$ transition. Physically, as n_e increases, more excited particles are created and the ionization from the ground state is supplemented by the ionization of the excited atoms, the latter requiring lesser electron energy. Therefore, when the stepwise ionization is important, the term $\Xi_{sources}$ in Eq.(3.2) grows with n_e steeper than $\nu_i n_e$, where ν_i is the rate of ionization of argon atoms from the ground state [154].

Real-time dynamics of the optical emission intensity depicted in Fig. 3.11 supports the importance of multistep atomic excitation/ionization processes in the discharge maintenance. One can note that the emission line at 840.82 nm originates due to

$$3s^23p^5(2P^0_{1/2})4s \rightarrow 3s^23p^5(2P^0_{1/2})4p$$

atomic transition. It is remarkable that in this case the 4s and 4p manifolds of argon atomic spectra that can contribute up to 30–40 % of the atomic ionization [91, 128] are involved. Hence, multiple states of argon atoms can be excited in relatively dense E-mode plasmas. The energy released through the de-excitation of the excited states can instantly contribute to the gas breakdown in the electromagnetic mode.

Meanwhile, optical emission spectra of the discharge before and after the $E \rightarrow H$ transition reveal that there is a notable difference between the relative steady-state amplitudes of different emission lines in the two established discharge modes [154]. For example, the emission intensities of most of the spectral lines of neutral (such as 420.7, 427.22, and 840.82 nm, Ar I) and ionized (such as 434.81 nm, Ar II) argon atoms are higher in the electromagnetic mode. Furthermore, relative intensities of two different emission lines in the same discharge mode also change after the transition.

In the electrostatic (E) mode, the intensity of 394.9 nm emission line is higher than that of the 420.7 nm line. Likewise, the intensity of Ar I 789.11 nm line is higher than that of Ar I 840.92 nm line. However, in the established H mode, the OEIs of 420.7 nm and 840.82 nm lines become higher than the intensities of 394.9 nm and 789.11 nm lines, respectively. This evidences that the excited states of argon atoms are involved in the mode self-transition and, more importantly, different atomic levels emit during the mode transition quite differently.

To conclude this section, we stress that the sources of high-density inductively coupled plasmas with internal RF coil configurations undoubtedly have an outstanding potential for the future use by a number of industries for a wide range of tasks related to synthesis and modification of surface and bulk properties of various advanced materials, including biomaterials and nanomaterials, as well as development of new generations of micro- and nanoelectronic integrated circuitry and devices.

However, there are still some issues to be resolved before this class of advanced plasma sources can be widely adopted in industry. One of the remaining issues is the optimization of the uniformity of the number densities and fluxes of the species over larger, than in the experimental prototype discussed in this chapter, surface areas and bulk volumes.

Another critical point is to avoid any undesirable factors that might result in unstable operation of the plasma source. One of them is the observed mode self-transition that originates due to the drift of the plasma parameters during the discharge operation in the electrostatic mode in the vicinity of the $E \rightarrow H$ mode transitions. However, this problem is most likely due to the overheating of the internal RF coil and can be eliminated by a better engineering design. Nonetheless, we believe that the IOCPS and other advanced inductively coupled sources with internal RF coil configurations should attract a wider international attention and become a focus of intense research efforts in the near future.

3.3
ICP-Assisted DC Magnetron Sputtering Device

In this section we describe a hybrid configuration with a high degree of flexibility in controlled generation of a large variety of working species by using gas phase and solid precursors. Originally designed to be a low-frequency ICP source (discussed in detail in Chapter 2), the system of our interest here is an upgraded LFICP version modified by inclusion of a DC magnetron electrode inserted through the bottom endplate of the vacuum chamber.

With this adaptation, this hybrid system shown in Fig. 3.12 can be used to sustain plasma discharges in three distinct modes:

- conventional DC magnetron sputtering mode;
- ICP mode; and
- ICP-enhanced DC magnetron sputtering mode.

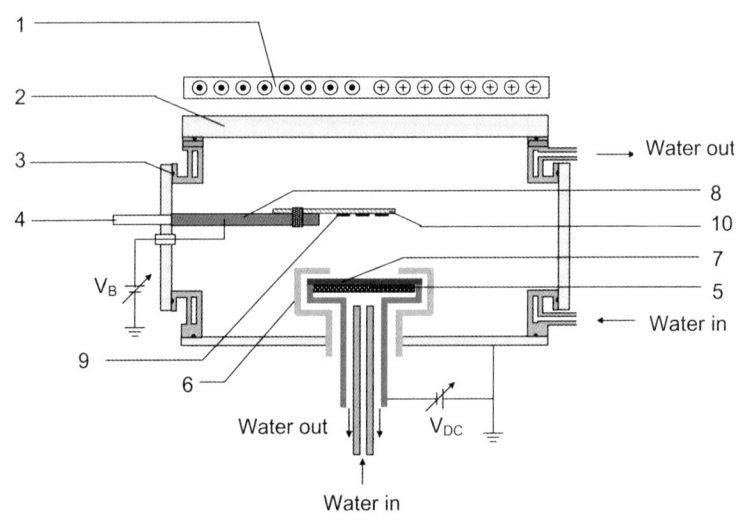

Fig. 3.12 Schematic of the ICP-assisted DC magnetron sputtering facility. Here, 1-RF planar coil; 2-quartz plate; 3-O-ring; 4-quartz rod; 5-magnet; 6-guard; 7-sputtering target; 8-optical detector; 9-substrate; 10-substrate holder.

This combination of the two different ways of plasma generation has been motivated by the acute necessity to maximize the yield of the species released from the magnetron sputtering target electrode. In conventional DC magnetron sputtering plasmas, the electrons are confined in the vicinity of the target by varying configured DC magnetic fields and draw closed trajectories in crossed **E** and **B** fields.

The electrons confined in such a way ionize working gas (such as argon) thus creating fluxes of positive ions that are driven to the target electrode (cathode) by the DC electric field maintained between the anode (target assembly) and the cathode. The electrons in this device are magnetized, which means that the period of their gyration $t_{ge} = 2\pi/\omega_{ce}$, where ω_{ce} is the electron cyclotron frequency [15], is much shorter than the time needed to traverse the near-target sheath. However, the ion sheath traverse time τ_i^s should be much shorter than the time of their gyration $t_{gi} = 2\pi/\omega_{ci}$, where ω_{ci} is the ion cyclotron frequency to enable the ions to accelerate to a reasonably high energy (when the sputtering yield is maximum) and impinge on the target surface.

Therefore, the magnetic field should be neither too weak to enable the electron confinement nor too strong so as not to magnetize the positively charged ions and reduce the efficiency of their interaction with the target surface. Another limitation of the efficiency of the target material sputtering yield is a consequence of the relatively low ionization rates and hence, number densities of the electrons and ions in DC magnetron-based discharges.

In the hybrid plasma device of our interest here (Fig. 3.12), the two independent means of the plasma generation are combined. In fact, the RF inductive coil placed on top of the processing chamber is used to generate inductively coupled plasmas in a usual manner, by using RF electromagnetic fields induced by the RF current in the coil. Depending on the operation regime, the plasma can either be confined near the coil (low-power *E*-mode discharges) or can occupy the entire volume of the reactor chamber (high-power *E*-mode or *H*-mode).

When the RF coil is not powered, the plasma can be created in the vicinity of the DC magnetron electrode. However, it is natural to expect that when both plasma generation channels are operated simultaneously, the ionization efficiency can be dramatically increased and will eventually result in the enhancement of the sputtering yield of the target material.

Moreover, one can also expect a fascinating variety of mode transition and hysteresis phenomena in the hybrid plasma source. Indeed, one should expect complex transitions and nonlinear hysteresis phenomena between multiple steady states of the ICP-assisted DC magnetron sputtering discharge. These phenomena will be discussed further in this section.

For consistency, the hybrid plasma facility should be termed as the ICP-Enhanced DC Magnetron Sputtering Plasma Source. However, for simplicity and for the sake of shorter abbreviations, this facility will be further referred to as the Plasma-Enhanced Magnetron Sputtering Facility (PEMSF).

The PEMSF comprises the following key components, such as the vacuum chamber, external flat spiral ("pancake") inductive coil, vacuum and gas handling systems, magnetron electrode, RF power supply with a matching network, magnetron electrode power supply, and a range of advanced diagnostic and data acquisition and processing instrumentation. Most of these components are the same is in the base version of the LFICP source and have been described in detail in Chapter 2.

Here, we describe in more detail the arrangement of the sputtering target assembly. Specifically, the target electrode consists of four major components: the target disc, the set of annular permanent magnets, the DC powered electrode, and the grounded aluminium guard. Sputtering targets of different sizes (up to 15 cm in diameter) are secured onto the top flange of the magnet container, which essentially acts as a cathode. To maintain the temperature of the magnetron target electrode constant and prevent the magnets from demagnetization due to rising gas temperatures during the discharge, a cooling flow of running water is driven through the interior section of the electrode as shown in Fig. 3.12. Surrounding the cathode is a grounded aluminium guard that acts as an anode.

It is important to mention that the spacing between the target surface and the top quartz window can be varied so as to adjust the RF power deposition into the plasma, number densities, and spatial profiles of the main working species (i.e., when the ICP is in use). A tuneable DC power supply provides the target electrode with a variable negative potential (150–300 V).

To summarize, due to the inherent difficulties in increasing deposition rate in conventional DC magnetron sputtering devices, a separate inductively coupled plasma source is introduced to enhance the plasma production and sputtering. In fact, this device configuration allows one to decouple the plasma generation and magnetron sputtering functions and distribute them between the two different parts of the PEMSF. The choice of the LFICP as the ionization enhancer owes to excellent plasma uniformity and high ionization degrees (see Chapter 2). It is hoped that the deposition rates in the Plasma-Enhanced Magnetron Sputtering Facility can be substantially higher than in the conventional DC magnetron sputtering devices. More importantly, one can also expect a superior quality of functional coatings, nanofilms, nanoassemblies, etc., over large surface areas, in particular, due to highly uniform plasma generation. Some relevant examples will be given in the following section.

3.3.1
Enhancement of DC Magnetron Sputtering by an Inductively Coupled Plasma Source

Here we show evidence that concurrent operation of the ICP and DC magnetron sputtering discharges results in a substantial gain in the net fluxes of positive ions onto the target electrode. This effect is quantified in terms of a drastic increase of DC currents through the magnetron circuit at the same DC bias voltage. When a high-density (up to 8×10^{12} cm^{-3}) plasma is generated by an external RF coil and occupies the entire chamber volume, including the areas adjacent to the magnetron sputtering electrode, it is reasonable to expect that the sputtering yield can be enhanced by positive ions additionally created in the chamber bulk as a result of RF power deposition.

Figure 3.13 shows the variation of the current flowing in the target electrode with the (negative) voltage applied to the cathode at different levels of RF power (0, 0.5, and 1.5 kW) supplied to the external inductive coil. Apparently, the RF power here is used to independently generate and sustain high-density inductively coupled plasmas. From Fig. 3.13, it can be seen that at a given DC voltage, the cathode current is much larger at higher levels of ICP-sustaining RF power.

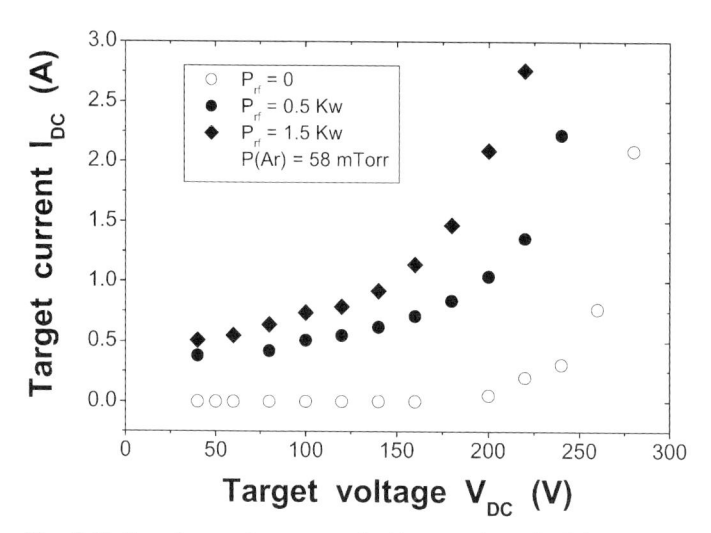

Fig. 3.13 Target current versus applied target voltage for DC magnetron sputtering only (a) (open circle), inductively coupled plasma assisted DC magnetron sputtering at RF power of 0.5 kW (b) (solid dot) and 1.5 kW (c) (diamond).

It is noteworthy that when a negative voltage is applied to the cathode, the current value quantifies the net ion flux incident on the target surface. In other words, additional bulk ionization in the RF fields can dramatically enhance the incident ion flux, even at small voltages between the anode and the cathode. Such unusually high ion currents are beyond the capabilities of a conventional DC magnetron sputtering device operating at the same voltages (see the open circle plot in Fig. 3.13).

It can be easily rationalized that an increase in the incident ion flux on the target surface can result in a higher sputtering yield, and eventually, deposition rates. Figure 3.14 shows the optical emission intensities of the main species originating in the plasma-assisted sputtering of Ti targets in DC-magnetron, inductively coupled, and ICP-assisted DC magnetron plasmas in nitrogen. The optical emission has been collected by using a collimated optical probe positioned 2–3 mm below the (downward facing) deposition substrates (see the schematic diagram of the PEMSF in Fig. 3.12). From the optical emission spectra, it is clear that the intensities of the emission peaks of the (gas source) nitrogen and sputtered (metal source) titanium species are much higher when the external RF coil is powered.

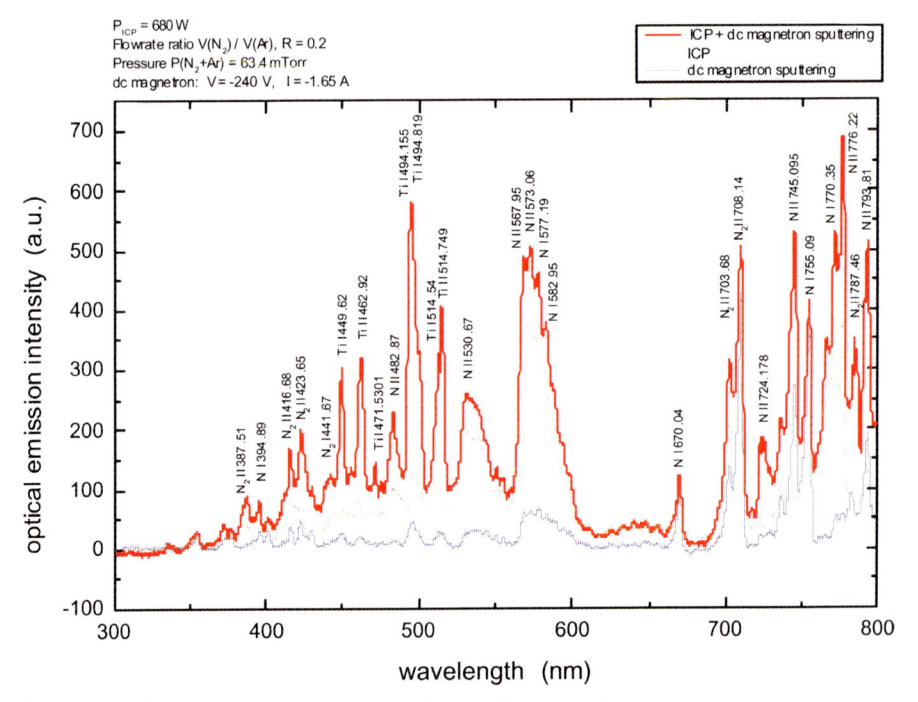

Fig. 3.14 Optical emission spectra observed under three conditions: DC magnetron sputtering only (bottom line), inductively coupled plasma (middle line), and an inductively coupled plasma assisted DC magnetron sputtering (upper line).

The results in Fig. 3.14 are of paramount importance to establish the correlation between the efficiency of the PEMSF and the eventual film properties. Indeed, an increased flux of the plasma species due to the inductively coupled plasma-enhanced DC magnetron sputtering can eventually lead to a substantial increase of the rates of micro- or nanofeature formation. These include nanofilm deposition rates, rates of nanostructure or nanocrystal growth, pore or trench development rates, etc.

The results of the original work [156] convincingly confirm this effect in the practically very important case of deposition of superhard TiN coatings by means of DC magnetron sputtering of Ti targets in reactive nitrogen-based plasmas. Similar conclusions are expected for a large number of processes utilizing different types of plasma-assisted sputtering processes. There are tantalizing prospects for the improvement of vacuum diode, DC/RF magnetron, pulsed magnetron, and other sputtering processes by continuously and independently operated RF plasmas. These possibilities are still awaiting their experimental and commercial realization.

3.3.2
Mode Transitions in ICP-Assisted Magnetron Sputtering Device

We now turn our attention to the fascinating variety and complexity of discharge mode transitions in the hybrid plasma setup shown in Fig. 3.12. As we have mentioned above, both main plasma-generating systems (ICP and DC magnetron) feature two stable discharge operating modes. The RF plasma device can be operated in the electrostatic (E) and electromagnetic (H) modes, whereas the DC magnetron plasmas can be generated in the D (vacuum diode) and M (magnetron) operation regimes. Thus, instead of two stable discharge states peculiar to each of the systems involved, one has four operation modes! More importantly, this opens an incredible opportunity to manipulate the hybrid discharge operation by varying the DC voltage applied to the magnetron electrode and RF power applied to the external inductive coil.

The details of mode transitions between the two operating modes of inductively coupled plasmas, which are quite similar to what has been discussed in Chapter 2 (with some minor differences in mode transition thresholds due to reduction of the plasma volume and redistribution of the electric potential and plasma species due to additional large electrode introduced into the chamber), will not be considered here. Instead, we will discuss the dynamics and physics behind the $D \rightarrow M$ transitions and the origin of complex hysteresis phenomena when both plasma-generating systems are operating.

At the beginning of mode transition experiments, when the voltage is gradually increased from the zero value, a dim discharge can be observed. When the voltage increases, the dark region between the plasma and the target surface decreases in size. During this period, the plasma luminosity is weak and the plasma bulk is almost uniformly located above the target surface. This operation mode is commonly referred to as the diode (D) mode.

When the voltage reaches a certain threshold denoted here as V_{DM}, the discharge suddenly brightens. Simultaneously, bright annular "plasma rings" hovering above the target surface are observed. When the voltage is further increased, the plasma luminosity also increases, with the annular rings still present. On the other hand, when the voltage is decreased from an initially high value, the plasma luminosity decreases. As long as V_{DM} is not reached, the plasma is still located within the annular zones. As the voltage is decreased until the turning point (threshold of the inverse $M \rightarrow D$ transition), there is an instantaneous dip in the luminosity. The plasma is freed from the annular zones and is once again almost uniformly distributed above the target surface. Further reduction in voltage only results in weakening the optical emission intensity of the discharge. For convenience, the process of increasing the cathode voltage will be referred to as the current-increasing phase, whilst the corresponding decreasing of voltage will be termed the current-decreasing phase.

The dynamic process, which leads to the brightening/dimming of the discharge and the formation of the bright annular plasma rings, can be explained as follows. Initially, when a DC voltage is applied to the magnetron electrode, because of the potential difference between the anode and cathode, the setup resembles a capacitor (i.e., $E = V/d$, where E is the electric field, V is the potential difference, and d is the distance between the plates). Therefore, an increase in voltage increases the electric field magnitude. This electric field is a primary source of energy needed to sustain the ionization and excitation processes in the discharge.

Thus, the ionization degree and proportion of the plasma species in excited states increases with the applied DC voltage. This in fact results in an increased discharge luminosity. At the beginning, when the voltage is relatively low, the plasma density is low and hence the plasma sheath thickness is large, i.e., the plasma bulk is seen relatively far from the target surface. In this case the electrons are not affected by the magnetic field, which is concentrated near the electrode surface. At this stage, the electrons behave just as in a conventional diode sputtering device. Hence, this discharge phase is termed as a D (diode)-mode discharge.

However, when the voltage is further increased to V_{EM}, the plasma sheath thickness is reduced, so that the plasma can be affected by the magnetic field strength. It is exactly at this point the electrons are suddenly driven (and confined) into the annular zones where the magnetic field magnitude is higher. In addition, the electrons are simultaneously subjected to a cycloidal motion due to the $\mathbf{E} \times \mathbf{B}$ drift. Thus, a Hall current is induced just above the target surface. With the emergence of the Hall current, the collision frequency between the electrons and neutrals increases significantly. Thus, the ionization is enhanced and the plasma becomes much brighter.

A subsequently rising cathode voltage further increases the excitation and ionization of the working gas, and hence, the plasma luminosity. At this stage, the discharge evolves into the magnetron (M) discharge mode. Hence, as the cathode voltage increases from a low value to magnitudes above V_{DM} (the mode transition point), the discharge transits from the D-mode to the M-mode. The same explanation can be adopted when the voltage decreases from a high value to a low value, eventually reaching V_{MD}—the threshold voltage for $M \rightarrow D$ mode transitions.

When the ICP discharge is in operation, the picture of mode transitions becomes a lot more complicated. If the magnetrons are switched off, one observes conventional $E \leftrightarrow H$ transitions described in detail in Chapter 2. However, when both discharges are run simultaneously, there is a significant distortion of the $D \leftrightarrow M$ and $E \leftrightarrow H$ mode transitions peculiar to the DC magnetron and ICP devices, respectively. In particular, one can observe complex hysteresis phenomena with multiple spiral loops, for example in the dependences of the cathode current (current on the magnetron sputtering target) and optical emission intensity of selected plasma species on the RF current in the external inductive coil.

The original investigations of the effect of the variation of the RF current on the cathode current (and also on TiN film deposition) have been conducted in the electromagnetic (H-mode) discharge mode by using a N_2(5 sccm)+Ar(50 sccm) mixture at a fixed cathode voltage of 260 V. The data have been collected using the optical emission spectroscopy system, together with the measurement of the target ion flux (cathode current), by varying the RF current [156].

Figure 3.15 shows the evolution of the cathode current (a) and the emission intensity of a selected sputtered target element, Ti I (λ = 494.8 nm; neutral titanium atoms) (b) when the peak-to-peak RF current is varied [156]. The data points of the cathode current values and the emission intensities are represented by sequentially numbered frames, with number "1" being assigned to the first data point collected. These points have been enumerated in sequence and thus a larger number corresponds to a data point collected at a later time moment.

It is not surprising that the measurements show that such a variation gives rise to the formation of distinct spiral loops when the cathode current is plotted against the RF current. Both figures exhibit similar spiral loops; each of them comprise two closed loops: one main loop and the other, a sub-loop. The closed sub-loop is within the main loop and its traces are close to the trace of the main loop in the RF current decreasing phase. The results in Fig. 3.15 reveal that the plasma generated by the ICP (H-mode)-enhanced DC magnetron sputtering is multistable. For instance, a single value of the RF current in the range 75–95 A can correspond to four different discharge states, i.e., four cath-

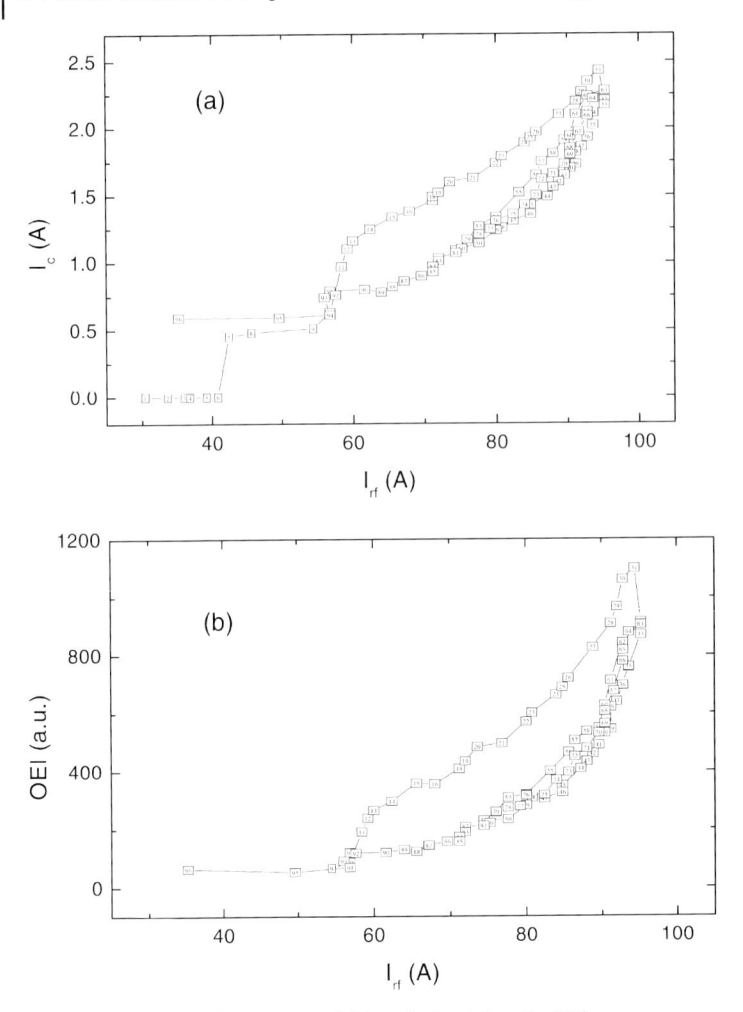

Fig. 3.15 (a) Cathode current and (b) emission intensity (Ti I 494.819 nm) *versus* peak-to-peak RF current for N_2+Ar discharge. The cathode voltage is -260 V [156].

ode currents and four emission intensities. It is remarkable that the plasma shows similar nonlinear characteristics for the two completely independent measurements [156].

The experimental discovery of the PEMS discharge multistability reveals the overwhelming complexity of hybrid discharges featuring multiple operation modes. Therefore, in applications, to ensure the acceptable reproducibility of the ion flux to the target electrode (reflected by the measurements in Fig. 3.15(a)) or number densities of working species in the vicinity of the substrate surface (reflected by the OEI measurements in Fig. 3.15(b)), one has to set the required level of the RF current consistently and in the same way from one experiment to another. For instance, the result of a direct increase of the RF current from 0 to 80 A will not be the same as its reduction from 85 A down to the 80 A if the required working point has been missed by 5 A while increasing the current from the initial zero value.

These fascinating hysteresis phenomena still await their conclusive explanations based on comprehensive discharge modeling. In the following section, we will consider a more advanced and more flexible arrangement of different means of source species generation. This facility is a further approach toward meeting the continuously rising demands of plasma-aided nanofabrication.

3.4
Integrated Plasma-Aided Nanofabrication Facility

A large number of functional nanoassemblies and nanofilms discussed in this monograph have been synthesized in custom-designed Integrated Plasma-Aided Nanofabrication Facility (IPANF); a photograph and schematic of which are shown in Fig. 3.16.

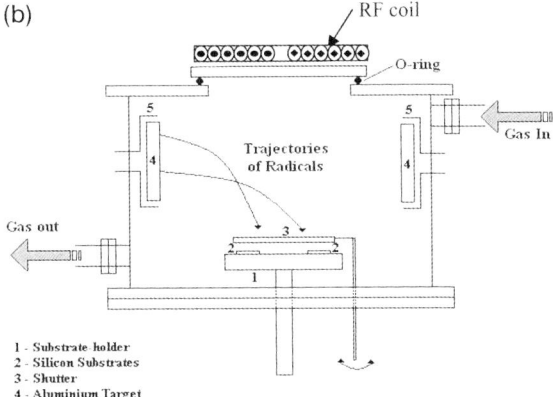

Fig. 3.16 Photograph and schematic of the IPANF [157].

This device encompasses the means of generation of highly uniform high-density plasmas by driving the external spiral (this configuration is sketched in Fig. 3.16(b); the coil is installed above the quartz top end plate, similar to the coil arrangement in the PEMSF (see Fig. 3.12)) or internal oscillating unidirectional RF currents (discussed in detail in Section 3.1), operation of low-pressure discharges in mixtures of inert and reactive gases, control of deposition processes by using the substrate stages with the temperature- and deposition area-control functions, multiple RF sputtering targets (two of them are shown in Fig. 3.16(b)), and advanced diagnostic instrumentation. The internal RF coil can be installed by replacing the top quartz window by the top assembly (similar to what is shown in Figs. 3.2 and 3.3) that houses the internal RF coil.

This facility is suitable for large-area plasma processing due to its large (50 cm) internal diameter. The RF current is driven in either antenna configuration by a 460 MHz RF generator via an in-house designed matching network. The exact ranges of the input power and parameters of the RF circuit to sustain the discharges in the required reactive gas feedstock and working pressure are estimated by using the results of our numerical simulations of the discharges sustained in reactive gases. To enable the efficient deposition of various thin films in the temperature-controlled regime, the moveable (in the vertical and azimuthal directions) substrate stage with a built-in thermocouple and external temperature control unit, is installed as shown in Fig. 3.16(b). An automated shutter (with or without any perforated pattern) enables one to partially cover any part of the surface being processed and control (e.g., focus) the ion/neutral fluxes from the plasma. Three equidistantly positioned (along the circumference of the chamber) 13.56 MHz RF magnetron sputtering electrodes enable the sequential and controllable release of the target material (e.g., Al, Fe, Ni, Co, In, Si, Ti, etc.) into the reactive gas environment. Optionally, the magnetron assemblies can be powered by using DC or pulsed power supply; thus, the DC/pulsed magnetron sputtering can be enabled.

This hybrid technique is similar to that adopted in the PEMSF of Section 2.3 and is particularly useful in the plasma-assisted synthesis of various semiconductor quantum confinement structures and biocompatible films, as discussed in other sections of this book. The advanced plasma diagnostic instrumentation includes Langmuir and magnetic probes, ultra-high-resolution optical emission spectroscopy, and quadrupole mass spectrometry (QMS). Further details of the plasma source operation, stability, plasma diagnostics, and parameters can be found elsewhere [157].

The Integrated Plasma-Aided Nanofabrication Facility can be operated in four main regimes:

- CVD mode (mode (i); no plasma, heated substrate stage);

- magnetron sputtering mode(s) (mode (ii); in an inert gas such as argon); with the DC, RF, and pulsed magnetron options;

- plasma-enhanced chemical vapor deposition mode (mode (iii); in mixtures of inert and reactive gases); with optional external and internal RF coil configurations; and

- combined PECVD and magnetron sputtering mode (mode (iv)); with the DC, RF, and pulsed magnetron and external/internal RF antenna options.

The variety of operating modes makes this device extremely flexible in the synthesis of numerous functional nanoassemblies, films, and coatings. The advantages of this hybrid (integrated) device become even more apparent when multistage processing is needed. We emphasize that the ability to perform multiple stages of complex nanofabrication processes in the same reactor chamber, without disrupting vacuum and/or exposing the samples to air, is one of the key requirements for successful commercialization of the plasma facilities and techniques.

For example, synthesis of the carbon nanotube/nanofiber/nanotip structures (considered in detail in Chapter 4) requires the modes (ii) and (iii) to be run consecutively. Indeed, mode (ii) is used to deposit nickel catalyst layers, and mode (iii) is required for pretreatment of the catalyst layer and actual PECVD of carbon-based nanostructures.

On the other hand, nanofabrication of various low-dimensional semiconductor structures (considered in detail in Chapter 5) requires plasma-assisted sputtering of one or more solid targets (mode (iv)) preceded, if necessary, by modes (ii) and (iii). For example, synthesis of SiC quantum dots on AlN buffer layers on Si(100) involves pretreatment of silicon surface (mode (iii)), deposition of AlN buffer interlayer by RF magnetron sputtering deposition in nitrogen-based plasmas (mode (iv)), followed by either concurrent sputtering of Si and graphite targets in argon (mode (ii))or PECVD in silane and hydrocarbon gas mixture (mode (iii)) or sputtering of either of Si (in hydrocarbon-based plasmas) or graphite (in silane-based plasmas) targets (mode (iv)).

In addition to the four main operation regimes, wherein solid and gaseous precursor species (building/working units) are generated, the facility can also be equipped with the liquid precursor feed system shown in Fig. 3.17. In particular, this system enables one to introduce reactive vapor, such as H_2O, much needed for successful hydroxylation of hydroxyapatite bioceramic coatings discussed in Chapter 6. Reactive gases and vapors also play a prominent role in reactive sputtering of magnetron targets; this technique is commonly referred to as the reactive magnetron sputtering. The way of introducing reactive vapor into the processing chamber is elucidated in Fig. 3.17(b). An inert gas (e.g., argon) passes, under pressure, through the water vapor feeding

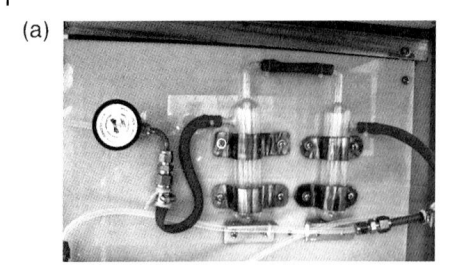

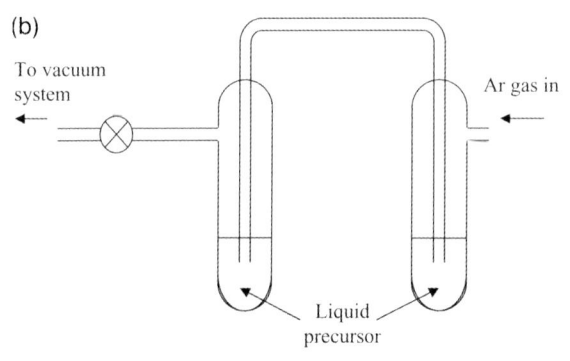

Fig. 3.17 Photograph (a) and schematic (b) of the liquid precursor feed system.

tubes, contributes to evaporation, and finally expels the vapor into the reactor chamber (Fig. 3.17(b)).

The outstanding flexibility of the Integrated Plasma-Aided Nanofabrication Facility is further evidenced by a large variety of different precursors (and hence fabrication processes) one can implement *in vacuo*. For example, in numerous applications considered in Chapters 4–6, low-to-intermediate pressure (0.07–26.6 Pa) discharges in the following main combinations of reactive gas mixtures and sputtering materials are used:

- $Ar+CH_4/C_2H_2+H_2$ for fabrication of various carbon-based nanostructures (Chapter 4);

- $Ar+SiH_4+H_2+N_2+O_2$ (with e.g., Ta, Ti metal precursors) for deposition of silicon-based quantum structures, barrier coatings/interlayers, and nanocrystalline films and nanostructures (Chapters 5 and 7);

- $Ar+N_2+SiH_4$ (with e.g., Al, In, Si, C, SiC, Ti, Ta, V, Er, etc. solid-state precursors) for the fabrication of various quantum dots, nanowires, nanostructured, and nanocrystalline films (Chapters 5 and 7);

- $Ar+Ti$ (metal precursor) + H_2O (liquid precursor) + hydroxyapatite (HA) for the fabrication of biocompatible calcium–phosphate-based coatings (Chapter 6);

in each specific process a large number of the above-mentioned combinations of the gaseous, liquid, and solid precursors can be used.

We emphasize that depending on the specific process requirements another reactive gases and sputtering targets can be used in the IPANF reactor. More details about the operation regimes and process parameters will be given in appropriate sections of this book.

3.5
Concluding Remarks

In this section, we have followed possible modifications and upgrades of the base version of the low-frequency inductively coupled plasma source of Chapter 2. First, by replacing the external flat spiral ("pancake") coil by an internal two-layered antenna that generates unidirectional oscillating radio frequency currents, one can improve the uniformity of the species number densities and eventually deposition fluxes, in particular, in the areas close to the chamber axis.

This nonuniformity is common to most of the ICP sources with external flat spiral coils. Work in this direction needs to be continued to elucidate general plasma stability criteria in the IOCPS plasma source and optimize the RF power deposition to increase the number densities of reactive species to above 10^{13} cm^{-3} at sub-kilowatt input powers, the level that has successfully been achieved in conventional LFICP sources [93].

Introduction of additional sources of gaseous, liquid, and solid precursor species by varying configuration and positioning of magnetron sputtering electrodes and using extra gas feed lines and liquid precursor feed systems significantly improve flexibility of the plasma reactors. Indeed, they become suitable for controlled delivery of the required building units to the nanoassembly sites, functionalization of the growth surfaces, complex surface processing, including simultaneous deposition of the desired building units and reactive etching of undesired species.

More importantly, installation of different source feeds, substrate processing features, and sophisticated diagnostic instrumentation in the same reactor chamber will ultimately enable one to implement multi-staged processes meeting rigorous industrial standards and specifications. The Integrated Plasma-Aided Nanofabrication Facility of Section 3.4 is the best candidate (among the plasma sources discussed in this chapter) to meet the demands of plasma-aided nanofabrication.

However, a lot of efforts are still warranted, in particular, to achieve a better stability of operation of hybrid magnetron and RF plasma sources and equip the plasma facilities in question with reliable diagnostic instrumentation. Finally, we believe that ICP-assisted pulsed magnetron sputtering systems, one of them currently being designed and commissioned at the School of Physics, the University of Sydney, holds very exciting prospects for processing of matter at nanometer and even sub-nanometer scales.

4
Carbon-Based Nanostructures

As has been discussed in the introductory chapter, one of the ultimate goals in the bottom-up approach to the self-assembly of nanostructured matter is to achieve a reasonable control and predictability in the size, chemical structure, architecture, and ordering of the nanostructures (NSs). Such nanostructures are among the main building blocks of a wide variety of mesoscopic functionalities and devices. Several applications, such as electron field emitters (EFEs) [158], require excellent ordering of the NSs both in the growth plane and in the growth directions [159]. Positional control in most of the existing nanostructure fabrication techniques is achieved by using prepatterned substrates and expensive nanolithography processes.

Thus, the apparent challenge is to develop a suitable technique for the growth of the ordered self-assembled nanoislands that promote bottom-up assembly of the functional nanostructures thereon without any prepatterning routines [160–162]. However, the efficiency of the specific envisaged application also depends on a number of other requirements, including shape, internal organization and chemical structure, phase and elemental composition, electrical/optical parameters, suitable mechanical and adhesive properties, and several others. From the manufacturing point of view, the number of steps required for the nanoassembly and subsequent device integration is to be minimized.

Self-assembled carbon nanotips (CNTPs) grown by chemical vapor deposition (CVD) on metal catalyst layers are promising for the electron field emission applications as efficient field enhancing structures with size-dependent electronic properties [163]. One of yet unresolved puzzles is to synthesize the ordered CNTP arrays under the process temperatures well below the metallic interconnect melting points to enable the efficient integration of the nanostructures into electronic/photonic devices [160, 161]. Most of the existing CVD techniques based on a thermal decomposition of hydrocarbon feedstock gases fail to meet the above requirement.

Another common requirement is to synthesize the field emitting bits (e.g., carbon nanotubes, nanofibers, or nanotips) with the optimized and balanced combination of their geometric and electron confinement properties. As was

Plasma-Aided Nanofabrication. Kostya (Ken) Ostrikov and Shuyan Xu
Copyright © 2007 WILEY-VCH Verlag GmbH & Co. KGaA, Weinheim
ISBN: 978-3-527-40633-3

already discussed above in this book, the best result for the electron field emission efficiency can be achieved when the CNSs involved have the highest possible aspect ratio and the thinnest possible width. The latter requirement is needed to enhance the electron confinement effects and ultimately reduce the work function, the energy needed for an electron to leave the material concerned. Meanwhile, the former requirement is crucial to ensure the strongest electric fields in the vicinity of the microemitter tips.

A reasonable combination of the above two factors should be used to optimize the electron emitter performance. However, better mechanical stability of high-aspect-ratio structures can be achieved by widening the base width of the CNSs, eventually ending up with the high-aspect-ratio conical nanotips, potentially outstanding candidates for the industrial carbon-based microemitters of the next generation. For this reason, Sections 4.1–4.4 are devoted to such carbon nanotip structures, and especially, the single-crystalline nanotips (Sections 4.3 and 4.4) that are extremely promising from the point of view of optimized electron confinement properties, structural integrity, chemical purity, electric field generation, and other factors. In particular, Sections. 4.1–4.3 report on various regimes of the plasma-aided nanofabrication of relevant carbon-based nanostructures, and Section 4.4 contains the results of the *ab initio* density functional theory computations of stable atomic structures of single-crystalline carbon nanotip microemitters.

Here, it would be instructive to recall the generic "cause and effect" approach that enables one to bridge the spatial gap of nine orders in magnitude by using proper atomic and radical building units [36]. There are experimental and modeling results on the generation, transport, and deposition of the building units required for specific nanoassemblies that support the above approach. However, this is not the main focus of this chapter. Instead, the primary aim of this chapter is to introduce the plasma-aided techniques used and comment on possible growth scenarios.

Section 4.5 is devoted to the discussion of the approaches and challenges of the plasma-assisted postprocessing of carbon nanotube-like structures. Examples of such postprocessing include but are not limited to the nanostructure coating with ultrathin functional films and control of electronic structure by introducing foreign elements (doping).

In the final section of this chapter (Section 4.6) we explore the PECVD of quasi-two-dimensional carbon nanostructures, such as nanowalls and nanoflakes of different sizes and intricate configurations. This section also contains our comments on probable plasma-generated building units of such nanostructures. Wherever practical, we also discuss some of the main issues related to the plasma-based methods for nanoassembly of relevant carbon-based nanostructures.

4.1
Growth of Carbon Nanostructures on Unheated Substrates

It is a common requirement for conventional thermal CVD-based CNS growth processes to maintain deposition substrate temperatures high enough to ensure metal catalyst melting and fragmentation into smaller nanoparticles. In some processes, surface temperatures of externally heated substrates can reach 800–900 °C and even higher. It is remarkable that many plasma-aided nanoassembly processes do not require any external substrate heating and yet yield the nanostructures of comparable and even superior quality compared to thermal CVD.

In this section, we consider the high-density inductively coupled plasma-assisted self-assembly of the ordered arrays of various carbon nanostructures for the electron field emission applications and show that the nanostructures in question can be grown without any external substrate heating [160, 161]. It is remarkable that various CNSs such as carbon-based nanoparticles, nanotips, and pyramid-like structures, with the controllable shape, ordering, and areal density can be assembled under remarkably low process temperatures (260–350 °C) and pressures (below 100 mTorr), on the same Ni-based catalyst layers, in a DC-bias controlled floating temperature regime [160]. More importantly, this technique enables one to achieve a reasonable positional and directional ordering, elevated sp^2 content, and a well-structured graphitic morphology without the use of prepatterned or externally heated substrates.

It is notable that several plasma-enhanced CVD methods have recently proved their efficiency in the low-temperature synthesis of various carbon-based nanostructures and the process temperatures as low as 120 °C have already been reported [75]. However, in many cases reporting such low substrate temperatures, one usually refers to the temperature underneath the substrate, measured by thermocouples built in the substrate stage and not the actual temperature of the substrate surface exposed to the plasma.

In reality, the temperature on the surface of the metal catalyst layer exposed to the plasma can be quite different owing to intense surface bombardment by impinging ions that transfer their energy into thermal energy of the catalyst. Recent experimental and numerical results suggest that in some cases the actual temperature of the catalyst layers can be 100–150 °C higher than the temperature of the underlaying substrate [36].

The required ordering of the carbon nanostructures in the growth direction can be achieved, e.g., by applying a DC bias to the substrate [73,74]. Nevertheless, most of the existing plasma-based methods still heavily rely on external substrate heating to activate the catalyst layer and promote the carbon nanostructure growth.

As will be discussed below in this chapter, one can manage the growth of carbon nanostructure arrays without prepatterning of the substrate in high-density reactive environments of low-frequency (∼460 kHz) inductively coupled plasmas (ICPs) and synthesize different architectures of the individual nanostructures in the low temperature range ∼260–350 °C on the same (composition and thickness) metal catalysts by varying the DC substrate bias [160].

Moreover, this technique does not require any external substrate heating and the necessary activation of the Ni-based catalyst layer is achieved by a combination of the reactive etching and surface heating by a hot neutral gas and energetic ions in the reactor chamber. In this way, it appears possible to control the site density and spacing, as well as to achieve excellent uniformity of the CNTPs and carbon pyramid-like structures (CPLSs) over the large areas. Likewise, the nanostructures grown in this way also feature an elevated content of sp^2-bonded carbon. The above-mentioned factors are particularly favorable for the development of new-generation carbon-based microemitters.

4.1.1
Process Details

Now, following the original work [160], we discuss the experimental details and main results relevant to the growth of carbon nanotip-like structures on unheated Ni-catalyzed substrates. The deposition reactor based on the source of low-frequency inductively coupled plasmas described in detail in Chapter 2 (see also Ref. [93]) on ∼30–40 nm-thick Ni/Fe/Mn catalyst layers predeposited on Si(100) substrates (with different doping levels and conductivity) in an ultrahigh vacuum sputtering reactor. The substrates are placed on the top surface of a DC-biased substrate stage positioned in the area of the maximal electron/ion density in the plasma reactor.

Working gases Ar, H_2, and CH_4 are introduced into the chamber sequentially. A 30 min wall/substrate conditioning in argon is followed by a 20 min catalyst activation in the Ar+H_2 mixture. Thereafter, a carbon source gas CH_4 is fed into the chamber for the entire duration of the 40-min PECVD process. The partial pressures in the Ar+H_2+CH_4 gas mixture are maintained at 60, 17, and 8 mTorr, respectively.

The high-density ($n_{e,i} \sim 10^{12}$ cm^{-3}) plasma is sustained with RF powers of ≈2 kW, which corresponds to the RF power density range ∼0.09–0.11 W/cm^3. We emphasize that in the floating (DC bias-controlled) temperature growth regime the substrates are heated internally by the hot working gas and intense ion fluxes.

4.1.2
Synthesis, Characterization, and Growth Kinetics

Figure 4.1 shows FE SEM micrographs of the nanostructures grown at different DC substrate biases V_s [160]. The corresponding variation of the substrate

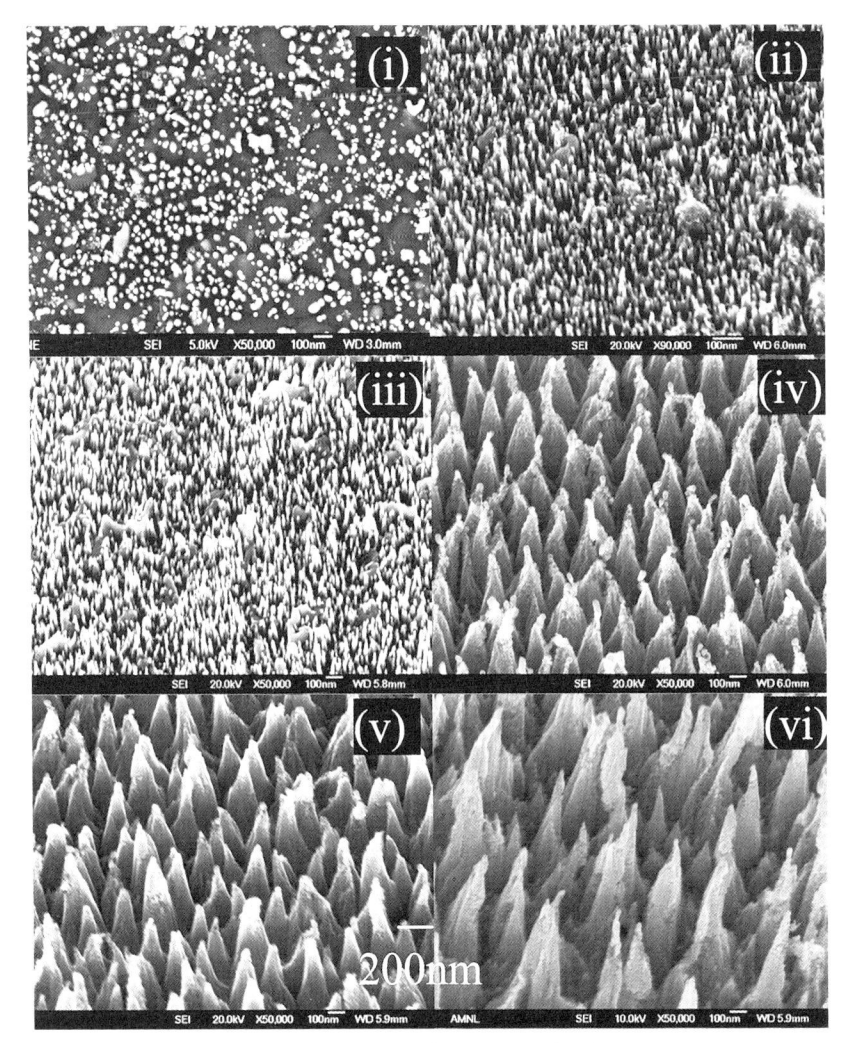

Fig. 4.1 Field Emission Scanning Electron Microscopy of carbon nanostructures grown at different DC biases. Micrographs (i)–(vi) correspond to $V_b = 0, -60, -100, -200, -300,$ and -400 V, respectively [160].

temperature T_s is given in Fig. 4.2, where and below the same numbering as in Fig. 4.1 is used.

At a zero DC bias (Fig. 4.1(i)), only nanoparticles and nanoparticle agglomerates partially cover the surface and there are no other nanostructures visible. One can observe that there is a minimum negative DC bias (in the 50–60 V range) that enables the growth (with an average growth rate of 2–5 nm/min) of small (typically 10–20 nm in width and 80–100 nm in height) carbon nanotips (Fig. 4.1(ii)). The minimum T_s for the nanotip growth appears to be

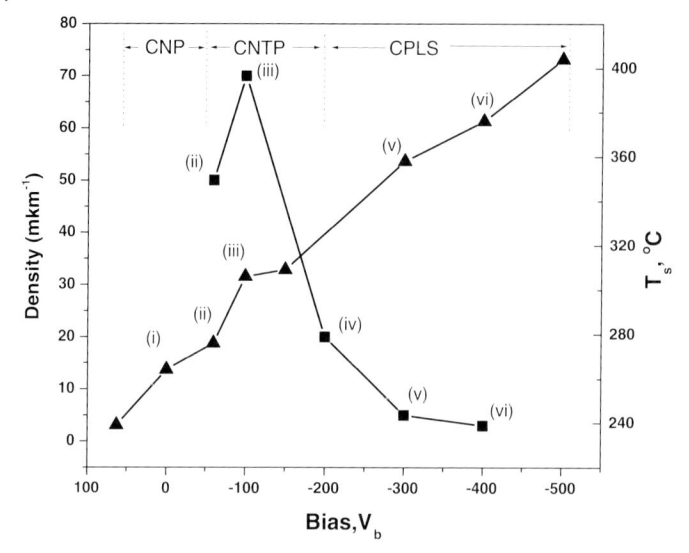

Fig. 4.2 Substrate temperature (triangles) and nanostructure surface density (squares) versus V_b for the same conditions ((i)–(vi)) as in Fig. 4.1 [160].

$\approx 270\,°\text{C}$ [160]. With an increase of the bias to -100 V (and the substrate temperature to $T_s \sim 310\,°\text{C}$), the nanotips grow in size (Fig. 4.1(iii)) and their linear density (the averaged number of individual nanostructures per linear micron) increases as shown in Fig. 4.2.

When V_s increases further to -200 V, the CNTPs disappear giving rise to the pyramid-like structures (typically 100–150 nm in width and 400–450 nm in height) shown in Fig. 4.1(iv). The minimum T_s that enables the growth of the CPLSs (with the average growth rate 10–13 nm/min) is estimated to be $\sim 320\,°\text{C}$ [160].

We note that the CNTPs are still observable at the substrate biases $V_s \sim 150$ V and temperatures $T_s \sim 300\,°\text{C}$. The assembly of the carbon pyramid-like structures becomes pronounced and peaks at $V_s = -300$ V (Fig. 4.1(v)), when the substrate temperatures rise to $\sim 350\,°\text{C}$. Meanwhile, the CPLSs areal density diminishes with DC bias, with approximately four structures per 1 μm at $V_b = -400$ V, as can be seen in Figs. 4.1(vi) and 4.2.

It is remarkable that all the resulting NSs are *aligned vertically and perpendicular to the substrate surface*. Physically, the direction of the DC electrostatic field is an energetically most favorable orientation of the one-dimensional CNSs [77]. This issue has already been briefly discussed in the introductory Chapter 1.

Thus, Figs. 4.1 and 4.2 reveal an excellent ordering and uniformity of the CNTPs and CPLSs both in the growth direction and over the large surface areas. Furthermore, even a modest change in the substrate bias (δV_s ∼50–100 V), results in a structural transformation of the CNTP arrays into CPLSs. Quite similar transformations are the case for a number of different CNSs [78]. We emphasize that the process temperatures and gas pressures of the experiments of Tsakadze et al. [160] are noticeably lower than in many relevant plasma-based methods [73, 74, 77–79, 158].

A detailed SEM analysis also reveals that the carbon nanotip-like structures structures in Figs. 4.1(ii) and (iii) do not grow on the nanoparticles and nanoparticle agglomerates (the latter also grow in size with bias and form large islands of irregular shapes) and follow tiny growth islands on the thermally and plasma-chemically activated catalyst surface. On the other hand, no more nanoparticle agglomerates can be seen when larger carbon pyramid-like structures start to self-organize into ordered surface morphology elements (Fig. 4.1(iv)). A further comparison of the surface morphologies of the catalyst-coated samples removed from the reactor chamber immediately after the hydrogen etching stage and those at the initial growth stages reveals a remarkable correlation between the widths of the CNTPs and the nanoislands created as a result of the reactive chemical etching (RCE) and thermal activation of the catalyst surface.

It is not clear, however, if the preferential growth of the CNTPs on the nanoisland edges [79] is the case in the experiments concerned [160]. Taken a small size of the growth islands, low temperatures of the process, and relatively large thickness of the catalyst layer, one can speculate that the reactive chemical etching rather than the plasma heating is a dominant surface activation mechanism here. However, this assertion still needs a detailed verification.

Figure 4.3 shows X-ray diffraction and Raman spectra of the nanostructured films of our interest here [160]. It is interesting that the Raman spectra shown in Fig. 4.3(b) exhibit two well-resolved G (at 1580 cm^{-1}) and D (at 1350 cm^{-1}) peaks suggesting the formation of a well-structured graphite-like morphology [160]. A notable photoluminescence background indicates the presence of the disordered nanotip structures and large amounts of amorphous carbon (a-C) [79]. With an increase of the DC bias, the ratio of the magnitudes of D and G peaks I_D/I_G grows. Thus, the amount of unorganized graphite nanocrystals and the number of nanosized sp^2 clusters increase. These clusters play an important role in the field emission from nanostructured carbons [164]. From Fig. 4.3(b), one can note that the D peak is quite high and its amplitude is comparable to that of the G peak, which suggests the presence of microscopic defects in the structure.

It is also remarkable that the structures grown at $|V_s| > 200$ V are not only morphologically but also structurally different from those grown at lower DC biases. As can be seen in Fig. 4.3(b), at $|V_s| > 200$ V the diffraction peaks (006)

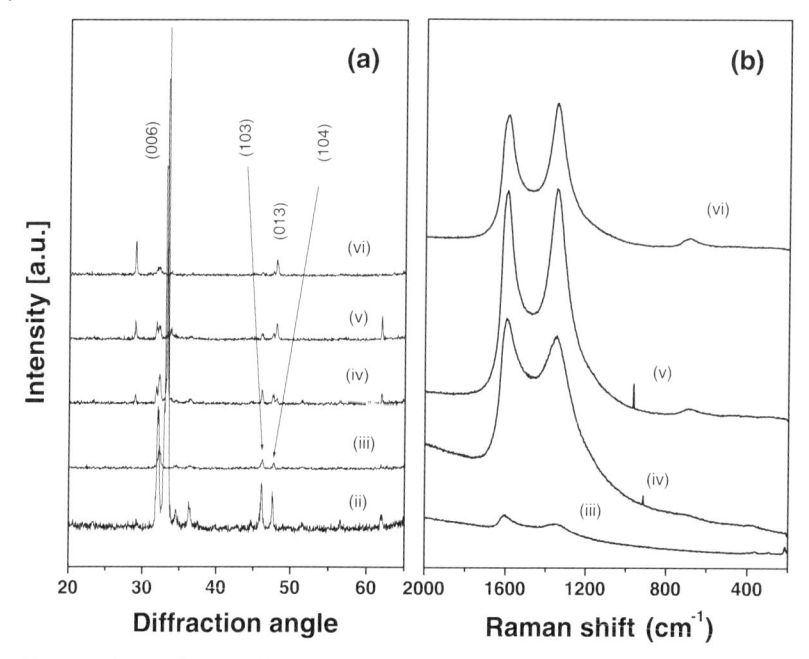

Fig. 4.3 X-ray diffraction (a) and Raman (b) spectra of the films grown at the same conditions ((i)–(vi)) as in Fig. 4.1. In case (iii), the amplitude is multiplied by a factor of 10 [160].

and (104) are split and a new peak at $2\Theta = 29°$ appears. Thus, the preferred orientations of the crystal growth change when the formation of the carbon pyramid-like structures is triggered [160]. This can serve as a qualitative indicator of the enhanced growth of the pyramid-like structures.

It is thus quite likely that the films concerned contain at least two (crystalline and amorphous) carbon phases and can be termed polymorphous in a manner similar to nanostructured silicon-based films [46]. One can also note that the resulting structures are strongly affected by the competition of the reactive chemical etching of the growth surface by hydrogen and PECVD of a new carbon material from the gas phase. In the low-temperature ($<350\,°C$) regime [160], a chemisorption of CH_3 radicals to hydrogen-terminated carbon surfaces is one of the most probable mechanisms of the carbon film growth [165]. In this case the CH_3 radicals stick to the surface as a result of a bias-controlled activation of the hydrogen-terminated carbon bonds by the impinging ions [165]. Therefore, activation of the nanostructured surfaces discussed here can indeed be controlled by the substrate bias voltage.

Apparently, the value of DC bias does affect the relative efficiencies of the competing RCE and PECVD processes. Generally, when the anisotropic etching prevails (at $50\,V < |V_b| < 150\,V$), the growth of high-aspect-ratio carbon nanotips is favored. At higher ($|V_b| > 200$ V) bias and substrate temperature, the reactive chemical etching cannot keep the pace with the elevated amount of carbon material deposited onto the growing surface and extruded through the metal catalyst, so that the pronounced lateral growth evolves giving rise to the pyramidal structures (Fig. 4.1) [160].

On the other hand, there is a minimum DC bias required for the nanostructures to start growing. Indeed, in the dissolution/precipitation growth mechanism, formation of NiC on the catalyst surface requires external supply of the energy (≈ 9.8 eV), which can be provided as a result of heavy particle collisions involving sufficiently accelerated (by the DC electric field) cations in the near-substrate sheath [158]. On the other hand, the observed minimum bias is needed to provide the minimum temperature of the surface catalyst through the energy transfer from the impinging ions to the growth surface.

Low surface temperatures for the CNTPs growth can be attributed to the outstanding properties of the LF ICPs such as very high plasma densities, absence of high near-substrate self-biases common for many parallel-plate plasma reactors, and externally controllable ion fluxes onto the substrate [93]. In the plasma of interest here, the near-substrate sheath (~ 1 mm) is smaller than the mean free path of CH_3 radicals, which is ~ 1.76 mm at 85 mTorr. Thus, contrary to the case reported by Shiratori et al. [158], the LF ICP sheath is nearly collisionless. Moreover, the bias-controlled cation fluxes onto the substrate are very strong and can exceed the diffusion fluxes of neutrals due to near-substrate density gradients [166].

An outstanding ability of the plasma-catalyst system to support the growth of the nanostructures at low surface temperatures ($T_s < 300\,°C$) can also be due to the excellent fragmentation of the Ni catalyst layer in this temperature range. Specifically, insufficient fragmentation of nickel films on silicon surfaces above 300 °C can be explained by the diffusion of Ni into the Si, leading to the formation of a silicide, $NiSi_x$ [79]. It is yet another advantage of this technique that no special barrier interlayers (such as SiO_2), adversely affecting the adhesion properties of the film, are required [160].

Therefore, the original results [160] suggest that high-density environments of reactive RF plasmas are indeed favorable for the low-temperature (below 350 °C) fabrication of ordered arrays of carbon nanotips and nanopyramid-like structures. In this process the ordering, surface density, and architecture can be controlled by a competition of the reactive chemical etching, plasma-enhanced chemical vapor deposition, and growth island self-organization processes.

However, better structural and positional uniformity is required before the structures concerned can be used in the electron field emitter technology. One of the reasons is the remaining uncertainty in the actual surface temperature of the catalyst layer, in particular due to its strong dependence on the applied

DC bias to the substrate. On a positive note, the experiments have shown fairly reproducible results and should be continued in the future. In the following section, we will consider another plasma-aided growth regime of the carbon-based nanostructures, wherein the surface temperature is externally controlled by using advanced temperature controllers.

4.2
Temperature-Controlled Regime

In this section, the main features of the temperature-controlled growth regime of vertically aligned carbon-based nanostructures in low-pressure RF plasmas are considered [162]. Electron field emitting properties of some of the structures concerned will also be discussed. The main details of the CNS synthesis are the same as in Section 4.1. However, in this set of experiments the substrate temperature is maintained at the preset level by means of external heating.

For this purpose, a custom-designed substrate stage equipped with an insulated heating element encapsulated in the substrate-holding assembly, digital temperature controller, and a thermocouple, is introduced in the LFICP-based plasma reactor as shown in Fig. 4.4. The substrate stage is electrically insulated from the grounded chamber and externally biased with a negative DC potential.

This growth regime of the carbon nanotips can thus be termed as a temperature-controlled growth (TCG) regime with the external control and stabilization of the substrate temperature. To enhance the ion fluxes onto the samples, the top surface of the substrate holder is positioned in the area of the maximal electron/ion density in the plasma reactor. Field emission scanning electron microscopy (FE SEM), X-ray diffraction (XRD, CuK$_\alpha$ source with 0.154 nm X-ray wavelength) and Raman (Ar$^+$ laser, 514.5 nm spectral line) spectroscopy are used to characterize the morphology, crystalline structure and chemical states of the nanostructures.

To find the optimum value of the substrate bias, the original experiments started with a preselected substrate temperature $T_s = 500\,°C$, and the bias was varied from 0 to -300 V. The best results for the area density of CNTPs can be achieved when the substrate bias is ≈ -80 V [162]. Moreover, some minimum DC bias value V_{min}, of ≈ -50 V is required to trigger the nanostructure growth. Remarkably, it is much lower than reported elsewhere under similar conditions [74]. At zero bias, only massive nanoparticle agglomerates that partially cover the surface can be seen and no other CNSs are observable. When the substrate bias is increased to $V_b = -50$ V, the surface morphology comprises irregular-shaped structures (ISS), presumably formed from the nanoparticle agglomerates, and individual carbon nanotips (typi-

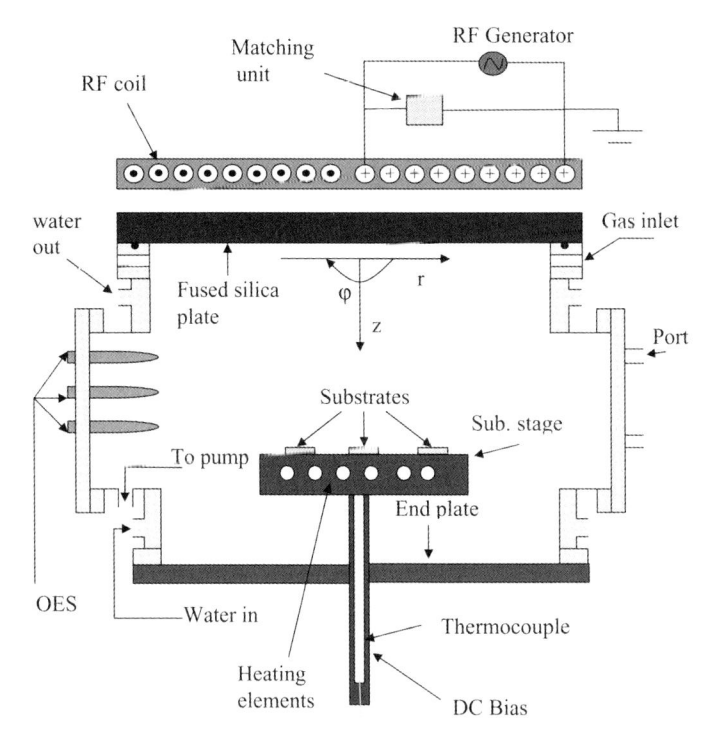

Fig. 4.4 Schematic of the LFICP plasma reactor with the substrate stage, external heating element, and built-in thermocouple [167].

cally 10–20 nm in diameter and 50–70 nm in height) rarely distributed over the surface between the ISSs. Further increase of the bias yields higher surface densities of the needle-like structures and reduces the amount of the nanoparticle agglomerates on the surface.

We note that at fixed substrate temperatures the density of the nanotips strongly depends on the bias. Indeed, a modest change of V_b to -60 V leads to a pronounced increase of the number of CNTPs per unit area [162]. From this point of view, the best results can be achieved by applying the DC bias $V_b = -80$ V. In this case the nanotips almost entirely cover the areas between the ISSs; however, their ordering still remains quite poor. At the same time, a strong photoluminescence background in Raman spectra suggests that there is a notable content of the amorphous carbon (a-C) phase in the films.

In an attempt to inhibit the deposition of amorphous carbon and minimize the CNTP deposition temperature, the nanostructure synthesis was conducted at lower T_s (300–400 °C). The results of the nanostructure growth in this temperature range and fixed DC bias of -80 V are shown in Fig. 4.5 [162]. One can notice that at $T_s = 400$ °C (Fig. 4.5(a)) the ISSs and self-organized nanotip "bundles" are still present in the film.

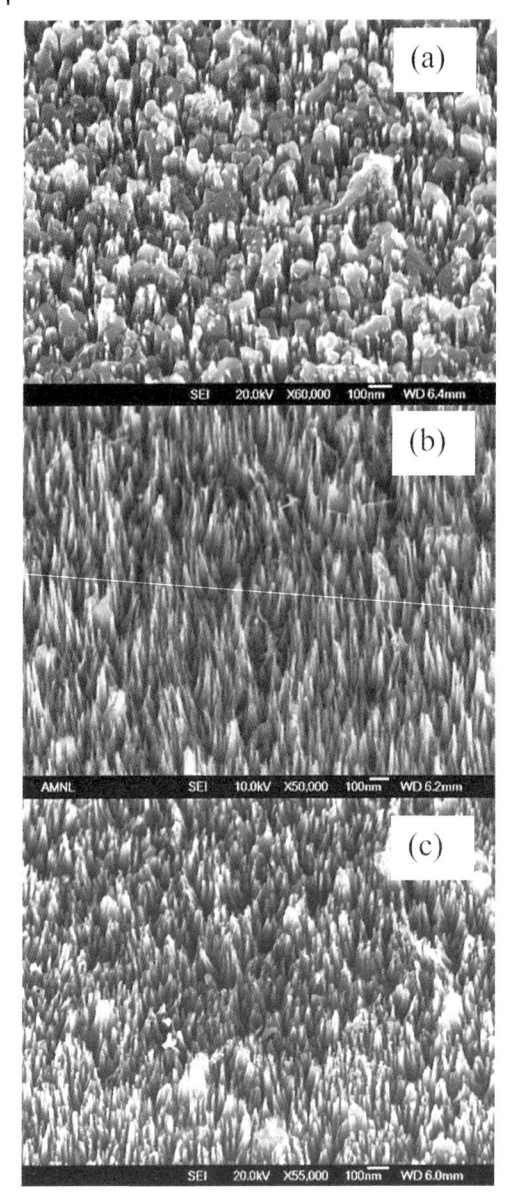

Fig. 4.5 FE SEM images of CNTP structures grown at $V_b = -80$ V and $T_s = 400$ (a) 350 (b) and 300 °C (c) [162].

However, a further decrease of the substrate temperature to 350 °C and 300 °C results in the growth of more regular and well-resolved nanopatterns with the carbon nanotips (20–30 nm in diameter and 140–160 nm in height) (Figs. 4.5(b) and (c)). More importantly, ordering and uniformity of the CNTP structures (critical factors in the development of electron field emission devices) significantly improves as compared to the cases with surface temperatures exceeding 400 °C.

The estimated value of the film thickness is ≈ 500 nm. The linear density (per linear micron) of the individual nanoassemblies is also affected by this change in the temperature, with the average number of CNTPs being ∼32–47 µm^{-1}. It is important to mention that the nanopattern shows an excellent surface uniformity over surface areas up to ∼10 cm^2 and even larger [162].

From Figs. 4.5(b) and (c) one can conclude that an excellent surface uniformity of the nanotip pattern can be achieved in the low-pressure (<13.2 Pa) range most suitable for microelectronic manufacturing. Furthermore, the ordering of the nanostructures is quite sensitive to the changes in the substrate temperature. It is notable that the process temperatures and gas pressures [162] are noticeably lower than in many relevant plasma-based methods of fabrication of various carbon-based nanostructures [73, 74, 77–79, 158, 168].

The films feature a high degree of crystallization, which is evidenced by well-resolved peaks in the XRD spectra [162]. More importantly, the intensities of most of the diffraction peaks vary with the decrease of the substrate temperature from 500 to 300 °C, which indicates on the changes in the preferential crystal growth direction with T_s. It is interesting that the XRD spectrum of the films grown at 300 °C features several broadened peaks in the range of diffraction angles 2Θ of 33–37° indicating on the presence of nanosized crystals [169] otherwise not present at other temperatures.

The origin of the nanocrystals can be attributed to the diffusion properties of the Ni-based catalyst layer [162]. Specifically, at the temperatures above 300 °C, the diffusion of Ni into Si (leading to the formation of a silicide NiSi$_x$) is inhibited, which results in an excellent fragmentation of the Ni-based layer into nanosized particles [79]. Thus, more efficient suppression of the interface diffusion at $T_s < 300$ °C results in better nanostructuring of the catalyst and hence to the XRD peak broadening. On the other hand, since the growth regime of interest here features quite low temperatures (300–400 °C), there is no need for any special barrier interlayer (e.g., SiO$_2$ or TiN [164]) that can adversely affect the adhesion properties of the nanostructured film.

Raman spectra of the nanofilms show two distinct peaks [162]. Specifically, the tangential C–C stretching mode corresponding to the G peak at ≈ 1580 cm^{-1} is attributed to the crystalline graphite, whereas the D peak at ≈ 1350 cm^{-1} is attributed to the disorder-induced (e.g., due to the presence of defects in curved graphite sheets) Raman scattering from sp^2-hybridized carbons. The location and well-defined shape of the D and G peaks suggest a well-structured graphitic morphology in the films. One of the Raman peaks, at ∼211 cm^{-1} is attributed to the radial breathing mode of the carbon needle-like structures and serves as an additional indicator of the presence of the nanostructures. It is remarkable that the photoluminescence background decreases at lower temperatures, which indicates on the reduced amorphous carbon content [162].

Interestingly, the ratio of intensities of the D and G peaks I_D/I_G becomes higher at lower surface temperatures T_s, which indicates on an increased number of nanosized sp^2 clusters that play an important role in the field emission from nanostructured carbons [164]. The observed shapes and frequency shift of the major Raman peaks can be described by the conventional phonon confinement model [170,171]. By using this model, one concludes that the crystallite size is the smallest at $T_s = 350\,°C$. We recall that the process temperatures and gas pressures of the nanostructure growth experiments [162] are lower than in many existing low-temperature plasma-based methods.

However, lower ($\sim 350\,°C$) substrate temperatures do not necessarily lead to the best results for the electron field emission properties. Figure 4.6 shows the electron field emission properties of the synthesized nanopattern and presents the I–V characteristics of the CNTPs grown at different substrate temperatures [162]. One can see that the needle-like structures grown at low substrate temperatures yield lower field emission currents. The field emission current becomes stronger with an increase of the substrate temperature. The field emission threshold fields are 16.6, 11, and 6.4 V/μm at $T_s = 350$, 400, and $500\,°C$, respectively. The threshold field is defined as the electric field magnitude when 0.1 μA emission current can be emitted by the surface area of 0.28 cm^2 [172].

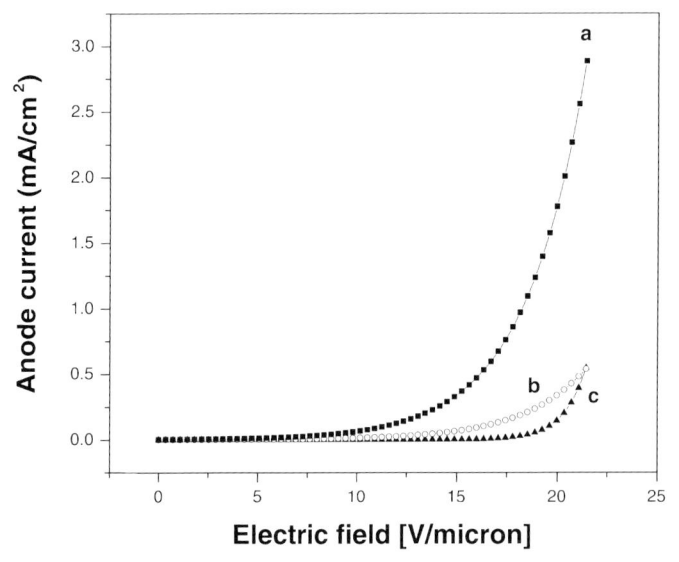

Fig. 4.6 Field emission properties of the nanopatterns grown at surface temperatures of (a) 500, (b) 400 (multiplied by 10), and (c) 350 °C (multiplied by 100) [162].

From Fig. 4.6, one can conclude that the rarefied "forests" of carbon nanotips produces much higher emission currents than denser patterns synthesized at lower substrate temperatures. These results suggest that the shielding of the electric field between closely packed carbon nanotips in densely populated patterns effectively decreases the field emission from the high-aspect-ratio nanotips.

On the other hand, in the rarefied nanotip "forest" the field shielding effects are apparently weaker due to the lower surface density of the nanostructures. It is also of interest to mention that the nanopatterns are remarkably stable in air over long periods of time. In fact, the field emission measurements made after several days or months after the deposition show the same results [162].

We now comment on the growth kinetics of the carbon nanotip structures of our interest here. Similar to the case discussed in the previous section, the resulting surface morphology is strongly affected by the competition of the reactive chemical etching of the growth surface by hydrogen and plasma-enhanced chemical vapor deposition of a new carbon material from the gas phase. Apparently, the value of the substrate temperature does affect the relative efficiencies of the competing chemical etching (combined with physical sputtering) and PECVD processes. At low substrate temperatures ($300\,°C < T_s < 350\,°C$) the anisotropic etching prevails and the growth of high-aspect-ratio CNTPs is favored [162]. Furthermore, a variation of the substrate temperature does affect the deposition rate of neutral particles from the ionized gas phase. When T_s increases, the gas heating is more efficient near the deposition surface. Since the gas density in the plasma bulk is higher than near the substrate surface, increasing the substrate temperature effectively elevates the pressure gradient (and hence the diffusion flow of neutrals) in the near-electrode area.

On the other hand, measurements show that the ion current on the substrate remains essentially the same at different temperatures [162]. We can thus relate an overall increase in the deposition rate at higher T_s to the neutral radicals. Meanwhile, higher near-substrate temperature gradients result in stronger thermophoretic (in our case repulsive) forces on the larger (typically in the 10–100 nm range) carbon-based nanoparticles grown in the ionized gas phase.

Similar to Section 4.1, the nanofilms grown in the temperature-controlled regime can also be termed polymorphous in a manner similar to nanostructured silicon-based films. Therefore, the successful growth of the carbon nanostructures does imply a preferential growth of the nanostructured crystal phase and a strongly inhibited growth of the amorphous phase, the latter being an unwelcome component in many device grade films [168].

Physically, in the gas/substrate temperature range 300–350 °C, the amorphous carbon phase predominantly grows as a result of a chemisorption of

CH$_3$ radicals to hydrogen-terminated carbon surfaces. In this case the carbon dangling bonds can be activated by intense fluxes of the impinging (with the main contribution from Ar$^+$) ions [165]. It is interesting that the presence of the amorphous carbon phase on the substrate surface can be minimized by a heavy dilution of the carbon-bearing gas (CH$_4$ or C$_2$H$_2$) in the etching gas (e.g., H$_2$ or NH$_3$) [168], which leads to more effective and faster etching processes.

Moreover, fabrication of carbon needle-like structures does require an optimized DC bias [162]. Specifically, the bias should not be too low to be able to sustain the minimal level of the reactive ion etching and physical ion sputtering to activate the catalyst layer. Meanwhile, the DC bias should be strong enough to support the preferential growth direction and vertical alignment of the CNTPs. On the other hand, V_b should not be too high to avoid unwelcome over activation of the carbon dangling bonds, which is consistent with several reports that a minimum (typically ≈ -50–100 V) CD bias is required for carbon material to be deposited on the substrate [74, 162].

Similar to the explanation given in the previous section, the minimum DC bias for the temperature-controlled growth of carbon nanotips can be explained by noting that in the dissolution/precipitation growth mechanism, carbonization of nickel on the surface requires external supply of the energy (~ 9.8 eV), which can be provided as a result of heavy particle collisions involving sufficiently accelerated (by the DC electric field) cations in the near-substrate sheath [158]. When the bias is too large, the PECVD of the amorphous carbon (as well as an unwelcome physical sputtering of the growth surface by argon ions) becomes dominant. On the contrary, when the bias voltage is low, the kinetic energy of cations is not sufficient to carbonize nickel and then the CNTP structures are not formed [162].

To conclude this section, we emphasize that the resulting carbon nanotip structures are aligned vertically and perpendicular to the substrate surface. The alignment direction is the same as the direction of the DC electrostatic field in the near-substrate area, and is an energetically most favorable orientation of one-dimensional carbon nanostructures [77]. Remarkably low substrate temperatures (300–350 °C) sustaining the growth of the carbon nanotip structures can be attributed to the high efficiency of the plasma-surface interactions in high-density inductively coupled plasmas. In addition to high ion densities ($n_i \sim 10^{12}$ cm^{-3}), the LF ICP reactor features very low near-substrate sheath potentials, which make it possible to efficiently control the ion fluxes on the substrate by the DC bias [93] (see also Chapter 2).

4.3
Single-Crystalline Carbon Nanotips: Experiment

Nanosized carbon-based structures are widely regarded among the most promising materials for nanotechnology applications. Recently, various types of low-dimensional carbon nanostructures have attracted enormous research interest because of their virtually unlimited technologically important applications, including electron field emission displays, molecular electronic device components, novel composite materials with unique, enhanced mechanical properties, hydrogen storage devices, bioimaging markers, and smart drug delivery.

In this section, we introduce an interesting sort of crystalline carbon nanostructures and discuss their main characteristics. Such structures are single-crystalline carbon nanocone-like assemblies with the preferred crystallographic orientation (002) and have been synthesized by using low-frequency inductively coupled plasmas in $Ar+CH_4+H_2$ gas mixtures [157, 167, 173, 174] discussed in detail in Chapter 2.

Typical dimensions of the nanoneedle-like conical nanotips are 60–100 nm in diameter (at base) and up to 1 µm in height. Interestingly, these graphite-layered, sharp-tip crystals are fundamentally different from all carbon nanostructures discussed so far. Scanning electron microscopy reveals that high-density, vertically aligned sharp-tip carbon crystal structures are oriented along the normal to the substrate as can be seen in Fig. 4.7. It is worth noting that the alignment direction is same as the direction of the electric field in the plasma sheath that separates the plasma bulk and the nanostructures.

On the other hand, high resolution transmission electron microscopy (HRTEM) shows that the nanotips are made up of continuous horizontal

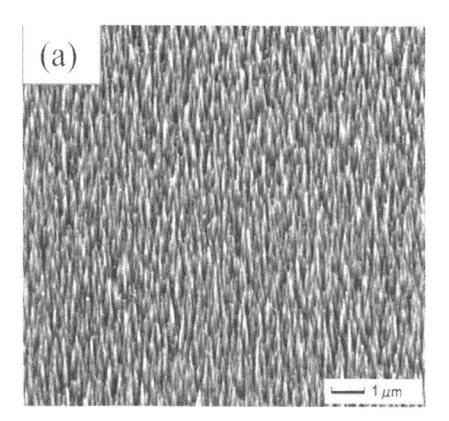

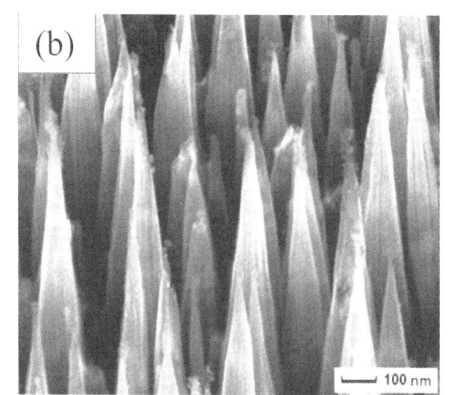

Fig. 4.7 FESEM images of carbon nanotips synthesized in low-frequency inductively coupled plasmas: (a) large-area view, (b) fine structure [173].

planes (up to several thousand layers) perpendicular to the growth direction (002) (Fig. 4.8). From the applications perspective, such sharp carbon nanostructures are ideal for electron field emission displays (FEDs). More details of the synthesis and characterization of such unique nanostructures follow.

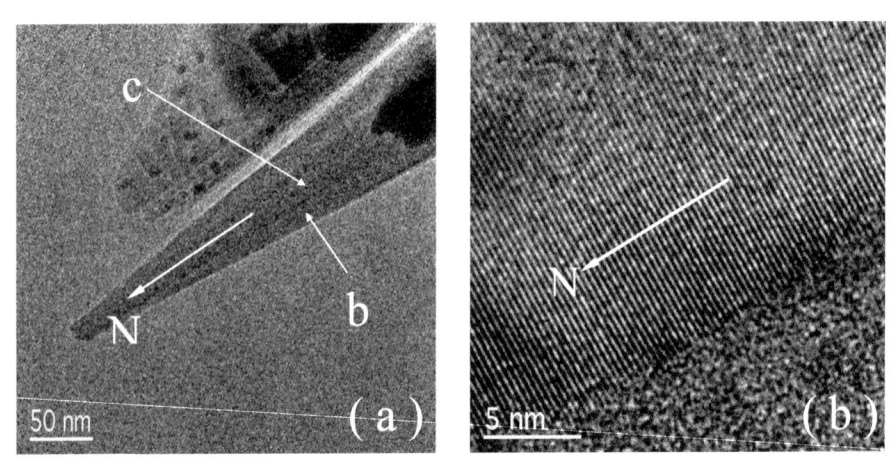

Fig. 4.8 TEM images of the single-crystalline carbon nanotip: (a) low-magnification; (b) high-resolution image taken from the lateral side of the nanostructure [173].

Carbon nanocone arrays can be fabricated by using a low frequency inductive coupled plasma reactor [93] (see also Chapter 2 for more details). Here we briefly recall the most important experimental details. The plasma is generated in a double-walled vacuum chamber with an inner diameter of 32 cm and height 20 cm. A 17-turn flat spiral coil is fixed 3 mm above a quartz window atop of the vacuum chamber. A 460 kHz RF generator drives an RF current in the spiral coil through a matching network to initiate a plasma discharge. The chamber is evacuated by a 450 l/s turbomolecular pump down to a pressure of 2×10^{-5} Torr. Other details of the plasma source are described elsewhere [93]. In the experiments, Long et al. used Si(111) wafers as deposition substrates.

high-purity Co catalyst layer was predeposited on the silicon substrate by using plasma-assisted RF sputtering deposition in the IPANF discussed in Chapter 3. Prior to the deposition, the substrates were first degreased in acetone bath in an ultrasonic cleaner and then blown dry with purified nitrogen. A gas mixture of high-purity ($\sim$99.9%) Ar, H_2, and CH_4 is used as working gas. During the deposition process, Ar is used as a diluent gas that facilitates the discharge ignition and stable run. Methane is used as a carbon feedstock gas and reactive hydrogen serves for the purpose of controlling the surface dangling states and removal of unwanted amorphous carbon from the surface. The following conditions are used for the film deposition: RF power

$P_{RF} = 2$ kW, working pressure $p_0 = 10$–80 mTorr, substrate temperature $T_s = 400$–600 °C and (negative) DC substrate bias $V_s = 50$–500 V.

The characterization of the thin films is performed via the analysis of microscopic surface morphology, bonding states, and crystalline structure. A JEOL JSM-6700F field emission scanning electronic microscope (FESEM) is used to observe the micromorphology of the grown thin film [173]. The bonding states of the carbon films are studied by using a RENISHAW Raman spectrometer. A JEOL transmission electron microscope is used to analyze the fine crystalline structure of individual single-crystalline nanoassemblies.

The resulting nanopatterns are very sensitive to the process parameters, such as partial pressures of working gases, bias voltage, and externally controlled substrate temperature. In particular, variation of the experimental parameters significantly change the morphology (such as shape and size) and distribution of the nanotips. Figure 4.7 shows the SEM images of the carbon nanotips grown on a Co-catalyzed highly doped Si(111) substrate in Ar+H_2+CH_4 reactive plasma environment. The experimental conditions are: $p_0 = 60$ mTorr, $V_s = -150$ V and $T_s = 500$ °C. The mole ratio between Ar, H_2, and CH_4 used to synthesize nanostructures in Fig. 4.7 is 3.5:1:60. It is observed that the vertically aligned carbon nanotips cover a large surface area (Fig. 4.7(a)). Moreover, the nanostructured films feature an excellent adhesion to the silicon substrate.

Figure. 4.7(b) shows a high-resolution SEM micrograph of the carbon nanocones taken at 30° tilt. It is observed that most of the cone-like structures have an apex angle ranging from 10 to 15°. The nanotip height varies from 0.8 to 1.2 μm. The sharp tips of the nanotips have a size of a few nanometers, and the bases are of up to 100 nm in diameter. The cones form large-area uniform nanopatterns with the surface density ∼32–35 nanostructures per square micrometer.

The bonding configurations of carbon atoms are studied by using the Raman spectroscopy. Figure 4.9 shows a representative Raman spectrum of a nanostructured carbon film containing conical nanotips. One can notice that two clearly resolved peaks located at 1350 and 1593 cm^{-1} appear in the spectrum. The 1350 cm^{-1} peak is the so-called D-band and is attributed to disordered carbon states. It arises from the breathing modes of sp^2 atoms in rings [175] via a double resonant Raman process [176]. On the other hand, the peak at 1593 cm^{-1} corresponds to the G-band (E_{2g} mode) of disordered graphite. This peak results from the C–C stretching vibration and evidences the graphitic structure of the carbon nanocones of our interest here. We note that similar peaks also appear in Raman spectra of other carbon nanofilms discussed in Sections 4.1 and 4.2.

High resolution transmission electron microscopy (HRTEM) (Fig. 4.8) shows that each individual carbon cone is composed of a single graphite crystalline grain that grows along the (002) direction. The HRTEM images have

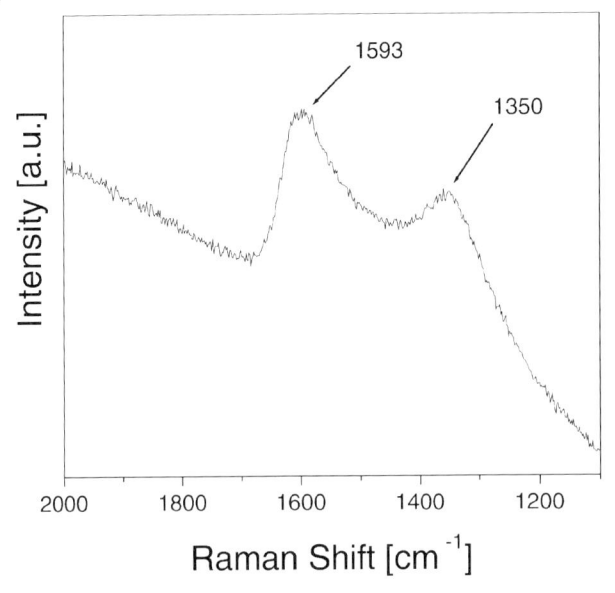

Fig. 4.9 Raman spectra of the as-grown carbon nanotips.

been made along the normal direction of the crystal plane (the N direction in Fig. 4.8) from the tip to the base of the cone-like structure. More importantly, the crystal structures appear to be exactly the same near the tip and the base of the CNSs. Figure 4.8(b) displays the HRTEM images taken from the lateral side of the carbon cone synthesized under conditions of Fig. 4.7. It is clearly seen that parallel crystalline planes cross the entire body of the nanotip. Thus, the conical carbon nanotips consist of a single crystalline grain. The regular interval between the planes is measured to be 0.34 nm. This interplane space corresponds exactly to the interlayer spacing between the (002) planes of graphite. These parameters of the crystal lattice will be used in numerical simulations in Section 4.4.

The carbon nanotips of our interest here can easily be mistaken as the hollow carbon nanocones originally discovered in 1994 by Ge and Sattler [177]. Subsequently, Krishnan and coworkers [178] successfully synthesized the cones with five opening angles. It was found that carbon could form nanosize cones or conical filaments, which can further be classified as closed or open cones. Pentagonal or heptagonal defects can be introduced into a graphene network to form nonplanar, conical structures. For instance, by removing a 60° wedge from a graphene sheet and joining the edges gives rise to a nonplanar cone tip with a single pentagon defect at the apex similar to the sketch structure in Fig. 4.10.

Previous work [179, 180] reveals that apex angles of closed cones take discrete values that are in a reasonable agreement with theoretically calculated

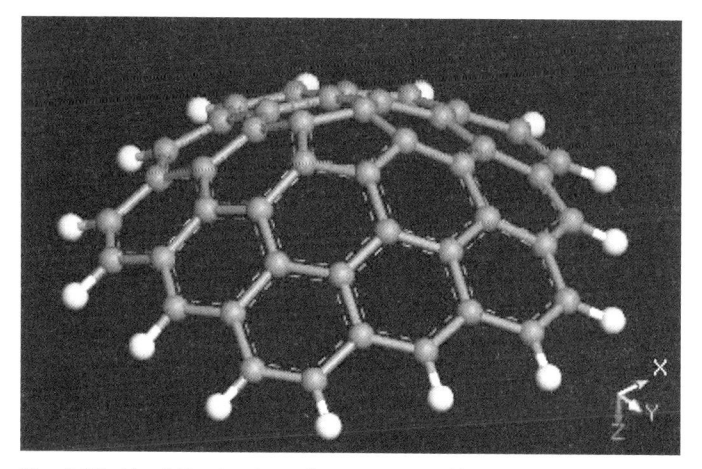

Fig. 4.10 Atomistic structure of a nanocone with a pentagon defect in the center.

values and they depend on the number of pentagonal defects incorporated in a graphene sheet. Unlike conventional carbon nanocones reported in the last decade, we deal with a different class of carbon nanostructures with the unique architecture and single-crystalline structure. The discovery of these exotic nanostructures motivated further studies of their atomic architecture, structural and electronic properties summarized in the following section. Last but not the least to mention, rigorous attempts of the team of the Plasma Sources and Applications Center of NTU, Singapore to synthesize similar nanostructures under plasma-off conditions, did not succeed. This is yet one more illustration that the plasma is crucial for the assembly of this sort of nanostructures. However, why exactly the low-temperature, thermally nonequilibrium plasma is so important for this process, still remains an essentially open question. Numerical simulations can shed some light on the nature and capabilities of this plasma-based process.

4.4
Single-Crystalline Carbon Nanotips: *ab initio* Simulations

In this section, following the original work [181] we show how state-of-the-art First Principles (*ab initio*) calculations based on the local density approximation to the density functional theory (DFT) can be used to explain unusual properties of single-crystalline carbon nanostructures of Section 4.3. A sophisticated package, DMol^3, which enables one to calculate variational self-consistent solutions to the DFT equations, is used to optimize and stabilize the geometrical atomic arrangements of the nanotip structures.

The atomistic models of such nanoassemblies (NAs, with significantly reduced numbers of atoms) are based on the SEM, HRTEM, and XRD analytical characterization. The geometrical arrangement of individual atoms in the nanotips can be optimized and the total energy of the nanostructure minimized. Remarkably, hydrogen termination of outermost carbon atoms located at the periphery edges can stabilize the atomic network. Furthermore, the molecular orbital analysis reveals that the carbon nanotips are very narrow bandgap (of the order of 0.01 to 0.05 eV) semiconductors and are ideally suited for electron field emission applications.

4.4.1
Theoretical Background and Numerical Code

The density functional theory (DFT), originally developed by Hohenburg, Kohn, and Sham [182, 183], provides a supreme technique for *ab initio* (First Principles) studies. A comprehensive review of this theory can be found elsewhere [184]. The main approximation used in this section is the local-density approximation (LDA) [183]. Despite its simplicity and lesser accuracy compared with some other *ab initio* approaches, the LDA turns out to be computationally convenient and surprisingly accurate, which is evidenced by numerous successful applications that prove this technique to be remarkably useful. Moreover, for many systems, including metals, transition metal compounds, organic and inorganic molecules, the LDA gives surprisingly good results, especially for the prediction of structural properties, which are of the most interest in this section.

$DMol^3$ is a state-of-the-art computational package based on the density functional theory that features a high accuracy while keeping the computational cost fairly low compared to other *ab initio* methods. This tool allows users to model electronic structure and energies of a wide range of systems including organic and inorganic molecules, molecular crystals, covalent solids, metallic solids, or infinite surfaces of a material [185]. The $DMol^3$ can be used to calculate variational self-consistent solutions to the DFT equations, expressed in a numerical atomic orbital basis, which uses single-electron functions or orbitals to approximate the full quantum mechanical wave function [186].

The solutions of these equations provide the molecular wave functions and electron densities, which can be used to evaluate the energetics and the electronic and magnetic properties of the system. Results obtained from $DMol^3$ provide a reliable predictive method for theoretical exploration of the properties of unknown compounds and nanoassemblies. In addition, this technique can explain various properties of the existing materials at the most fundamental atomic level. The primary advantage of the DFT approach is its relatively low computational requirements, which enables one to study reasonably larger atomic assemblies impossible by other *ab initio* methods.

4.4.2
Geometrical Stability of Carbon Nanotip Structures

In this subsection we build and optimize atomistic models of carbon nanotip structures shown in Fig. 4.11. The atomic structures consist of a small number of atoms that are arranged in stacking, layered graphite sheets. The number of carbon atoms along the vertical direction (axis of the conical needle-like structure) decreases toward the apex. This structure has been chosen consistent with the HRTEM analysis of Section 4.3.

For convenience, the layers of such model pyramid-like CNTP atomic assemblies are labeled A to E from the top to the base. For example, the numerical symbol 5 in the 5ABCDE refers to the total number of layers in the atomic structure, whereas A, B, C, D, and E denote the top layer, the second, the third, the fourth, and the fifth layer, respectively.

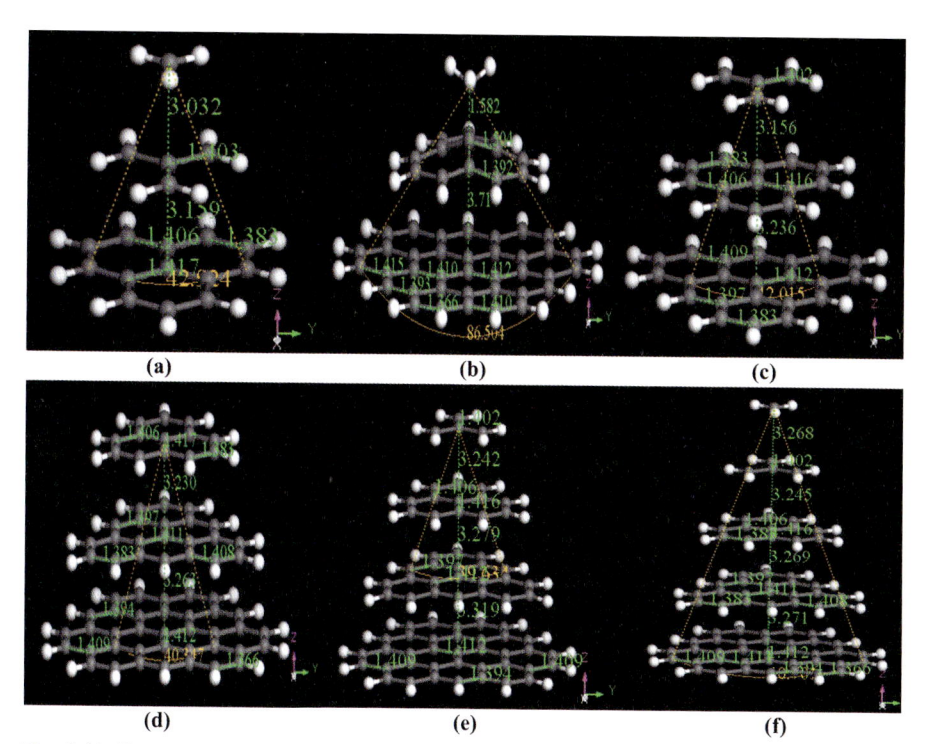

Fig. 4.11 Geometry-optimized atomic structures (a) 3ABC(H); (b) 3ABC(H)W; (c) 3BCD(H); (d) 3CDE(H); (e) 4BCDE(H); and (f) 5ABCDE(H) computed by using a local density approximation to the density functional theory.

Before one can study the electronic properties, it is essential to ensure the overall stability of the structure by means of total energy minimization. This procedure is called the geometry optimization. The objective of the geometry optimization is to find an atomic arrangement, which makes the nanoassembly most stable. The nanostructures are most stable when their total energy is the lowest. A series of iterations is needed to adjust the locations of individual atoms to reduce the total energy of the structure to the lowest possible value. Ideally, the chemical structure of a geometry-optimized model structure should closely resemble that of the real nanoassembly.

In the case considered, termination of dangling bonds of carbon atoms located at the periphery edges stabilizes the entire atomic assembly. For example, a three-layered hydrogen-terminated model carbon nanotip 3ABC(H) is stable, whereas the energy of the 3ABC structure cannot be minimized.

The computational routine for the *ab initio* investigation of chemical structure of model carbon nanoassemblies shown in Fig. 4.11 involves the following sequence. First, the atomic assembly is built and terminated on its lateral surfaces by hydrogen atoms. Thereafter, the geometry optimization (energy minimization) is performed. If the energy can be minimized, then the energetic characteristics are computed. Otherwise, the structure is re-built and the cycle is repeated until the geometric stability criteria are met.

We now consider the energetic characteristics of various stable atomic carbon nanotip-like structures and comment on the possibility of their synthesis in the plasma environment. Figure 4.11 shows six different geometry-optimized and hydrogen-terminated carbon-based structures derived from the five-layered structure 5ABCDE(H). The structures of Fig. 4.11 consist of three to five layers of different combinations; for instance, the 3ABC(H) structure consists of only the top three layers, whereas 3BCD(H) comprises the second to the fourth layers of the 5ABCDE(H) structure. Likewise, the 3ABC(H)W structure is similar to 3ABC(H) but has a larger apex angle and a larger number of atomic building units.

Table 4.1 summarizes the most essential parameters that characterize the geometrical stability of seven representative carbon nanotip structures. All the energies are expressed in Hartree (Ha) units, where 1 Ha = 27.212 eV. Here, E_{tot} is the total energy, N_a is the total number of atoms ($N_a = N_C + N_H$), N_C and N_H are the numbers of carbon and hydrogen atoms, respectively; N_e is the total number of electrons, and $\epsilon_e = N_{tot}/N_e$ is the energy per electron. The structure's cohesive energy E_c is defined as $E_c = E_{tot} - (N_C E_C^a + N_H E_H^a)$, where E_C^a and E_H^a are the total energies of a single carbon or hydrogen atom, respectively. The structures are ordered in Table 4.1 according to their total energy per electron, from the highest (−5.48 Ha) to the lowest (−5.86 Ha) values. One can notice that the 3CDE(H) structure has the lowest energy per electron and is thus the most stable nanoassembly from the min-ϵ_e point of view.

Tab. 4.1 Main characteristics of atomic structures.

Structure	Formula	N_a	N_C/N_H	$|E_{tot}|$	N_e	ϵ_e	E_c
3ABC(H)	$C_{18}H_{18}$	36	1:1	690.3	126	−5.48	−8.35
4ABCD(H)	$C_{40}H_{30}$	70	1.33:1	1528.5	270	−5.66	−17.54
3BCD(H)	$C_{39}H_{27}$	66	1.44:1	1489.1	261	−5.71	−16.88
5ABCDE(H)	$C_{77}H_{45}$	122	1.71:1	2935.4	507	−5.79	−32.43
4BCDE(H)	$C_{76}H_{42}$	118	1.81:1	2896.0	496	−5.82	−31.78
3ABC(H)W	$C_{51}H_{27}$	78	1.89:1	1942.7	333	−5.83	−21.2
3CDE(H)	$C_{72}H_{36}$	108	2:1	2741.5	468	−5.86	−29.47

The most exciting observation from Table 4.1 is that the ordering of the nanoassemblies according to the ratios N_C/N_H of carbon and hydrogen atoms in them is exactly the same as that according to the energy per electron. Thus, higher carbon elemental concentrations relative to hydrogen correspond to lower ϵ_e and vice versa! Therefore, Table 4.1 serves as a qualitative indicator of the amount of carbon and hydrogen atoms one needs to deliver to the growth site to synthesize any specifically desired nanoassembly.

For instance, to synthesize the 3CDE(H) structure, one should maintain the ratio of carbon to hydrogen atoms 2:1, whereas the largest nanotip 5ABCDE can be synthesized when the carbon and hydrogen atoms are dosed 1.71:1. Another important conclusion made by following the min-ϵ_e principle is that the structures with broader topmost layers (e.g., 3CDE(H) of 4BCDE(H)) and a larger number of layers feature a better geometrical stability.

On the other hand, the nanoassemblies are more stable when the absolute values of their cohesive energies E_c are larger. The results of Table 4.1 reveal a similar trend: carbon nanotips with broader bases and a larger number of the layers have higher values of $|E_c|$ and thus feature a better geometrical stability.

We now compare the two similar structures 3ABC(H) and 3ABC(H)W (Fig. 4.11(a) and (b)) with the apex angles of 43° and 87°, respectively. The data in Table 4.1 suggest that a larger apex-angle carbon nanotip has a greater absolute cohesive energy and a larger total energy per electron. Therefore, larger apex-angle nanotips are more geometrically stable than sharper nanotip.

To summarize, the DFT numerical simulations predict most important structural characteristics from the total energy per electron and cohesive energy points of view. To this end, the three major factors affecting the geometrical stability of the carbon nanotip-like atomic structures are the number of layers in the structure, the broadness, and diameter of the top-most layer, and the apex angle. As the results of this subsection suggest, the most geometrically stable nanotip is the one with the largest number of layers, the broadest diameter tip and the larger apex angle.

4.4.3
Electronic Properties of Carbon Nanotips

Prior to considering the details of numerical simulations of the electronic properties of carbon nanotips, we recall the basics of the electron field emission and its relevance to the Fermi level. Field emission, also known as Fowler–Nordheim tunneling, occurs when a strong electric field applied to the surface of a metal lowers its potential barrier to the point that electron emission becomes possible via quantum mechanical tunneling. The presence of the electric field deforms the barrier and enables the electrons to tunnel through the deformed potential barrier at the surface of the metal.

This type of electron emission is distinct from photoemission and thermionic emission, in which electrons are given sufficient energy to overcome the local potential barrier, otherwise termed as the work function. For this reason, the advantages of field emission devices are in their operation at low temperatures and their very short activation times. This makes the field emission widely usable as a current source across thin barriers in metal-semiconductor junctions, field emission microscopy, femto-second cameras, and field emission displays (FEDs).

As was mentioned above, the efficiency of the field emission can be greatly enhanced by using sharper emitter tips, which intensify the applied electric field. For instance, carbon nanotubes, nanofibers, nanotips, and nanoneedle-like structures were found to be excellent field emitters due to the durability and sharpness of the tips. The height of the barrier is equal to the work function of the material ϕ, which is defined as the energy required to remove an electron from the Fermi energy level E_F, from the surface of the material. A smaller work function means that less energy is required to remove an electron from the Fermi energy level. Therefore, the field emission is closely related to the work function of a material, and in other words, also depends on the Fermi level energy.

The classical textbook definition tells us that the Fermi level is defined as the energy of the topmost filled level in the ground state of the system of N electrons at absolute zero. Electrons are fermions and, according to the Pauli's exclusion principle, different electrons cannot exist in identical energy states. Thus, at absolute zero, they pack into the lowest available energy states and build up the so-called "Fermi sea" of the electron energy states.

In DMol^3 computation package, the total energy is computed with respect to the vacuum reference level; therefore, the Fermi energy of this section corresponds exactly to the work function of the model atomic carbon nanotip structures. Hence, using Fermi energy as an indicator of the field emission efficiency, the electronic properties of different atomic structures can be compared and evaluated. If the absolute value of the (negative) Fermi level energy decreases (E_F becomes less negative), the work function decreases, implying

Tab. 4.2 Main parameters of the electronic structure of carbon nanotips (all energies are in eV).

Structure	Formula	E_F	LUMO	HOMO	Bandgap
3ABC(H)W	$C_{51}H_{27}$	−3.818	−3.786	−3.917	0.131
3CDE(H)	$C_{72}H_{36}$	−3.728	−3.721	−3.784	0.063
3BCD(H)	$C_{39}H_{27}$	−3.758	−3.754	3.782	0.028
4ABCD(H)	$C_{40}H_{30}$	−3.77	−3.767	−3.793	0.026
4BCDE(H)	$C_{76}H_{42}$	−3.711	−3.7	−3.712	0.012
3ABC(H)	$C_{18}H_{18}$	−3.819	−3.818	−3.828	0.01
5ABCDE(H)	$C_{77}H_{45}$	−3.718	−3.717	−3.721	0.004

a better field emission. Tabulated in Table 4.2 are the Fermi level energies for each atomic structure considered in the previous subsection.

From Table 4.2 one can work out the following sequence 3ABC(H) → 4ABCD(H) → 3BCD(H) → 3CDE(H) → 5ABCDE(H) → 4BCDE(H) of the atomic structures according to an increased field emission efficiency. Examining the Fermi level energies in Table 4.2, one can see that the values of E_F of all the atomic structures are consistently very close. The average Fermi level energy is ≈ −3.76 eV. This implies that applying 3.76 eV of electric energy to the downscaled carbon nanotips considered in this section will result in electron field emission.

Table 4.2 also contains valuable information about the highest occupied molecular orbitals (HOMO) and the lowest unoccupied molecular orbitals (LUMO), respectively. The HOMO is the molecular orbital of the highest energy that has at least one electron in it, which can act as an electron donor. On the contrary, the LUMO is the molecular orbital of the lowest energy that does not have electrons in it, which can act as an electron acceptor. Hence, the HOMO–LUMO gap can be envisioned to be similar to the energy bandgap between the valence band and conduction band in solids.

Careful examination of the results in Table 4.2 reveals that the HOMO–LUMO gaps are very small, of the order of 0.01 to 0.05 eV. Such small HOMO–LUMO gaps indicates that, by their atomic structures, the carbon nanotips are narrow-bandgap semiconductors. Examining the energy levels of the HOMO, the LUMO, and the Fermi level in the 5ABCDE(H) atomic structure, one can find that there is some overlap between the HOMO and the LUMO. Therefore, the 5ABCDE(H) structure is metallic.

Thus, one can anticipate that larger atomic structures approaching in size to bulk graphite, feature metallic structures. Indeed, the DFT computations of our interest here suggest that the largest molecular structure 5ABCDE(H) exhibits metallic properties, whereas smaller molecular structures are narrow-bandgap semiconductors.

We now consider the electron density distributions in the atomic carbon nanotip structures. Physically, the electron density is the distribution of the probability of finding an electron at a certain position. Regions of high probability have high electron density as it is implied that the electron spends a greater proportion of time in these regions.

The DFT asserts that the ground state energy is determined by the electronic density, therefore throwing out the fearsome complexity of a multidimensional wave function. Electron density plots of the optimized 3ABC(H) atomic structure generated from solutions of the Schrödinger equation are illustrated in Fig. 4.12, which is the constant-density surface plot. This means that every point on the surface has the same electron density. The electron density varies between the layers. Higher electron density regions are located closer to the center, whereas lower electron density regions are smeared outward. It is obvious that there is distinct separation between the intense density fields, which results from the lack of molecular bonds between the layers. This further elucidates the existence of Van der Waals bonding that keeps the separated graphene planes together.

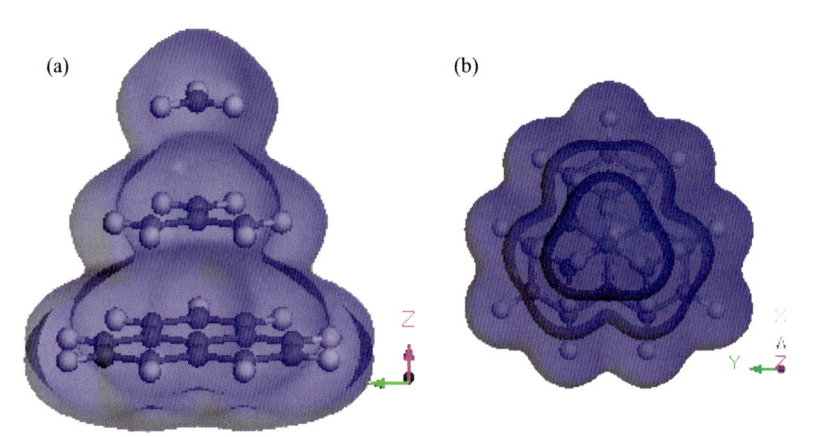

Fig. 4.12 Density plots of optimized atomic structure 3ABC(H) in (a) side view (b) top view.

Therefore, the density functional theory computations presented in this section convincingly suggest the existence of stable nanotip-like atomic structures with the different number of layers, atoms, and apex angles. The amount of computational effort required to perform the geometry optimization and energy calculation stages increases dramatically with the number of atoms building up the structure. The maximum number of atoms in the structures shown in Fig. 4.11 is 122 for 5ABCDE(H).

It is important to note that such structures are downscaled versions of the real carbon nanotip structures. One such structure with a height of 300 nm and the width (at the base) of 60 nm would contain $\approx 1.5 \times 10^6$ carbon and 2×10^5

hydrogen atoms. Therefore, it is a futile attempt to compute the geometric configurations of real "large" nanostructures by using the DMol^3 package.

This approach can be adopted to compute the equilibrium structural and energetic states of smaller nanoassemblies, such as smaller size quantum dots considered in Chapter 5. Nonetheless, appropriate scaling considerations can be used to work out the numbers of carbon and hydrogen atoms, and atomic layers needed to create real carbon nanotip structures. Further comparison with the best matching (by the ratio of carbon to hydrogen atoms, for example) atomic structure can enable one to estimate the main parameters of the real carbon nanotips. In the following section, we discuss the benefits of doping carbon-based nanostructures and the relevant experiments that use various plasma environments.

4.5
Plasma-Assisted Doping and Functionalization of Carbon Nanostructures

In the previous section, density functional theory computations have predicted that small atomic structures, downscaled versions of real carbon nanotips, exhibit a narrow band-gap semiconductor nature. The process of adding impurities, commonly referred to as doping, is important for fabricating devices and semiconductors having well-defined regions of different conductivities. It is well known that elements from Group III and V in the Periodic Table are very important dopants in semiconductor devices.

The effects of impurities in carbon nanotips are of great interest to experimentalists as structural or electronic properties can be changed with intentional introduction of impurities during the synthesis process. In this section we detail the effects of substitutional boron (group III) and nitrogen (group V) dopants on geometrical stability and electronic properties of the 3ABC(H) atomic structure. We also discuss relevant experiments on doping and functionalization of most common carbon nanostructures.

4.5.1
Doping of Carbon-Based Nanostructures: Density Functional Theory Considerations

We begin with a short summary of what happens when certain dopants or impurities are added to semiconductors. First and foremost, the band structure of the semiconductor is modified [187]. If a semiconductor is doped with atoms containing three outer-shell electrons, such as boron and/or aluminium, the three electrons form covalent bonds with neighboring semiconductor atoms, leaving an electron deficiency—a hole—where the fourth bond would otherwise be if an impurity atom electron were available to form it. The energy level of this hole lies in the energy gap, just above the valence band. An electron from the valence band has enough energy at room temperature to fill this impurity level, leaving behind a hole in the valence band. As a trivalent atom accepts an electron from the valence band, such impurities are referred to as acceptor atoms. Thus, a semiconductor doped with trivalent impurities is known as a p-type semiconductor because the majority of charge carriers are positively charged holes.

On the other hand, when an atom containing five outer-shell electrons, such as nitrogen or arsenic, is added to a semiconductor, four of the electrons form covalent bonds with atoms of the semiconductor and one is left over. This extra electron is nearly free of its parent atom and has an energy level that lies in the energy gap, just below the conduction band. Such a pentavalent atom donates an electron to the structure and hence is referred to as a donor atom. Due to the small spacing between the energy level of the electron of the donor atom and the bottom of the conduction band (typically 0.05 eV), only a small amount of thermal excitation is needed to cause this electron to move into the conduction band. Semiconductors doped with donor atoms are called n-type semiconductors as the majority of charge carriers are electrons, which are negatively charged.

Table 4.3 summarizes the effect of addition of one, two and three boron or nitrogen atoms to the 3ABC(H) atomic structure representing a downscaled single-crystalline carbon nanotip. Two modified atomic structures obtained by addition of one boron and two nitrogen atoms are shown in Figs. 4.13(a) and (b), respectively.

It is interesting that only certain atomic sites can accommodate the dopants. Tabulated in Table 4.3 are the number of electrons, total energy per electron (in Ha/e), Fermi energy, LUMO and HOMO (in eV) of the base 3ABC(H) $C_{18}H_{18}$ structure and its derivatives +1 B (one boron atom added), +2 B (two boron atoms added), +3 B (three boron atoms added), +1 N (one nitrogen atom

Tab. 4.3 Main characteristics of atomic structures with substitutional defects (all energies are in eV except for ϵ_e).

Structure	N_e	ϵ_e	E_F	LUMO	HOMO	Bandgap
3ABC(H)	126	−5.48	−3.819	−3.818	−3.828	0.01
+1 B	125	−5.42	−3.97	−3.95	−4.65	0.7
+2 B	124	−5.35	−4.33	−4.3	−4.56	0.26
+3 B	123	−5.29	−4.38	−4.33	−4.51	0.18
+1 N	127	−5.56	−3.1	−2.24	−3.1	0.86
+2 N	128	−5.65	−2.54	−2.28	−2.55	0.27
+3 N	129	−5.73	−2.45	−2.31	−2.49	0.18

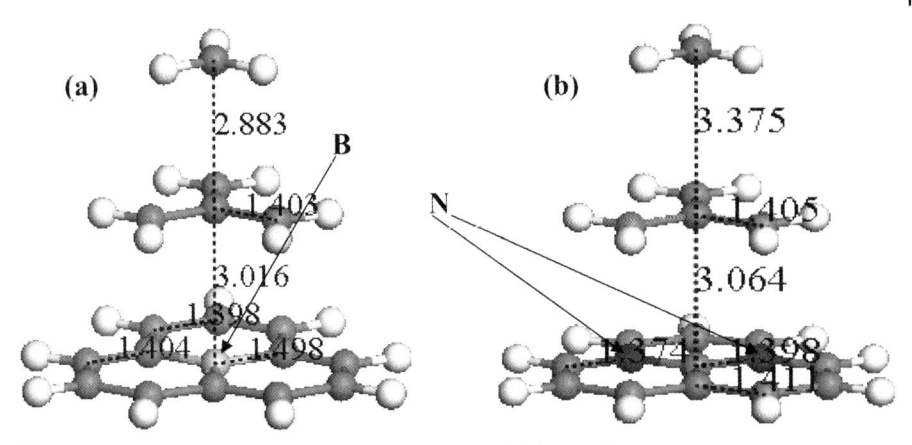

Fig. 4.13 Geometry optimized atomic structure 3ABC(H) with (a) one substitutional boron atom and (b) two substitutional nitrogen atoms.

added), +2 N (two nitrogen atoms added), and +3 N (three nitrogen atoms added).

A comparison of the values of their total energy per electron indicate that substitutional boron atoms added to the 3ABC(H) decrease the overall geometrical stability of the nanoassembly. The boron-doped 3ABC(H) structure becomes less stable when a larger number of boron atoms are incorporated into it. On the contrary, addition of nitrogen dopants increases the geometrical stability of the 3ABC(H) structure. Indeed, the total energy per electron decreases when a larger number of N atoms is incorporated.

Table 4.3 also shows the changes in the electronic structure of the downscaled carbon nanotip structures with the addition of 1–3 boron or nitrogen atoms. When boron dopants are introduced, the Fermi energy is shifted downward, toward the HOMO edge (i.e., the valence band), as can be seen from Table 4.3. This shift in the Fermi level energy is quite similar to the Fermi level shift of trivalent-doped semiconductors.

However, when nitrogen impurities are added to the 3ABC(H) structure (e.g., Fig. 4.13(b)), the Fermi energy is shifted upward, toward the LUMO edge (i.e., the conduction band), as the last three rows in Table 4.3 suggest. This shift in the Fermi level energy is quite similar to that of pentavalent-doped semiconductors. Furthermore, the Fermi level shift also indicates the enhanced field emission as the Fermi energy level becomes closer to the vacuum level. It is also instructive to note that because of quite similar trends observed, the semiconductor theory of bulk materials is fairly applicable to the atomic carbon nanotip structures.

4.5.2
Postprocessing of Carbon-Based Nanostructures: Experiments

In this section we highlight some of the issues related to postprocessing of carbon nanotube arrays by using plasma-based processes [24] and discuss some representative experimental results. Compact arrays of vertically aligned carbon nanotubes (CNTs) [188] such as nanotube forests [76, 189] and bundles [190] hold an outstanding promise for various applications owing to the many unique properties of the CNTs [191]. Such arrays find numerous applications in electron field emitters [192], chemical sensors [193], high-frequency transistors [194], reinforcement materials [195], nanoelectronic devices [173], nano-oscillators [196], nanoswitches [197], and several others.

Recent research suggests that postprocessing (e.g., coating, doping, or functionalization) of nanotube surfaces can significantly improve several structural, electronic, mechanical and other CNT properties and dramatically expand the field of carbon nanotube-related applications. This opens new avenues for the use of CNTs in various advanced devices that require controlled electric capacitance [198], thermal resistivity [199], hydrophobic properties [189], or interconnection between different nanotubes in nanoelectronics [200]. Relevant examples include coating of carbon nanotube surfaces by amorphous SiO_2 [201] for better integration in silicon-based ULSI microelectronic technology, deposition of nonwetting PTFE layers [189] for biodevice applications, or tungsten disulfide films for the development of new-generation light-emitting devices [202].

Meanwhile, functionalization of nanotube surfaces is commonly achieved, e.g., by using neutral fluxes of atomic hydrogen [203], organic molecules [204] or fluorine-based species [205]. Despite a remarkable recent progress in postprocessing of individual carbon nanotubes and low-density nanotip patterns, a similar treatment of dense nanotip arrays (which result from the most widely used porous template techniques [192]) still remains a major challenge [24].

The main issue is to achieve a high level of control and selectivity in the coating or doping/functionalization of specific surface areas of the nanostructures. For example, few-monolayer-thin films uniformly covering the entire surface of every nanotube in the array are of special interest. However, using liquid reagent solutions for such a purpose often appears problematic (even in the case of relatively low-density patterns) because of significant limitations of the reagent penetration into the internanotube space due to the surface tension forces [24].

The use of the neutral and ionized gas-based processes, commonly adopted for synthesizing CNT arrays would be a possible alternative. In this regard, it would be prudent to recall that plasma-aided methods have several important advantages, such as better vertical alignment and ordering in the pattern [79], control of the growth cites on the surface via ion implantation [206], deterministic shape control [23], and several others [36].

As was mentioned before, many of the existing and potential applications of carbon nanotubes require their sidewall functionalization with molecular or radical groups. For example, fluorination of single-walled nanotubes can dramatically change their electronic properties from insulating to conducting [207]. Moreover, chemical processing of the lateral surfaces of the nanotubes in many cases provides a "velcro" effect, leading to stronger and more uniform integration of nanotube patterns into polymer/composite matrices.

However, approaches that rely on wet chemistry or high-temperature chemical vapor may damage the nanotubes [203]. This is why the development of efficient low-temperature plasma-based approaches is nowadays of paramount importance. The plasma-based functionalization techniques mostly rely on controlled generation and delivery of reactive species in a nanofabrication environment [?4, 36].

A glow-discharge approach has been used for functionalization of single-walled carbon nanotubes (SWCNTs) [203, 208]. In these works, a microwave discharge in H_2 is used to generate atomic and molecular radicals. Interestingly, a very short (only ∼30 s long) exposure to the plasma glow made it possible to almost completely saturate carbon bonds on carbon nanotube surfaces with atomic hydrogen, in other words, achieve a surface coverage by H atoms approaching to 100 %. A microwave plasma in hydrogen is created in a 10 mm in diameter tube inserted in a microwave cavity operated at 2.45 GHz [203, 208]. An almost 500-μm Hg pressure difference across the chamber makes it possible to direct neutral atomic hydrogen species toward prefabricated SWCNT array on a CaF_2 substrate. Exposure of the samples to hydrogen plasma for 30, 60, and 90 s lead to drastic differences of the FTIR spectra of the plasma-functionalized nanotube specimens from those of reference (unprocessed) samples.

The most striking observation is the appearance of a strong band at 2924 cm^{-1}, which is due to the C–H stretching mode. Careful analysis of the FTIR and UV-VIS-NIR spectra enabled Khare et al. to unambiguously state that the carbon nanotube surfaces were indeed functionalized with atomic hydrogen. More importantly, the intensity of the FTIR peak corresponding to the C–H stretching mode remains almost unchanged after 30 s of exposure of the nanotube pattern to the hydrogen plasma. Therefore, rapid coverage of the nanotube surface with atomic hydrogen can be achieved in 30 s. It is also worthwhile to note that the cold plasma approach can be used for grafting of F, Cl, NH_2, and other functional groups to carbon nanotube surfaces [203].

Chan et al. reported on a successful doping of multiwalled carbon nanotubes (MWCNTs) by using microwave plasma-enhanced chemical vapor deposition (PECVD) [209]. Similar to other semiconductor materials (e.g., carbon nanotip structures with substitutional defects considered in the previous subsection), extrinsic doping can alter the electronic properties and binding configurations of carbon nanotubes. In the original work [209], both nitrogen and boron atoms are separately doped into MWCNTs by microwave PECVD. A typical exposure of MWCNT patterns to the plasma is ∼30 min.

Figure 4.14 shows high-resolution TEM microphotographs of the unprocessed and nitrogen-doped multiwalled carbon nanotubes. In the reference cylindrical MWCNT shown in Fig. 4.14(a) basal graphene planes are straight and parallel. After doping of the MWCNTs in nitrogen plasmas, the originally parallel layers of the nanotube walls become disturbed and highly curved as can be seen in Fig. 4.14(b). In this figure, one also see fullerene-like structures and the overall TEM image becomes looking similar to human fingerprints.

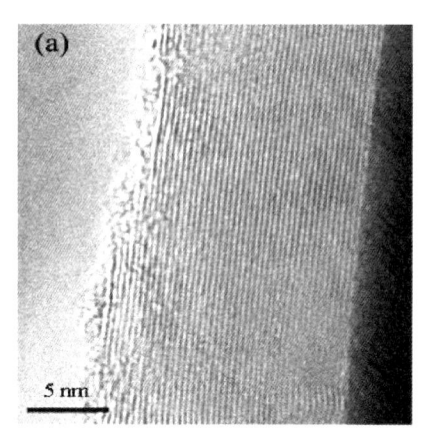

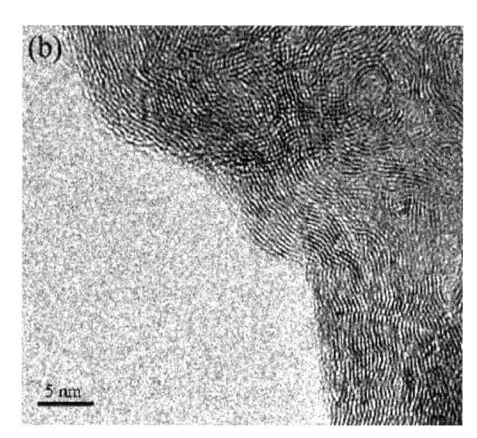

Fig. 4.14 High-resolution TEM images of (a) MWCNT, and (b) N-doped MWCNT by MPECVD [209]. The graphene layers of nitrogen doped MWCNTs appear noticeably distorted and twisted with some fullerene-like structures inside [209].

In this case one could expect that doped nitrogen atoms replace carbon atoms in the nanostructures, in a way similar to what was considered in the previous subsection, and therefore increase the electron density, which in turn is a decisive factor for the enhancement of the electron field emission. Such predictions have been confirmed by the measurements of the electron emission current from undoped, nitrided, and boronized multiwalled carbon nanotube patterns. For example, at the same applied voltage of 6 V/μm, nitrogen-doped (with the N/C atomic ratio ∼22%) nanotubes showed the emission current density of ∼0.85 mA/cm^2, which is more than two times larger than that from the undoped specimens.

On the other hand, boron-doped (with the B/C atomic ratio ∼24%) multiwalled carbon nanotube samples have shown emission current densities ∼0.15 mA/cm^2, which are more than two times lower than the currents emitted by the unprocessed nanotubes [209]. In the latter case, structural incorporation of boron into the carbon network apparently increases the concentration of electron holes that become electron traps and eventually suppress the electron field emission. It is also worthwhile to note that the field emission onset volatges for unprocessed, N-doped, and B-doped MWCNTs are 3.2, 2.5, and 4.6 V/μm, respectively [209]. As one can see, improving metallic properties of the carbon nanotubes by the plasma-assisted functionalization also leads to a significant (∼21 %) reduction of the electron emission onset voltage.

Another example of a successful application of plasma-based tools is fluorination of carbon nanotubes in CF_4 plasmas [210]. The main finding of this work is that carbon nanotubes gain more disordered sp^3 bonds as a result of functionalization and the degree of this disorder increases with the increase of the CF_4 gas inflow rates and the exposure times to the plasma. This is evidenced by the Raman spectroscopy, which shows that even though the shapes of the spectra remain unchanged after the plasma exposure, the ratio of intensities of the D and G peaks changes. For the details of the D and G peaks in Raman spectra of carbon nanostructures one can be referred to Sections 4.1 and 4.2 of this monograph.

As has already been mentioned in Section 4.1, higher ratios (which increased up to ∼50 % in the experiment concerned) of the intensities of the D and G peaks give an indication of an increased degree of disorder in the carbon films. It is remarkable that intense ion bombardment due to DC bias voltages as high as −300 V did not cause any significant damage to the nanotubes [210].

In fact, this finding refutes a common concern that ion bombardment can cause significant damage to carbon nanotube arrays. It has been demonstrated [211] that ion energies of the order of ∼100 keV are actually required to cause a significant damage to the nanotubes. We also note that the obvious advantage of carbon nanotube functionalization using plasma-based tools are very short periods of time (which typically range from 30 s to 2 min in the experiments concerned [210]) needed and the ability to run the process at low temperatures as opposed to other existing methods, which often require low process times and high-temperature treatment.

Larger size (up to 200–300 nm in diameter) nanotubes have also been uniformly coated with ∼20 nm-thick carbon fluorine films by using a 13.56 MHz RF plasma polymerization treatment [212]. In this work, C_6F_{14} was used as the monomer for plasma polymerization and owing to larger nanotube sizes, longer (∼10 min) plasma treatment times were required. Interestingly, surfaces of untreated nanotubes are usually rough, whereas the surface morphology of carbon fluoride-coated carbon nanotubes appears to be smooth. The polymeric films are approximately 20 nm thick, are tightly bound to and uniformly coat the surfaces of vertically aligned carbon nanotubes. Such conclusions have been confirmed by the TEM, SIMS, and FTIR analytical tools [212].

To conclude this section, we mention that reactive plasmas turn out to be a very efficient tool for coating various nanotubes and nanoparticles with ultrathin polymeric films [213–215]. For many envisaged applications, surfaces, structure, and other properties of nanotubes are far from ideal. To this end, the ability of uniformly coat, dope, or functionalize CNTs by using highly versatile low-temperature plasma methods would offer a wide range of exciting technological opportunities, based on intentional changes to both the physical and chemical properties of the nanoassemblies [212].

4.6
Synthesis of Carbon Nanowall-Like Structures

In previous sections of this chapter we mostly considered nanoassembly processes that involve methane as a carbon source gas. In this section, it will be shown that by replacing methane by acetylene (C_2H_2), one can synthesize quite different, quasi-two-dimensional carbon nanowall-like structures (CNWLSs) termed "nanoflakes" in the original work [173]. These nanostructures can also self-assemble in reactive environments of $C_2H_2+H_2+Ar$ low-frequency inductively coupled plasmas to form uniform patterns over large surface areas. Most interestingly, such quasi-2D structures can reach in height up to 3–4 μm remaining only a few nanometers thick, thus featuring height-to-width aspect ratios of the order of 10^3. Moreover, Raman spectroscopy reveals several higher-order peaks in addition to the usual G and D peaks (which can be seen, e.g., Fig. 4.3(b)), therefore suggesting that the nanostructures are mainly composed of disordered crystalline graphite.

Below, we will detail the process conditions and, following the original work [173], show that the morphology of the nanostructures can be controlled by the plasma-based synthesis conditions, in particular when an external DC bias is applied to the growth substrate (Si(111)). Prior to the CNWLSs growth, Ni catalyst is predeposited on Si(111) by using an ICP-assisted magnetron sputtering discussed in Section 3.3. The substrates are externally preheated up to 550 °C prior to the deposition. Other growth conditions are as follows: RF power $P_{in} = 2$ kW and working pressure $p_0 = 20$–40 mTorr. During the deposition, a negative bias of $V_b = 50$–200 V is applied to the substrate holder to draw the ionized carbon-containing species from the plasma bulk to the deposition substrates.

Carbon nanowall-like nanopatterns can have various and intricate morphologies that strongly depend on the plasma process parameters. Figures 4.15(a) and (b) show typical carbon nanowall-like structures synthesized in Ar(5 sccm)+H_2(25 sccm)+C_2H_2(100 sccm) reactive gas mixture by applying a bias voltage of -100 V to the substrate. The quasi-2D structures in this case are very thin (a few nm in width) and very tall reaching a few microns in height.

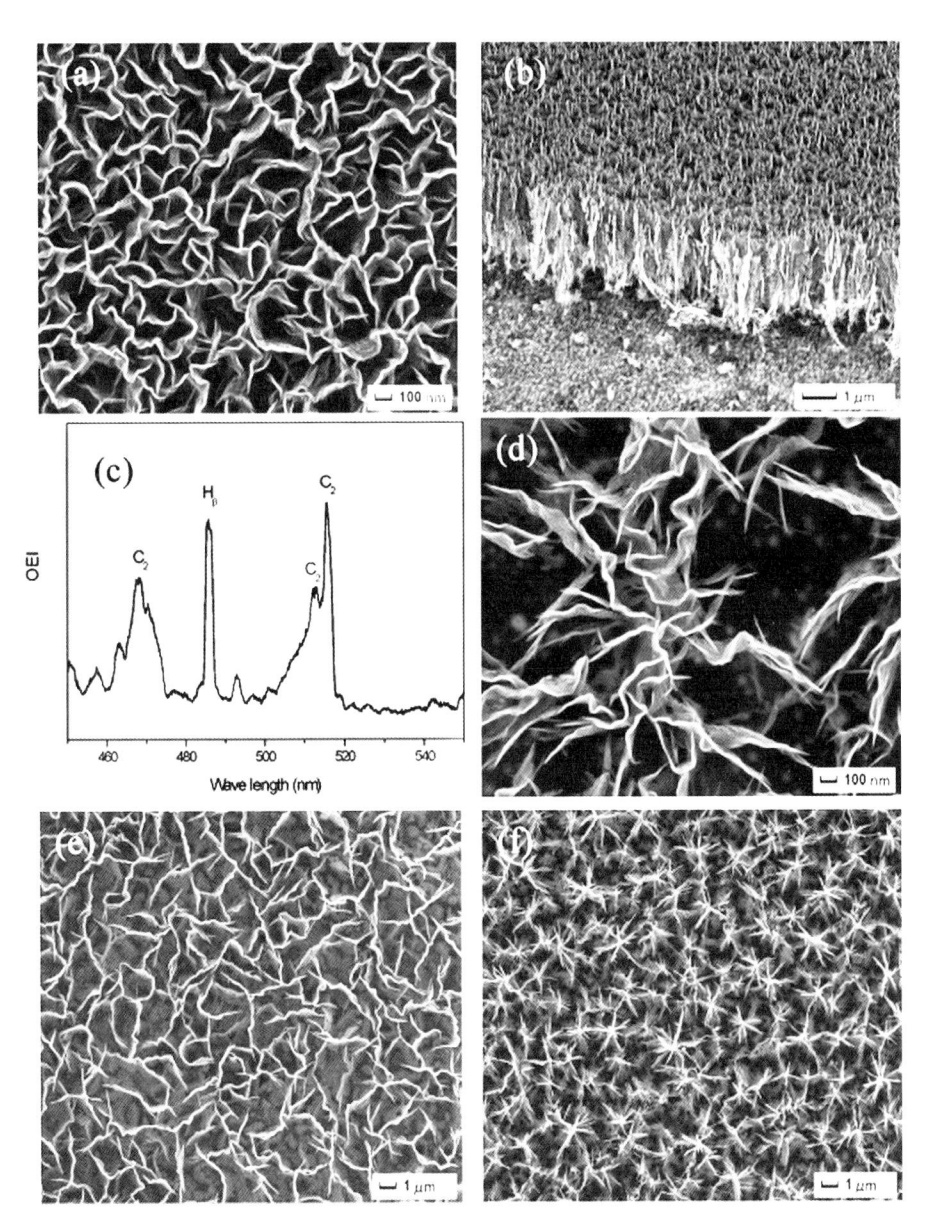

Fig. 4.15 Carbon nanowall-like structures synthesized in Ar + H_2+C_2H_2 plasmas [36, 173]. Electron images ((a), top view) and ((b), 30° tilt view) of the structures synthesized at gas ratio of Ar:H_2:C_2H_2 = 5 : 25 : 100, and −100 V bias. The optical emission spectra (c) recorded from the Ar + H_2+C_2H_2 plasma. Carbon nanostructures grown with −150 V bias and the same gas ratio (d). Electron micrographs (e) and (f) show carbon nanowall-like structures grown on (e) Ni–Fe–Mn alloy and (f) Ni catalyst layers.

Investigation of bonding states of the films by means of high-resolution Raman spectrometry suggests the presence of a major peak located at 2700 cm^{-1} (second-order G'-band), in addition to the two commonly observed D and G peaks at 1350 cm^{-1} and 1576 cm^{-1}, respectively. This implies that the CNWLSs have the attributes of disordered graphite with crystalline graphite lattices inside the structures.

It is notable that optical emission spectra of acetylene-based plasmas feature very strong emission peaks from carbon dimer C_2 that can clearly be seen in Fig. 4.15(c). Indeed, under prevailing experimental conditions, formation of C_2 molecules from acetylene source gas is very efficient. It is reasonable to assume that carbon dimer is the main building unit of carbon nanowall-like structures. As was discussed elsewhere [36], the C_2 dimer is highly reactive and is ideal for synthesizing carbon nanotubes and nanocrystalline diamond. The role of carbon dimer in the fabrication of such nanoassemblies has been confirmed by *ab initio* molecular dynamics simulations of the growth kinetics. Plasma discharges used for relevant growth processes also feature quite similar emission spectra in the same wavelength range as in Fig. 4.15(c). However, a conclusive answer about the microscopic growth mechanism of the CNWLSs can be made after detailed investigations of the atomic structure by means of high-resolution transmission electron microscopy, *ab initio* density functional theory modeling and numerical simulation of the growth kinetics by using first principles molecular dynamics or simplified fluid-on-fluid growth simulations.

Furthermore, the morphology of carbon nanowall-like patterns is strongly affected by the bias, catalyst, and discharge parameters. Figure 4.15(d) shows the quasi-2D structures grown with a -150 V bias and the same other conditions as in Figs. 4.15(a) and (b). It is seen that when the bias increases, the nanostructures become larger and the pattern becomes less dense. Figures 4.15(e) and (f) display the CNWLSs grown on Ni–Fe–Mn alloy and pure Ni catalyst layers, respectively. In this case the gas flow rates are 85 sccm (Ar), 10 sccm (H_2) and 15 sccm (C_2H_2) and the substrate bias is -100 V. It can be seen that the shapes of the CNWLSs are quite different on Ni–Fe–Mn alloy and pure Ni catalysts. Specifically, the CNWLSs in most cases look like individual nanoflakes on NiFeMn (Fig. 4.15(e)) and grow radially away from nickel catalyst islands (Fig. 4.15(f)). Finally, the growth kinetics of the carbon nanowall-like structures warrants focused and coordinated numerical and experimental efforts in the near future.

5
Quantum Confinement Structures

In this chapter we consider plasma-aided nanofabrication of semiconductor nanostructures that provide strong quantum confinement due to their reduced size in one, two, or three dimensions. Recently, synthesis of such nanoassemblies has become a matter of outmost importance and attracted a continuously increasing interest or microelectronic and optoelectronic industries because of their unconventional and extraordinary quantum confinement properties that make them invaluable for the development of integrated nanoelectronic and nanophotonic functionalities and devices.

It is commonly known that in semiconductor nanostructures, quantum confinement leads to an increase in the optical bandgap compared to the corresponding values of bulk materials, thereby leading to new possibilities for controlling photoluminescence effects and generating narrow emission spectra tunable over a wide range of wavelengths. Widening of bandgaps can be achieved by reducing at least one of the structure sizes to become comparable with the de Broglie wavelength of an electron. This size reduction results in remarkable changes in the density of states, as depicted in Fig. 5.1.

In a bulk crystal, there are two clear and nonoverlapping valence E_v and conduction E_c energy bands as can be seen in Fig. 5.1(a). If only one dimension is made small (Fig. 5.1(b)), an electron can only be partially confined; it still behaves as a free electron in the remaining two large dimensions. In this case one can observe the step terrace formation at the edges of the valence and conduction bands. When the electron motion is restricted in two dimensions but remains essentially unrestricted in one direction (Fig. 5.1(c)), the valence and conduction bands split into still overlapping sub-bands. As the object size decreases in the two "restricting" dimensions, the subbands become narrower and the overlap between them becomes less pronounced. When the electron motion is restricted in all three dimensions, the subbands evolve into well-defined, delta-function-like energy levels that resemble the energy levels of individual atoms or molecules, as can be seen in (Fig. 5.1(d)).

Quantum confinement structures can thus be classified by how many dimensions provide the electron confinement or inversely, how many dimensions allow free-electron behavior. In the examples shown in Fig. 5.1, the elec-

Plasma-Aided Nanofabrication. Kostya (Ken) Ostrikov and Shuyan Xu
Copyright © 2007 WILEY-VCH Verlag GmbH & Co. KGaA, Weinheim
ISBN: 978-3-527-40633-3

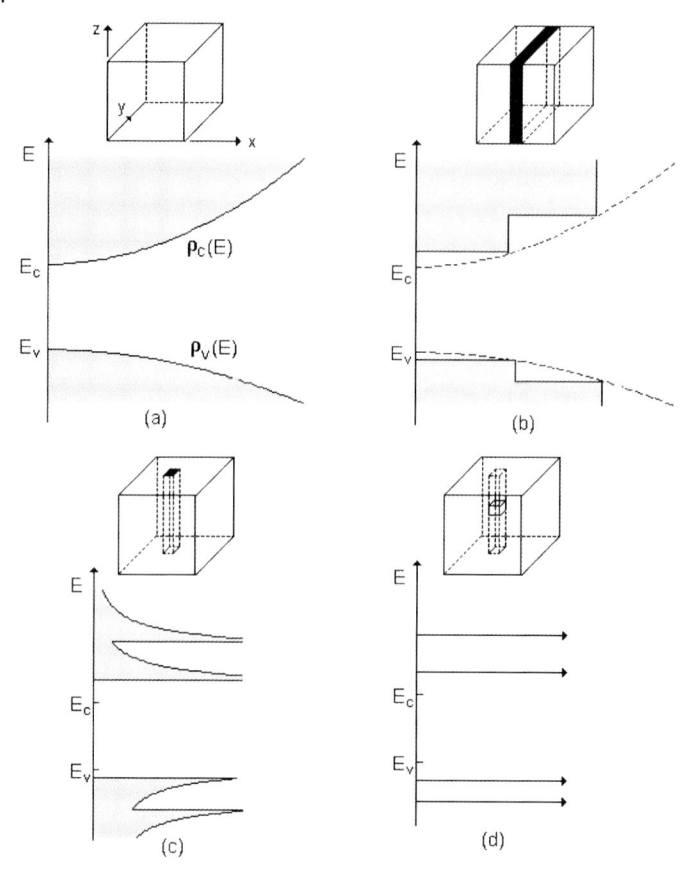

Fig. 5.1 Change in the density of states when the system dimensionality is reduced from (a) a three-dimensional bulk crystal to (b) a two-dimensional quantum well, to (c) a one-dimensional quantum wire, and to (d) a zero-dimensional quantum dot.

trons can be confined in one, two, or three dimensions. The corresponding low-dimensional semiconductor structures are commonly known as quantum wells (two-dimensional (2D) quantum structure) (Fig. 5.1(b)), quantum wires (one-dimensional (1D) quantum structure) (Fig. 5.1(c)), and quantum dots (zero-dimensional (0D) quantum structure) (Fig. 5.1(d)), respectively.

The main attention in this chapter is paid to the semiconductor quantum dot structures, whose electronic response is very similar to that of a single atom and manifests its discrete energy spectrum. Due to the quantum confinement effects, zero-dimensional quantum dots (QDs) exhibit a range of novel and exciting properties. The size changes other material properties such as the electrical and nonlinear optical properties of a material, making them very different from those of the material's bulk form. For example, quantizing allowable electron energies can make an electronic device more efficient and enable it to operate at lower voltages and higher speeds. Highly oriented nanodots can also form a template for preparing high-density arrays of a variety of self-assembled nanostructures.

In this chapter, we consider the semiconductor quantum confinement structures in order of increasing dimensionality, from zero-dimensional quantum dots to two-dimensional quantum well and superlattice structures. The chapter begins with Section 5.1, where the magnetron sputtering plasma-assisted fabrication of AlN quantum dots is considered. Intentional incorporation of additional functional building units, namely indium (In) atoms, enables one to synthesize in the reactive plasma environment, ternary $Al_xIn_{1-x}N$ quantum dots with tunable electronic properties. This possibility is considered in details in Section 5.2. Plasma environment also proves instrumental in the synthesis of SiC quantum dots directly on silicon substrates and on lattice-matching AlN interlayers (Section 5.3). The assembly process of ultra-high-aspect-ratio AlN and Si nanowires is discussed in Section 5.4. Section 5.5 introduces the plasma-aided fabrication technique and the most important properties of two-dimensional AlN/SiC superlattices. This chapter concludes with an overview of other relevant results and issues related to the use of plasma-based environments for nanoassembly of low-dimensional semiconductor quantum confinement structures (Section 5.6).

5.1
Plasma-Assisted Fabrication of AlN Quantum Dots

For the past decade, the group III nitrides semiconductors —aluminum nitride (AlN), gallium nitride (GaN), and indium nitride (InN), as well as their ternary and quarternary alloys, have been the subject of intense research [216]. These compounds are promising candidates for their potential use in optoelectronic devices in the UV/blue/green/yellow range for full color displays, semiconductor lasers, high-density information storage, and underwater communication. They are also used for high-temperature and high-power transistors, which are needed for automobile engines, future advanced power distribution systems, electric vehicles, and avionics [217]. This is due to their strong bond strengths and direct bandgaps spanning a wide range of energies from 1.9 to 6.2 eV [217,218].

Among them, AlN is the most interesting compound with many attractive physical properties. AlN is a direct-bandgap semiconductor and is an ideal material for UV light emission. For many device applications, AlN films must be single crystalline with a smooth surface. High-quality epitaxial heterostructures are usually required to fully realize these microelectronic applications.

To fabricate high-performance AlN quantum structures, techniques such as vapor phase epitaxy [220, 221], molecular beam epitaxy [221–224], chemical vapor deposition [225], ion beam nitridation [226, 227], reactive sputtering [219, 228–230] and pulsed laser deposition [219, 231] have been used. Although molecular beam epitaxy (MBE), chemical vapor deposition (CVD), and metallorganic chemical vapor deposition (MOCVD) have been known for their good control in the preparation of minimum lattice-mismatched compound semiconductor nanostructures [232], they are generally synthesized at high temperatures [228]. In order to achieve nanostructured AlN films, low temperature processes are generally considered more favorable [219]. In particular, low temperature reactive sputtering has the advantage in the ease of fabrication [229, 230].

Meanwhile, RF sputtering provides a highly reliable and reproducible method of film growth [233] and is particularly suitable for producing AlN thin films with a smooth surface at a lower oxygen content [229]. In order to maximize the deposition rate of a film, a magnetron electrode governed by the $\mathbf{E} \times \mathbf{B}$ drift is often employed.

Epitaxial growth of nitride semiconductors uses silicon (Si), 6H-silicon carbide (6H-SiC), α-aluminium oxide (α-Al$_2$O$_3$), or gallium arsenide (GaAs) substrates [217]. However, Si(111) is generally used as a substrate for deposition because the thermal expansion coefficient of AlN matches well with it and in addition, Si is widely used as a substrate in electronic applications [219].

In this section, following the original work [234], we discuss highly efficient synthesis of self-assembled and highly oriented AlN quantum dots on Si(111) substrates. The technique used to grow the high quality QDs is the plasma-enhanced radio frequency magnetron sputtering deposition in the Integrated Plasma-Aided Nanofabrication Facility (IPANF) described in detail in Chapter 3.

The morphology of self-assembled AlN dots is examined by using the field emission scanning electron microscopy (SEM) and atomic force microscopy (AFM). The results show that AlN nanodots are uniformly distributed throughout the substrate. Likewise, the X-ray diffraction peaks reveal the formation of c-oriented (002) QDs; the size of the dots varies from ∼5–20 nm, depending on the growth conditions. The dot sizes agree remarkably well with the SEM and AFM measurements. Raman and Fourier Transform Infrared Spectroscopy evidence the presence of strong Al–N bonding. Furthermore, the X-ray photoelectron spectroscopy analysis suggests that the QDs feature a chemically pure and near stoichiometric AlN. It is interesting that the AlN QDs with quite similar properties can be grown by the DC magnetron sputtering technique.

Before we present the specific details of the plasma-assisted assembly of AlN quantum dots, we summarize the main properties of the AlN material.

Tab. 5.1 Process conditions for the synthesis of AlN QDs [234].

Parameter	Condition
Base pressure	$< 1 \times 10^{-3}$ Pa
Reactive/sputtering gases	N_2, Ar
N_2 gas flow rates	10–85 sccm
Ar gas flow rates	2–30 sccm
Si substrate temperature	150–350 °C
RF power	200–400 W
Deposition pressure	0.4–1.4 Pa
Deposition time	2–20 min

Most importantly for optoelectronic applications, AlN is a direct-bandgap semiconductor, which makes it a very efficient source of light in the ultraviolet range. The wavelength corresponding to the bandgap of AlN (6.2 eV) is ~200 nm.

Among other properties of AlN, we mention its high dielectric constant (~8.5) that favors its applications in metal-insulator-semiconductor (MIS) devices; high thermal conductivity (3.2 W/cmK) makes it useful as efficient heat sinks; high refractive index (2.15) makes AlN an ideal device-quality material for surface acoustic wave devices and antireflection coatings; outstanding hardness (~12 GPa) for wear resistant coatings; high electrical resistivity (~10^{12} Ω/cm) attractive for the integrated circuit packaging applications; and very high melting point (>2000 °C).

Above all, a hexagonal wurtzite structure with $a = 3.112$ Å and $c = 4.982$ Å has a very small lattice mismatch with GaN ($a = 3.189$ Å, $c = 5.185$ Å) and SiC ($a = 3.081$ Å, $c = 5.039$ Å) [217, 218]. As such, AlN can be regarded as an ideal buffer layer [235, 236] for high-quality growth of these compounds, including the ternary semiconductor AlGaN [222].

We now briefly discuss the QD synthesis process and the deposition parameters. As was mentioned above, the synthesis process is conducted in the IPANF described in detail in Chapter 3. The precleaned Si(111) substrates are placed on the substrate holder and the shutter (denoted 3 in Fig. 3.16) is closed.

Prior to deposition, the vacuum chamber is evacuated till the base pressure reaches at least 10^{-3} Pa, after which nitrogen (N_2) gas is introduced into the chamber. Depending on the deposition conditions, argon (Ar) gas may also be introduced. Here, nitrogen gas simultaneously serves as a reactive and sputtering gas.

A circular high-purity aluminium (Al) target of diameter 12 cm is located on the RF-powered electrode, 10 cm vertically above and 20 cm horizontally away from the geometric axis of the substrate stage (see Fig. 3.16). An RF power at 13.56 MHz is delivered to the electrode via an L–C impedance matching network to initiate the RF magnetron plasma discharge.

The gas flow-rate(s) and chamber pressure are kept constant during each QD synthesis process. The target is conditioned (presputtered) for 20 min prior to each deposition, after which the shutter is opened. The growth of self-assembled AlN QDs on the substrates usually takes place under various deposition times up to 20 min. Other process parameters such as the substrate temperature, RF power, deposition time, working gas pressure, and nitrogen concentration ($[N_2]/[Ar]+[N_2]$) are also independently varied. Unless specified otherwise, the total working gas pressure is set at 0.8 Pa.

The deposition conditions of the AlN quantum dots are summarized in Table 5.1. The deposited samples remain in the chamber until they are completely cooled. During the cooling process, the chamber is continuously evacuated to avoid surface oxidation. Thereafter, nitrogen gas is fed into the chamber via the venting valve until the pressure inside the chamber is balanced against the atmospheric pressure. The sample loading port is finally opened and the samples are then removed.

Figure 5.2 shows the surface morphology of AlN samples deposited at a sputtering pressure of 0.8 Pa, RF power of 400 W, substrate temperature of 350 °C, and 3:1 N_2/Ar ratio. For 2 min growth duration, no particles are visible. As the deposition time increases (>5–7 min into the deposition), smooth-shaped AlN particles emerge. Moreover, the particles turn out to be homogeneous in size and are also uniformly distributed throughout the entire substrate surface. The average grain sizes at 20 min deposition time is ~18.5 nm; this is an indication that the AlN quantum dot arrays have been self-assembled on the Si(111) substrate. The QD pattern also reveals a densely close-packed structure, interspersed with voided grain boundaries. In most cases the dots follow a quite similar ordering pattern.

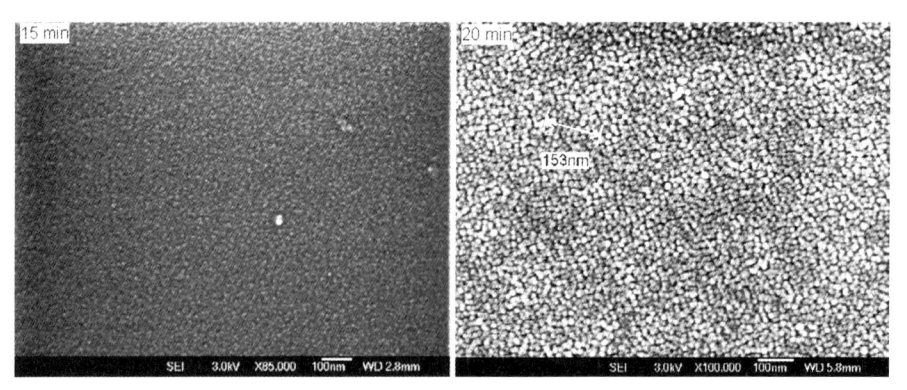

Fig. 5.2 SEM imaging of surface morphology of AlN QDs synthesized in a 15 and 20 min long deposition processes [234].

Cross-sectional SEM micrographs of AlN QDs (not shown here) reveal clear boundaries between the substrate and the AlN film indicating that the nanodots grow directly on the Si(111) surface. The films thicken with a reasonably high rate (2–3 nm/min) and reach ~40–50 nm in a 20 min long processes. However, at longer deposition times the nanodot structures evolve into elongated columnar structures discussed in detail in Chapter 7.

To verify the details of the QDs nanopattern, the deposited AlN samples have been further investigated by the atomic force microscopy in a tapping mode. The AFM technique gives a very high vertical resolution of 0.1 nm. The AFM scans over surface areas 0.5×0.5 µm of the films grown for 20 and 30 min are displayed in Figs. 5.3(a) and (b), respectively. It is evident that after 20 min of deposition a nanodot pattern uniformly covering the surface is formed. Besides that, the size of the quantum dots increases with deposition time, in agreement with the estimates made by using other techniques.

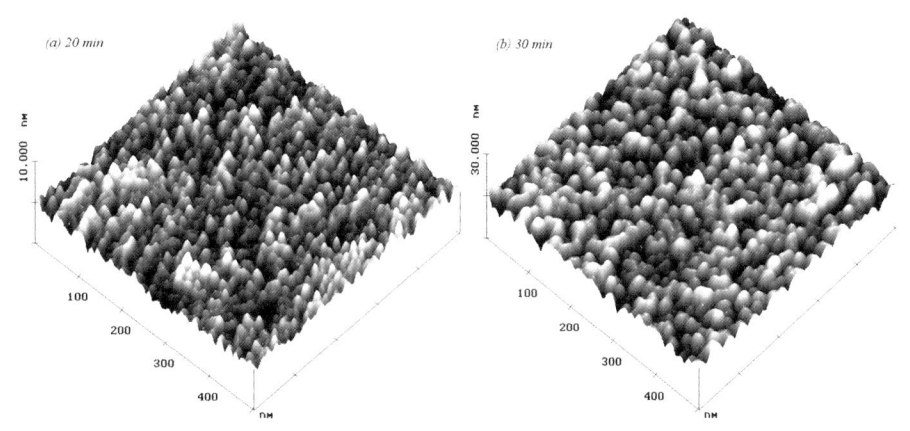

Fig. 5.3 AFM photographs of AlN films deposited for (a) 20 and (b) 30 min [234].

Moreover, the surface covered by the QDs is remarkably smooth. Indeed, the root mean square (rms) roughness of the samples deposited for 20 min is approximately 4.7 Å. The roughness of the AlN films grown for 30 min is approximately three times larger, which heralds a transformation of the ultrafine nanodot structures into more coarse nanorod-like columnar structures discussed in detail in Chapter 7.

The X-ray diffraction spectra reveal a single diffraction peak at 36.1°, which corresponds to the hexagonal AlN(002) reflection and suggests that AlN quantum dots are crystalline, with the preferential orientation along the direction of the c-axis. This diffraction peak appears in the XRD spectra of the films whose synthesis took more than 5 min and becomes more narrow at longer deposition times. The crystallite sizes can be determined by using the broadening of the X-ray peak according to the well-known Scherrer's equation

$$B = 0.9\lambda / d \cos \Theta$$

where B is the FWHM of the (002) peak in radians, λ is the wavelength of the incident X-ray radiation, d is the crystalline diameter, and Θ is the diffraction angle for the AlN(002) reflection.

The average size of the AlN dots as a function of the growth time is tabulated in Table 5.2. Standard fitting routines suggest that the lateral size of the nanocrystallites follows the cubic root dependence. Such time dependence is common to various theories of crystal growth via the so-called Ostwald ripening, which includes the coarsening, coalescence, and ripening processes that result in the growth of larger particles at the expense of smaller particles [237]. Here, the larger particles are the established AlN quantum dots and the smaller ones are the initial growth islands that do not survive in the nanopattern development process.

Tab. 5.2 Average QD size as a function of growth time estimated by the XRD technique [234].

Time (min)	2	5	10	15	20
Diameter (nm)	–	3	9	14	17

The estimates of the nanodot sizes from the XRD results also appear to be in a sound agreement with the images taken using both the SEM and the AFM techniques. The results of estimates of the AlN quantum dot sizes, by using the three different techniques, are summarized in Table 5.3. It is remarkable that the AlN quantum dot arrays of our interest can also be synthesized via the plasma-assisted DC magnetron sputtering deposition in the same nanofabrication facility. In this case, the XRD patterns also show a single AlN(002) peak pertinent to a highly c-axis oriented crystalline structure.

Tab. 5.3 Sizes of AlN QDs estimated by three different techniques [234].

Time	15 min	20 min
SEM	15 nm	18.5 nm
AFM	–	18 nm
XRD	14 nm	17 nm

In both cases, the results from the FTIR analysis show a 682 cm^{-1} peak that corresponds to the transverse-optical (TO) vibration mode of the Al–N bond. However, the XPS analysis evidences a rather high oxygen content of about 30 % in most of the specimens fabricated by using the DC magnetron sputtering deposition. Thus, the RF magnetron sputtering-based process results in a better quality AlN nanodot patterns.

Therefore, we can summarize that self-assembled AlN QD patterns can be synthesized at low substrate temperatures ~350 °C by using both the plasma-aided RF and DC magnetron sputtering deposition processes. Investigations into the elemental composition of the nanodots using both methods show the presence of Al and N with O as an impurity. The AlN QDs grown by the RF magnetron sputtering have a substantially lower oxygen content, suggesting that the RF magnetron sputtering technique is better than the DC magnetron sputtering one to obtain chemically pure and stoichiometric AlN. The SEM and AFM results reveal the presence of AlN QDs uniformly distributed throughout the substrate. Grain boundaries between the dots indicate that electrons can exhibit the quantum mechanical tunneling phenomena.

More importantly, the XRD spectra of the QDs depict a single hexagonal AlN(002) peak, indicating that the preferential growth direction of the nanodots is perpendicular to the substrate. One can thus expect that by increasing the deposition time, one can synthesize high-aspect-ratio columnar structures aligned perpendicular to the substrate. Successful growth of such rod-like structures is considered in Chapter 7.

The size of the dots varies from ∼5 to 20 nm, depending on the growth conditions. The growth dynamics of the dots follows the cubic-root-law size dependence. Thus, binary AlN semiconductor QDs can be successfully synthesized using the plasma-assisted magnetron sputtering deposition techniques and their lateral sizes can be predicted according to the cubic root growth curve. Knowledge of the growth behavior allows control of the quantum confinement of the dots, thereby allowing tuning of the energy levels and emission wavelengths.

Raman spectroscopy suggests the presence of the characteristic AlN E_2 (high) phonon mode, which is complemented by the FTIR spectroscopy depicting the characteristic E_1 (TO) mode of the Al–N bonding in both the RF- and DC-sputtered grown dots. Their values are slightly shifted due to residual compressive stresses present in the film and also due to the agglomeration of dots with deposition time. To conclude this section, we stress that highly oriented nanocrystalline AlN films also serve as lattice-matching interlayers in the nanofabrication of SiC quantum dots on silicon substrates considered in this chapter.

5.2
Nanofabrication of $Al_xIn_{1-x}N$ Quantum Dots: Plasma-Aided Bandgap Control

We now consider what difference can one make just by adding one additional element to the AlN quantum dot system considered in the previous section. Following the original report [238], we discuss how plasma-assisted RF magnetron concurrent sputtering of two solid targets can result in the formation of a ternary $Al_xIn_{1-x}N$ nanodot system. These quantum dots appear to be embedded in amorphous AlN films covering Si(111) deposition substrates.

It is interesting that the assembly and distribution of the $Al_xIn_{1-x}N$ quantum dots are affected by the substrate temperature, gas flow rates, partial pressures of reactive gases, surface morphology of the AlN buffer layer, and the relative rates of sputtering of In and Al targets (the latter is mainly controlled by the RF power applied to the magnetron sputtering electrodes). More importantly, quite separated nanodots emerge at substrate temperatures as low as 200 °C. Room-temperature photoluminescence (PL) of the $Al_xIn_{1-x}N$ quantum dots is directly related to relative elemental ratios of Al and In with a clearly observable PL maximum shift towards the UV domain for smaller QDs [238].

Let us recall here that the usefulness of quantum dot structures and their patterns in the envisaged applications critically depends on how well their size distributions can be understood and controlled. Self-assembled QD arrays can be grown epitaxially, with different possibilities emerging depending on lattice mismatch and bonding energies of the interface-forming elements involved. It is commonly known that in lattice mismatched systems, spontaneous formation of nanoislands relieves some of the strain energy.

There is a huge amount of data on the resulting distributions for many semiconductor systems. For example, InAs/GaAs(001) exhibits well-defined distributions [239, 240]; InP/GaInP [241, 242] and GaN/AlN [243, 244] show characteristic bimodal size distributions, with at least two types of islands; the InGaN/GaN family remains to be explored in more details.

Recently, $Al_xIn_{1-x}N$ has received considerable attention owing to strong bandgap tunability by the value of x. Due to the difficulty in the growth of $Al_xIn_{1-x}N$, its basic properties such as the value of its *size-dependent* fundamental bandgap, are still a subject of intense debates in the literature [238, 239, 243, 245–248].

It is well known that self-organizing growth techniques, based upon controlled strained-layer epitaxy, leads to the production of nanometer-scale dots with relatively homogeneous size distribution [238, 249–251]. MBE and MOCVD techniques are frequently adopted to produce quantum dots. Huang et al. reported the formation of $Al_xIn_{1-x}N$ QDs by the plasma-assisted magnetron sputtering technique, which is, in fact, seldom used to form quantum dots but suit to deposit uniform and large-area nanostructures [252].

The $Al_xIn_{1-x}N$ QDs have been grown on Si(111) substrates using the $Ar+N_2+H_2$ plasma-assisted RF magnetron co-sputtering technique. Pure Al and In discs are used as targets, the sputtering gas is a mixture of argon, nitrogen, and hydrogen. Base pressure of sputtering chamber is 8.3×10^{-4} Pa. Notably, the whole deposition process of the AlN interlayer and $Al_xIn_{1-x}N$ quantum dots has been implemented in a single vacuum cycle and the deposition parameters of the AlN buffer layer were the same [238]. The surface morphology and elemental ratios of Al, In, and N in the $Al_xIn_{1-x}N$ films can

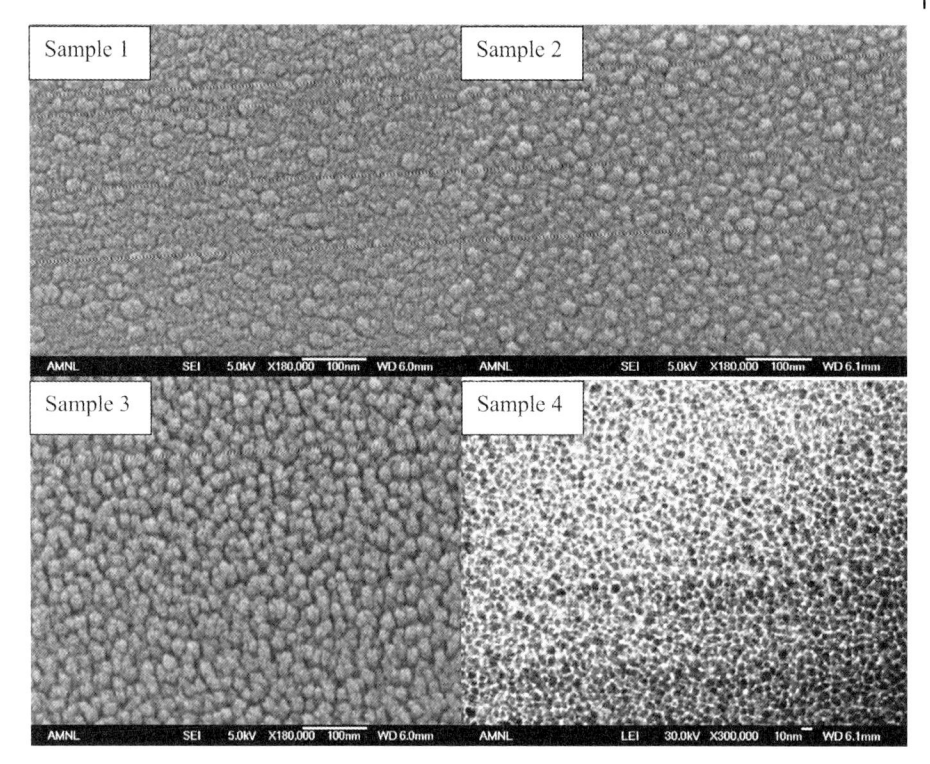

Fig. 5.4 SEM micrographs of the three $Al_xIn_{1-x}N$ QD samples and STEM micrograph of Sample 4 [238].

be effectively controlled by varying the RF power supply to the RF magnetron sputtering electrodes and other process parameters such as partial pressures of argon, nitrogen, and hydrogen gas feedstocks.

Let us consider the structure and photoluminescence (PL) properties of $Al_xIn_{1-x}N$ quantum dots embedded in amorphous AlN films grown on Si(111) substrates by using the plasma-assisted co-sputtering technique. Similar to the processes considered in other sections, field-emission scanning electron microscopy (FE-SEM) and X-ray photoelectron spectroscopy (XPS) measurements are used to characterize the structure and composition of the $Al_xIn_{1-x}N$. The photoluminescence was measured by a Raman spectroscope (514.5 nm Ar^+ laser) at room temperatures.

Figure 5.4 shows high-resolution FESEM images of three $Al_xIn_{1-x}N$ QDs samples (tilted 45°) grown on AlN buffer layers under different sputtering parameters. It turns out that the densities and mean sizes of the nanodots can be effectively controlled by the process conditions. For example, deposition time of Sample 1 is shorter than that of Sample 2, the substrate temperature of Sample 1 is lower than that of Sample 2, whereas the other deposition parameters are the same [238]. The density of nanodots in Sample 1 is smaller than in Sample 2, the mean heights of nanoislands in Sample 1 are somewhat smaller than that in Sample 2. However, the mean diameters of the QDs of the two samples are fairly similar.

On the other hand, the hydrogen partial pressures and the ratio of the RF power applied to the Al and In targets are different for Sample 1 and Sample 3, with all other process parameters remaining the same. From Fig. 5.4 one can notice that the surface density of nanodots in Sample 1 is smaller than in Sample 3, whereas the mean QD size in Sample 1 is smaller than in Sample 3. Sample 4 has been synthesized at the same deposition conditions as Sample 3 but with shorter process durations.

As the STEM micrograph in Fig. 5.4 suggests, the mean nanodot diameter in this case is only 10 nm. Typical surface densities of QDs in all four samples range from 2 to 5×10^{10} cm^2, which is a typical surface density of quantum dot systems [30]. On the other hand, the mean nanodot diameters typically range from 28 to 35 nm [238].

X-ray photoelectron spectroscopy (XPS) was used to determine the elemental composition in the $Al_xIn_{1-x}N$ films; the error did not exceed ± 0.05 at.%. For example, the XPS spectra of N1s, In3d, and Al2p of Sample 1 (Fig. 4 [238]) suggest that In and Al atoms are bonded with N and the value of x is 0.72.

Therefore, the field emission scanning electron microscopy and the XPS measurements reveal that the nanoassemblies in question are indeed $Al_xIn_{1-x}N$ nanodots, without any traces of contamination by carbon or other elements. Meanwhile, the x value of Sample 2 is 0.75. The elemental presence of aluminium in Samples 3 and 4 is in fact ~ 0.1.

It is remarkable that the relative elemental compositions of Al and In in the $Al_xIn_{1-x}N$ films can be easily controlled by adjusting the RF power applied to each magnetron sputtering electrode and other process conditions. The thickness of the AlN interlayer also plays a significant role: the minimum thickness has been reported to be ~ 15 nm [238]. Interestingly, when the x value exceeded 0.85, it turns out quite difficult to synthesize $Al_xIn_{1-x}N$ quantum dots on AlN.

It is instructive to note that the energy bandgap of $Al_xIn_{1-x}N$

$$E_g(\text{eV}) = 1.75 + 2.2x - 6.9x^2 + 9.1x^3 \tag{5.1}$$

can be expressed in terms of x [253]. We note that the excitation source (Ar$^+$ laser with $\lambda = 514$ nm) is not suitable to measure photoluminescence spectra of semiconductor materials with bandgaps exceeding 2.412 eV. Figure 5.5 shows room-temperature photoluminescence spectra of Samples 3 and 4; the peak positions are 1.793 and 1.796 eV, respectively.

Therefore, the difference in their bandgaps originates from the quantum confinement effect among the QDs, not from the relaxation of chemical bonds,

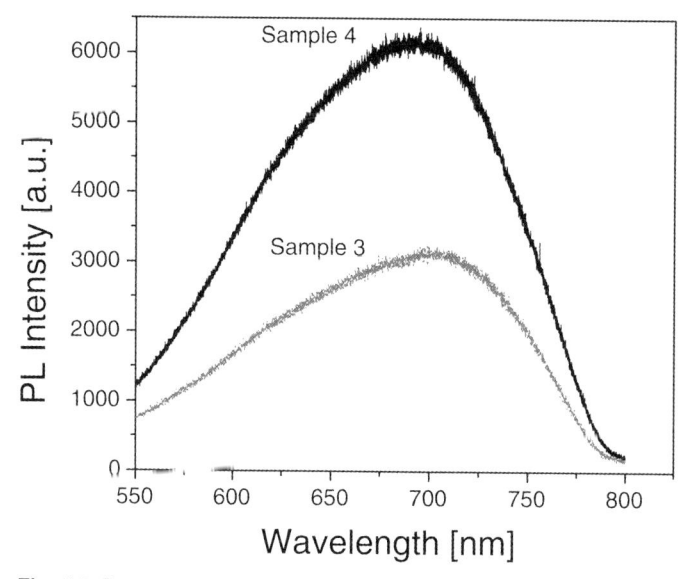

Fig. 5.5 Room temperature photoluminescence from samples 3 and 4 [238].

and the existence of defects required for the thermodynamic stability of the particles [254–257]. From Fig. 5.5 one can deduce that the full width at half maximum (FWHM) in the above two cases is approximately 0.3–0.4 eV. Since the energy gap of Samples 1 and 2 (about 3.358 eV) was larger than 2.412 eV, the peak positions of PL spectra could not be acquired by the exciting laser. According to Eq. (5.1), the energy gap of Samples 3 and 4 appears to be 1.791 eV, which perfectly fits the experimentally measured peak position (1.796 eV) in the PL spectrum.

To conclude this section, we recall that $Al_xIn_{1-x}N$ QDs embedded in an amorphous AlN matrix can be fabricated at very low temperatures on Si(111) substrates by using the RF magnetron sputtering technique [238]. The mean diameter of the QDs can be effectively controlled by adjusting the deposition conditions. The nanostructured films show strong photoluminescence excited by the Ar ion laser ($\lambda = 514$ nm). More importantly, Huang et al. successfully related the photoluminescence peak position, intensity, and the FWHM to the mean sizes and density of the $Al_xIn_{1-x}N$ nanodots, which are, in turn, controlled by the plasma process parameters [238].

A possible application of $Al_xIn_{1-x}N$ QDs grown on Si(111) substrates at low-substrate temperatures is light emitters and detectors operating in extremely wide spectral regions covering from deep ultraviolet to infrared. This can be made possible by the very efficient control of the QD bandgap energy (E_g) by varying the plasma-based process conditions, which in turn affect the density, mean diameter, and elemental composition of Al, In, and N in $Al_xIn_{1-x}N$ nanodot arrays.

5.3
Plasma-Aided Nanofabrication of SiC Quantum Dot Arrays

5.3.1
SiC Properties and Applications

In recent years, wide bandgap semiconductors (e.g., with the bandgap energy exceeding ~2.0 eV) like silicon carbide (SiC), gallium nitride (GaN), and some other Group III-nitride binary and ternary compounds [258, 259] have found many useful optoelectronic and microelectronic applications. The compounds have an outstanding potential for operating as high efficiency emitters and detectors in parts of the optical spectrum. In particular, SiC is an indirect wide-bandgap semiconductor whereas many III–V nitrides are photo-conductive and are suited for operation in the blue–green, blue, and ultraviolet regions. This makes it possible to use these materials in color displays, semiconductor lasers, and also in high-capacity optical storage media.

There is also a constantly rising need for electronic devices that can operate at high power levels, high frequency, high temperatures and in chemically hostile environments [260]. However, electronics based on existing semiconductor technologies of Si and GaAs cannot tolerate such hostile environments. As such, a suitable material is required to integrate existing semiconductor capabilities, light emitting properties and the ability to operate in hostile environments.

SiC has emerged as a suitable candidate for high-power and high-temperature device applications such as jet and automotive engines, deep well drilling, satellite systems, and advanced power distribution systems. The strong SiC chemical bond (bond enthalpy = 451.5 kJ/mol) makes it resistant to chemical attack and radiation hence ensuring its stability at high temperatures.

Moreover, a wide bandgap gives SiC some extra benefits, such as (i) an optical sensitivity in the blue and UV range; (ii) tolerance to strong electric fields, and (iii) ability to respond to high frequency signals [261]. As such, SiC is suitable for devices and integrated circuits operating at high voltage, high frequency, high power, and high temperatures.

Our choice of continuing our discussion of plasma-based methods of quantum dot synthesis by considering SiC is in part motivated by the fact that SiC and AlN bulk materials have closely matched parameters ($a = 3.112$ Å, $c = 4.982$ Å, for AlN; $a = 3.073$ Å, $c = 10.053$ Å for 4H SiC; $a = 3.081$ Å, $c = 15.117$ Å, for 6H SiC). Moreover, SiC and AlN have similar physical properties such as mechanical hardness [262], crystal structure [263] and thermal expansion [264]. Furthermore, since the bandgap of AlN (∼ 6.2 eV) is substantially larger than that of SiC (∼ 3 eV), it is therefore possible for the bandgap energy of AlN to act as a potential barrier to confine the electron wave function and hence allow quantum effects to manifest.

The possibility of manipulating optical properties of semiconductors through various degrees of dimensional or quantum confinement has attracted considerable attention during the last decade as it makes them potentially attractive for applications in nanodevice manufacturing. As mentioned in Section 5.1, in semiconductor nanostructures, quantum confinement effects lead to a modification in the electronic band structure, the vibronic states and the optical emission compared to the bulk material [265–268]. This leads to new possibilities for controlling photoluminescence effects of SiC quantum dots considered in this section.

In general, SiC is an indirect bandgap semiconductor with the energy bandgap of ∼3 eV at 300 K. As such, electrons can be excited from the valence band to the conduction band by absorption of reasonable amount of optical energy. Moreover, as an indirect semiconductor, the maximum of the valence band does not coincide with the minimum of the conduction band. Wave vector conservation then requires that the transition of an electron from the conduction band to the valence band requires a change in the electron momentum.

As a result of this indirect transition, energy is given up to the lattice as heat rather than as an emitted photon. Despite this, SiC can be made to photoluminesce by introducing discrete states in the bandgap by means of defect states (dopants) or by size-depending quantum confinement effects. The effect of size of SiC quantum dots on their photoluminescence properties is also considered in this section.

5.3.2
SiC Growth Modes: With and Without AlN Interlayer

There are various ways to fabricate SiC nanostructures. One common way is to grow SiC directly on Si substrates. Techniques such as plasma-enhanced chemical vapor deposition (PECVD) [269] and sputtering [270] can be employed. It is also desirable to grow crystalline SiC layers on substrates as this enhances carrier mobility and subsequently improves device performances [271]. Techniques adopted to prepare crystalline SiC film include hot filament chemical vapor deposition (PECVD) [272], laser ablation [273], and molecular beam epitaxy (MBE) [274].

AlN has a very small lattice mismatch with SiC and it is suitable as a buffer layer for the deposition of SiC quantum dots. Si(100) is commonly used as a substrate to deposit AlN because the thermal expansion coefficient of AlN matches well with it, and also because Si is widely used as a substrate in electronic applications [275]. The role of the AlN buffer layer is to allow heteroepitaxial growth of semiconductors with the minimum lattice mismatch [276]. We recall that AlN has an energy bandgap of ~6.2 eV. Comparing this to that of SiC (~3 eV), it is then possible for the bandgap energy of AlN to provide a potential energy barrier to confine the electron wave function as shown in Fig. 5.6.

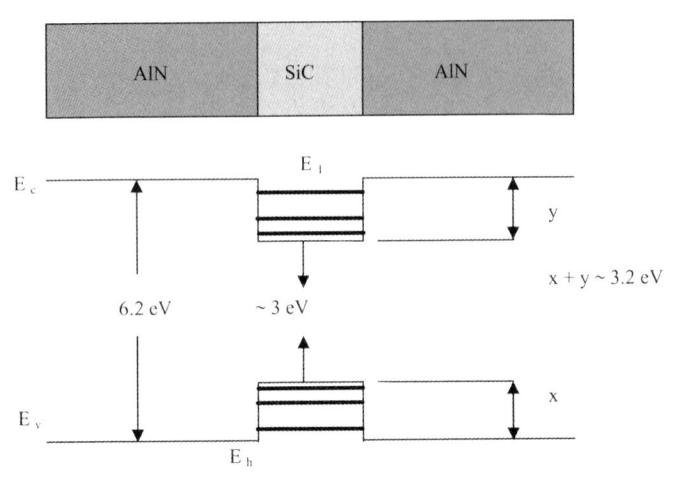

Fig. 5.6 Schematic of a one-dimensional quantum well composed of SiC sandwiched between AlN layers.

In the nanofabrication process discussed here, a wetting layer of AlN is first grown on the Si substrate. This is followed by the growing of an overlayer (SiC) on the wetting layer. Then a structural transition toward the formation of three-dimensional QDs occurs because it is energetically favorable over the planar growth [277]. These QDs are essentially defect-free and their zero-dimensional nature has been confirmed [278]. Similar to AlN nanodots of Section 5.1, SiC QDs also experience an Ostwald ripening process, in which larger islands grow at the expense of smaller islands when the growth is interrupted after the structural transition [279]. *In situ* real-time observation of quantum dot coarsening has demonstrated the importance of the ripening process for a more uniformly distributed QD ensemble [280].

However, only a few works report on the successful growth of SiC quantum dots. In most cases the SiC nanodots are directly grown on Si substrates, which inevitably lead to a lattice mismatch with SiC. If the interface strain exceeds a critical value, it is relieved via the production of dislocations or cracks, drastically reducing the material quality. These then lead to the necessity of an AlN buffer layer for the growth process.

Here we show the details of the successful use of the plasma-assisted RF magnetron sputtering technique to synthesize SiC QDs on AlN buffer interlayers [282]. More importantly, this technique allows large area nanodot fabrication at temperatures much lower than the melting point of SiC ($\sim 2800\,°C$) [261]. Furthermore, the resulting nanofilm structure can be tailored to closely match the composition of the sputtering target material [281]. In fact, this is a novel and unprecedented way of preparing SiC quantum dots. We also discuss the effects of the process control parameters on the structure, morphology, and composition of the SiC nanodot patterns.

The growth conditions of the SiC nanodot patterns are summarized in Table 5.4 and are used to form the required layers on unprocessed single-crystalline Si(100) wafers. Briefly, plasma-assisted RF magnetron sputtering in the IPANF plasma reactor (described in detail in Chapter 3) is used to obtain self-assembled SiC quantum dots on an AlN buffer interlayer. Generally, for both deposition processes, N_2, Ar, and H_2 are introduced into the chamber. The sputtering and reactive gases for AlN synthesis consist of Ar and N_2.

Tab. 5.4 Process conditions for the synthesis of SiC QDs on AlN buffer layers [282].

Parameter	AlN condition	SiC condition
Base pressure	2×10^{-3} Pa	2×10^{-3} Pa
Reactive/sputtering gases	N_2, H_2, Ar	N_2, H_2, Ar
Gas flow rates, sccm (Ar:H_2:N_2)	21.4:3.2:32.1	35.2:9.6:0–6.4
Substrate temperature (°C)	350	100–400
RF power (W)	150	100–500 W
Deposition time (min)	60	2–120

Alternatively, for SiC deposition, the working gas feedstock consists mainly of Ar, while N_2 serves as a dopant. In both processes, hydrogen is introduced to passivate the surface of the layer formed. SiC QDs on lattice-matched AlN buffer layers are grown according to the complex scenario, which combines the Stranski–Krastanov (SK) and nanoisland growth modes. In this two-stage growth process, deposition of a buffer layer of AlN is followed by the plasma-aided nanoassembly of SiC nanodot patterns. All the samples concerned have the same AlN buffer layer unless specified otherwise. On the other hand, when SiC QDs are directly synthesized on silicon substrates, the growth proceeds via the Volmer–Weber (VW) nanoisland growth scenario in lattice-mismatched systems.

We now show an example of the effect of the process conditions (with the main focus on the effect of AlN buffer interlayer) on photoluminescence (PL) characteristics, and surface morphology of the SiC quantum dot arrays. A Renishaw micro-Raman System 1000 with PL capability is used to measure room-temperature (~ 300 K) photoluminescence. Similar to the previous section, the system uses an Ar$^+$ laser excitation source with an incident beam with the 514.5 nm wavelength. The laser beam is focused (magnification of 50×) onto a surface area of 1–2 μm in diameter, with a spectral resolution of ~1.5 cm^{-1}.

Figure 5.7 shows the results of comparison of the PL measurements of (a) silicon substrate, (b) aluminum nitride grown on the substrate, and (c) silicon carbide deposited on the AlN interlayer on top of Si. The deposition conditions of the AlN layer are listed in Table 5.4. The SiC layer is grown with these conditions: deposition time of 5 min, RF power of 500 W, substrate temperature ~400 °C, and mass flow rates Ar:H$_2$:N$_2$ = 35.2 : 9.6 : 6.4 sccm. One can see that with or without AlN, the PL intensities of wafers are almost the same, indicating that AlN is transparent in the optical region concerned. Moreover, AlN, being a direct-bandgap semiconductor, is able to produce photoluminescence in a deep UV range.

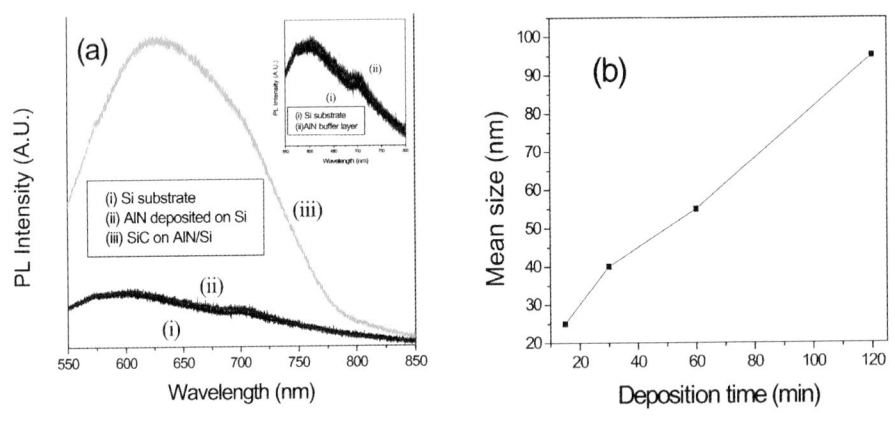

Fig. 5.7 Comparison of PL intensities (a) and plot of the SiC QD mean size against the deposition time (b). The inset in graph (a) displays PL emissions from Si and AlN [282].

On the other hand, Si is an indirect semiconductor but since the Si substrates are n-doped, it is able to generate photoluminescence albeit with a lower intensity. The energy of the incident radiation 2.41 eV is much higher than the bandgap of bulk Si and much less than that of AlN. Weak photoluminescence peaks at 575 nm, 600 nm, and 700 nm can be attributed to that of the Si substrate. A significant increase in the PL intensity is observed when a layer of SiC is deposited for 5 min. Subsequent photoluminescence measurements reveal a PL peak position at 630 nm, which cannot be routinely detected for either Si or AlN. Hence, the PL emission here is due to the presence of SiC.

The effect of a AlN buffer layer on photoluminescence properties of SiC film grown for 40 min is shown in Fig. 5.8. The deposition conditions for the AlN buffer layer are listed in Table 5.4. For the sample with the PL spectrum

(a) SiC was directly deposited on the Si substrate. The sample with PL spectrum (b) had SiC grown after AlN was deposited. The SiC deposition conditions for both samples are the same: deposition time of 40 min; RF power of 400 W; substrate temperature ~300 °C; and the ratio of the mass flow rates of Ar:H$_2$:N$_2$ are 35.2:9.6:6.4 sccm, respectively.

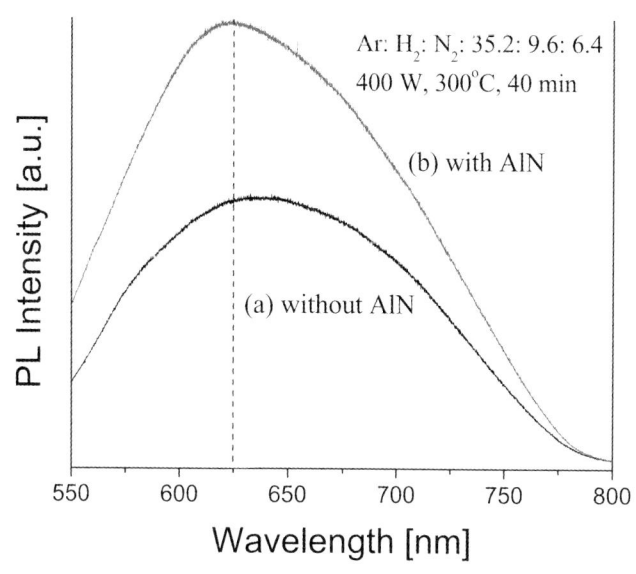

Fig. 5.8 Influence of AlN buffer layer on photoluminescence of SiC QDs deposited for 40 min [282].

The sizes of SiC nanodots grown on Si(111) substrates and AlN interlayers are 25–40 nm and 45–60 nm, respectively. The sample with AlN buffer layer shows a PL peak position at 625 nm and a full width at half maximum (FWHM) of 161 nm while the sample synthesized without a buffer layer had a peak centered at 639 nm and a FWHM of 166 nm. The photoluminescence spectra give evidence that the presence of an AlN buffer increases the PL intensity. This finding agrees well with the case of GaN films featuring an improved PL intensity when it was grown on an AlN buffer layer [283].

Moreover, a comparison of the PL peak position indicates a slight blueshift (from 639 nm to 625 nm) due to the presence of the AlN buffer layer (Fig. 5.8). This observation agrees well with the report of Liu et al. [284], which indicates on a quite similar blueshift of photoluminescence from InAs QDs as the thickness of InAlAs layer in InAlAs–InGaAs strain-reducing layer is increased. The observed blue shift of the photoluminescence maximum can be attributed to the modification of the potential barrier surrounding the quantum dots as the thickness of InAlAs increases [284]. It is thus possible that the AlN buffer layer significantly modifies the potential barrier surrounding the SiC QDs. Representative SEM micrographs of SiC quantum dots grown with and without

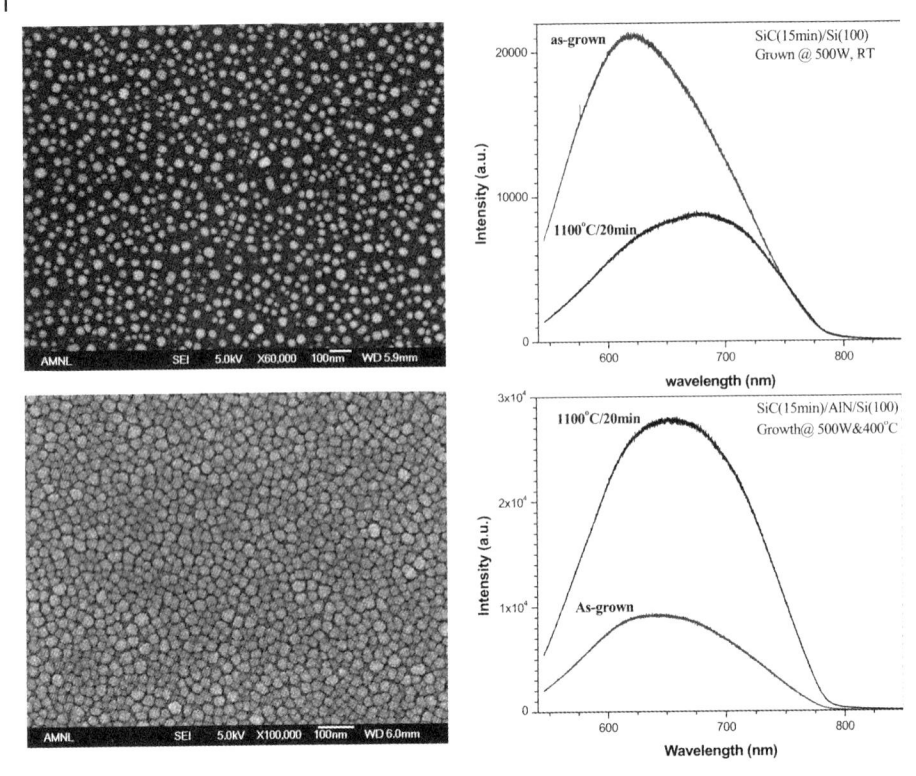

Fig. 5.9 SEM images and photoluminescence spectra of as grown and annealed (for 20 min at 1100 °C) SiC QD nanopatterns synthesized by 15 min deposition (RF power 500 W) on unheated Si(100) substrates (top panel) and AlN interlayer (grown at 400 °C for 60 min) on silicon (bottom panel) [282].

AlN interlayer are shown in Fig. 5.9. Interestingly, SiC QDs grown with an AlN buffer layer have a greater tendency to agglomerate and form clusters, a phenomenon reminiscent of Ostwald ripening.

This observation agrees well with the report of Hiramatsu et al. [285], which evidences that a fairly similar situation occurs for coalescing GaN nanoislands on AlN. We emphasize that SiC quantum dots deposited on an AlN buffer layer are observed to be generally larger in size (which can be seen in Fig. 5.9), possibly due to the effects of coalescence. From Fig. 5.9 one can also notice a quite different effect of postannealing on the room-temperature photoluminescence of SiC QDs synthesized without (top row) and with (bottom row) AlN buffer layer. Specifically, in the former case the PL intensity decreases after 20 min annealing of the specimen at 1100 °C. On the contrary, photoluminescence of SiC nanodot patterns fabricated on AlN buffers becomes even stronger after postannealing under the same conditions.

It is noteworthy that the sizes of quantum dots grown on AlN interlayers appear to be larger and the maxima of the PL intensity are smaller ("blue shift") than in the case of direct growth on silicon substrates. This observation is different from commonsense expectations that the emission wavelengths of smaller QDs must necessarily be shorter due to size dependence of the bandgap. However, in the case considered, we deal with a more complex system, where electron confinement in SiC quantum dots is controlled not only by the sizes of the nanodots themselves but also by the properties of AlN "sandwich" layers.

We now consider a simple model of confinement of an electron in a one-dimensional well of length a_x with the energy

$$\mathcal{E} = \frac{\pi^2 \hbar^2 n^2}{2 m_* a_x^2} = \frac{hc}{\lambda}$$

where λ is the peak emission wavelength, m_* is the effective electron mass, n is an integer, and $\hbar = h/2\pi$ is Planck's constant. If a_x is the lateral nanocrystal size, then the photoluminescence from larger nanocrystals (as is the case for SiC QDs on AlN buffers) should be red-shifted. This is not observed in the experiments discussed in this section. Thus, larger agglomerated nanocrystals is not the main factor that controls the peak emission wavelength. Therefore, some other mechanisms are more important in this case.

It is instructive to note that a one-dimensional quantum well structure shown in Fig. 5.6 can be considered to account for the quantum effects that manifest. This figure shows a nanometer-sized SiC layer sandwiched between two AlN layers. Due to electron quantum confinement, discrete energy states are formed in both the valence and conduction bands. For this structure, the quantum well has a finite height; hence only a finite number of bound states exist and the wells can interact [286].

When SiC quantum dots are deposited directly onto the AlN substrate, the SEM images in Fig. 5.9 reveal that the internanodot gaps are smaller than in the absence of the buffer layer, which is quite expected owing to the two different growth scenarios (SK and VW). By considering one of the dimensions of the QDs as a one-dimensional quantum well (Fig. 5.9), two closely packed QDs can be treated as a double quantum well. The latter can be formed when the lateral dimensions of AlN layers become comparable with the nanodot sizes. In this case the interaction between the individual wells would lead to a greater separation of the discrete energy states already existing in single quantum dot structures. Hence, any radiative transitions between the conduction and valence bands correspond to larger energy gaps and, hence, shorter emission wavelengths. This eventually results in a blue shift of the photoluminescence maxima.

Therefore, one can conclude that the effect of the AlN buffer layer on both photoluminescence properties and surface morphology of SiC quantum dot patterns is very strong. In particular, interaction between closely packed individual nanodots strained by the AlN nanolayer may lead to the reversal of the size dependence of the photoluminescence maximum on the QD size. It is interesting that quite similar observations have been made for dense nanopatterns of SiC QDs grown on silicon substrates by plasma-assisted magnetron sputtering deposition exceeding 30–40 min in duration.

To conclude this subsection, SiC QDs can be synthesized by using RF magnetron sputtering on an AlN buffer layer which is predeposited on Si(100) substrates. Sputtering of an aluminum target in a mixture of N_2 and Ar results in a thin AlN buffer layer. A subsequent sputtering of the SiC target can then be used. The above discussion involves the influence of the AlN buffer layer on the photoluminescence properties and morphology of SiC QDs. Regardless of the SiC deposition time, the buffer layer significantly enhances the PL intensity of the SiC quantum dots. The SEM analysis also indicates an increased coalescence (and hence, larger sizes) of SiC QDs when they are deposited on the AlN buffer layer. The influence of deposition time on PL intensity of SiC QDs deposited directly on the Si substrate is in agreement with that of SiC QDs grown on the buffer layer. For SiC QDs deposited directly on the substrate, SEM and STEM (Fig. 5.10) results reveal that SiC QDs are fairly uniformly distributed throughout the substrate. Grain boundaries between the dots are clearly seen in the STEM image in Fig. 5.10 and indicate that electrons can be subject to quantum tunneling phenomena. Last but not the least, the analysis of elemental composition by the energy dispersive X-ray spectroscopy confirms that the QDs of our interest in this section are made of SiC.

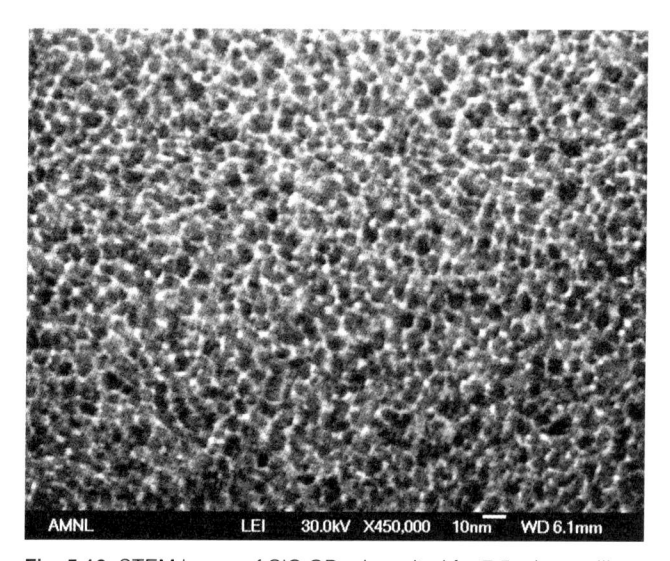

Fig. 5.10 STEM image of SiC QDs deposited for 7.5 min on silicon substrates. Quantum dots with an average size of 6 nm are uniformly distributed along the imaging area [282].

5.3.3
Quest for Crystallinity and Nanopattern Uniformity

Recently, Cheng et al. [287] have demonstrated the possibility of deterministic synthesis of highly uniform patterns of crystalline and stoichiometric SiC quantum dot patterns on p-Si(100) by using the inductively coupled plasma-assisted RF magnetron sputtering deposition process. The nanodots are highly stoichiometric ([Si]/[C]=1) and have a typical size of 20–30 nm at surface temperatures exceeding 250 °C. More importantly, an increase of the Ar+H_2 plasma density results in the improvement of crystallinity of the SiC nanopatterns.

Before we proceed to further discussion of the nanofabrication and properties of crystalline SiC QDs, we recall that bulk SiC has been commonly synthesized by using a variety of wet chemical, chemical vapor, and neutral beam epitaxy techniques [288]. However, the present day quest for tunable photoluminescence and other properties demands highly controlled nanofabrication of crystalline SiC QDs with a strong size-dependent response.

Despite a large variety of existing SiC-based nanostructures with different dimensionality, luminescent nanocrystals, nanoparticles, nanocrystalline films, heterojunctions, there has been no convincing evidence of the possibility of synthesizing highly crystalline SiC QDs on Si (further denoted as SiC/Si QDs) at low deposition temperatures compatible with the present day ULSI microelectronic process specifications. A very limited number of relevant reports indicates that nanodevice-quality SiC/Si QDs still remain elusive despite notable recent advances in *ab initio* modeling of chemical stability and optical bandgaps [289], synthesis of quasibulk (∼1 μm thick) nanocrystalline SiC [290] and SiC nanostructures of different dimensionality [291], molecular beam epitaxy of submonolayer quantities of SiC [292], and study of luminescence from colloidal SiC nanocrystals in different solvents [293].

This might seem as a bit of surprise especially because of much earlier advances in the synthesis of Si QDs on SiC (Si/SiC QDs) [294,295] and a common belief that SiC/Si QD is a merely inverted Si/SiC QD system. Indeed, both SiC/Si and Si/SiC systems feature a ∼20% lattice mismatch (the lattice constants of 3C–SiC and Si are 4.355 and 5.431 Å, respectively) and thus sustain the nanoisland growth modes. However, from the point of view of fabrication techniques and participating building units [36], the complexity of SiC/Si QD system is substantially higher, as the number of Si- and C-based building units to be delivered (from the nanofabrication environment such as chemical vapor or reactive plasma) to and redistributed (via surface migration) over the silicon surface need to be matched to achieve perfectly stoichiometric SiC QDs.

Due to much higher, compared to carbon, probability of silicon atoms to epitaxially recrystallize on Si surfaces, it is extremely difficult to independently (separately) deliver adequate number of silicon and carbon atoms. Furthermore, *ab initio* simulations [289] show that SiC (this also applies to *a*-SiC) can be synthesized under equilibrium conditions only in very narrow ranges of process parameters allowing one to avoid epitaxial recrystallization of Si or, alternatively, growth of carbon nanoislands.

More specifically, SiC can be formed, e.g., when the ratio of delivered Si and C species is low. It is amazing that delivering equal number of Si and C to the growth surface would turn counterproductive since Si and C adatoms are subject to very different conditions on a silicon surface. Above all, similar to many common nanodot systems, SiC/Si QDs are extremely sensitive to the rates of supply of building units and nanoisland crystallization rates. Indeed, the delivery rates should be reasonably low (which is extremely difficult to achieve in many existing fabrication techniques), whereas the crystallization rates should be high.

To this end, it is imperative to keep the substrate temperatures high enough (usually in excess of 600 °C) to achieve any appreciable crystallization of SiC, as is frequently the case in reactive magnetron sputtering [291]. Alternatively, special crystallization agents such as atomic hydrogen or carbon- or silicon-based reactive radicals are needed to increase the crystallization rates. It is notable that the above problems are in most cases fabrication technique-specific.

Therefore, there is a vital need to introduce a reliable nanofabrication technique suitable for the synthesis of stoichiometric and crystalline SiC/Si QDs. Cheng et al. [287] proposed the inductively coupled plasma (ICP)-assisted RF magnetron sputtering technique which made deterministic synthesis of SiC/Si QDs with the required properties, including ultrasmall size, required stoichiometry, crystallinity, and excellent uniformity over large surface areas, possible. In this way, it became possible to involve SiC compounds alongside with Si and C atomic units sputtered (with very different rates!) from sintered SiC targets in low-temperature, low-pressure Ar+H_2 plasmas independently created by an external inductive coil, to achieve the desired properties of the SiC/Si nanodot patterns.

The IPANF plasma reactor described in detail in Chapter 3, has been operated in the plasma-assisted RF sputtering mode. In this regime, the plasma densities are usually 1–2 orders of magnitude higher than in the pure RF magnetron sputtering regime used to synthesize SiC nanodots in the previous section. We emphasize that the plasma production and magnetron sputtering are essentially decoupled and can be controlled independently, by the inductive power P_{ind} (also for simplicity referred to as the ICP power) and RF magnetron sputtering power P_m. The inductive power (400–1000 W at 460 kHz) is delivered via a quartz window on the top of the chamber.

The RF magnetron power (50–150 W at 13.56 MHz) is applied to a sputtering electrode holding a high-purity, near-stoichiometric ([Si]/[C]=1), sintered 12 cm in diameter SiC target. An externally heated (in the 250–600 °C range)

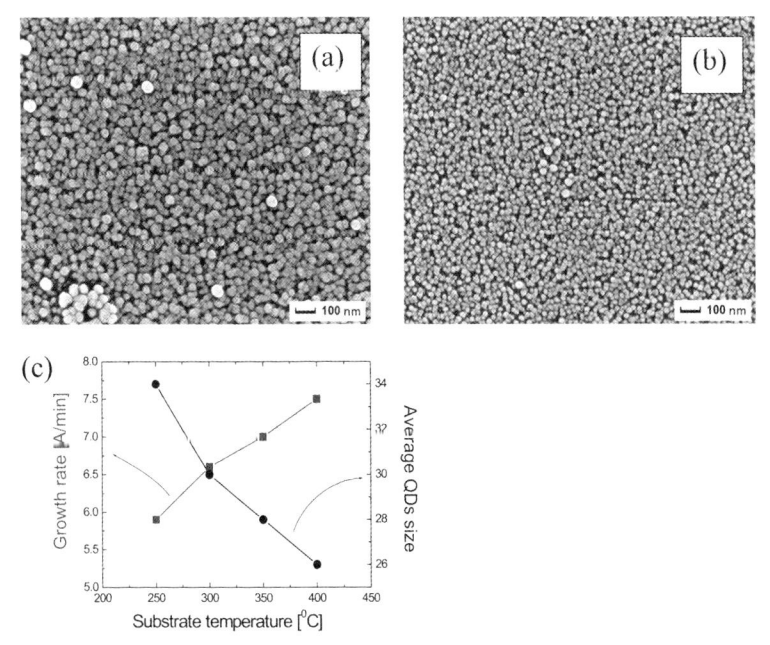

Fig. 5.11 SEM images of SiC QDs deposited at different substrate temperatures ((a) 250 °C; (b) 400 °C); (c) variation of growth rate and average SiC QDs size with T_s [287].

substrate stage is positioned in the chamber mid-plane approximately 15 cm beneath the quartz window. A low-pressure ($p_0 = 0.25$ Pa) mixture of high-purity Ar and H_2 gases with flow rates of 10 sccm, is let in the chamber, pre-evacuated to a base pressure of $\sim 10^{-5}$ Pa by using a combination of rotary and turbo-molecular pumps. SiC nanodot patterns are deposited on p-Si(100) substrates in the processes that lasted from 20 to 120 min. The surface morphology of SiC QDs is studied by standard analytical tools described in detail elsewhere [287] (see also other sections of this book).

At earlier growth stages the nanodot size d follows the cubic root-law dependence on deposition time t_D ($d \sim t_D^{1/3}$) according to the commonly accepted mechanism of Ostwald ripening [296]. Moreover, the quality of developed nanodot patterns is most sensitive to the substrate temperature T_s and inductive power. Figure 5.11 shows the surface morphology of the SiC/Si QD nanopatterns fabricated in a 60-min deposition process at $P_m = 100$ W, $P_{ind} = 800$ W, $p_0 = 0.25$ Pa, $p_{Ar}/p_{H_2} = 1:1$, and two different substrate temperatures (250 and 400 °C), where p_{Ar} and p_{H_2} are partial pressures of argon and hydrogen, respectively.

From Figs. 5.11(a) and (b) one can see that nanosized SiC grains very uniformly cover Si(100) surface. At $T_s = 250$ °C, the average size of SiC QDs is ~ 35 nm and the nanopattern development (growth) rate is ~ 0.58 nm/min

(Fig. 5.11(c)). As the temperature is increased to 400 °C, the average size of the grains reduces to ∼26 nm, whereas the growth rate rises to ∼0.75 nm/min (Fig. 5.11(c)). In this case, the uniformity of surface coverage by smaller SiC QDs becomes better as can be seen in Fig. 5.11(b).

Uniform coverage of large areas by ultrasmall SiC quantum dots also remains a striking feature of the experiments conducted at different inductive power and same other conditions as in Fig. 5.11 (Figs. 5.12(a) and (b)) [287]. At lower RF input powers (e.g., ∼400 W), QDs appear to be very small (∼20 nm) and cover large surface areas without any significant voids observable as is seen in Fig. 5.12(a). However, when P_{ind} increases, the average QD size increases in ∼50 % to reach ∼29 nm at 1000 W. The QD growth rates, on the other hand, decrease in almost two times, from 1.2 nm/min at P_{ind} = 400 W to just over 0.6 nm/min when the inductive power reaches ∼1000 W.

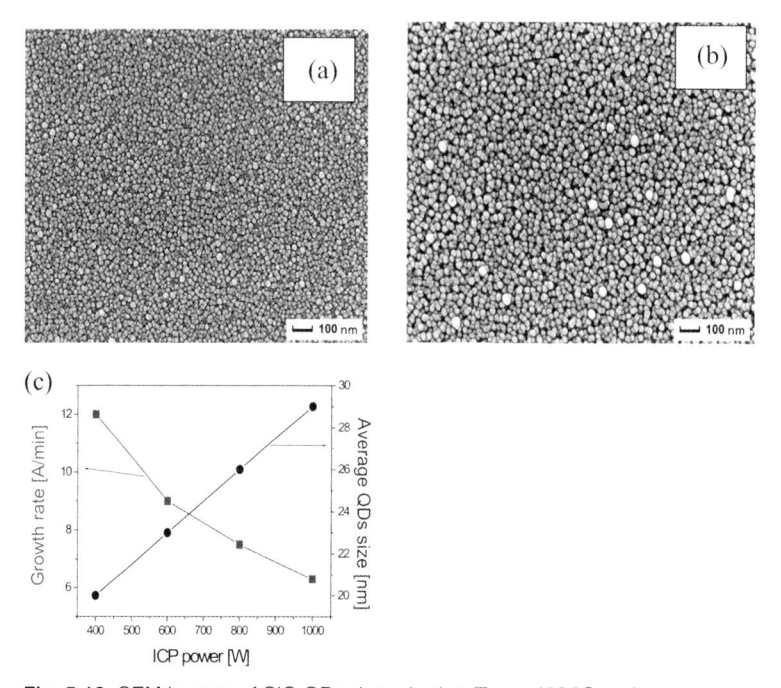

Fig. 5.12 SEM images of SiC QDs deposited at T_s = 400 °C and different ICP powers ((a) 400 W; (b) 1000 W); (c) variation of growth rate and average QDs size with the ICP power [287].

Elemental bonding states in the films have been characterized by using the XPS and FTIR techniques [287]. Figure 5.13 shows typical narrow scan XPS spectra of Si 2p (a) and C 1s (b). From the width and asymmetry of the XPS curve, one can infer the existence of multicomponent peaks. The peak fitting in the Si 2p spectrum yields four peaks, located at binding energy (BE) = 99.2, 100.5, 101.6, and 102.9 eV, respectively. These binding energies are attributed to Si–Si/Si–H, Si–C, O–Si–C, and Si–O$_x$ bonds [297, 298], respectively and are labeled in Fig. 5.13(a). Similarly, two deconvoluted peaks in the C1s spectrum in Fig. 5.13(a) correspond to binding energies 283.2 and 284.6 eV, respectively. These two binding energies are attributed to C–Si and C–C/C–H bonds, respectively [297–299].

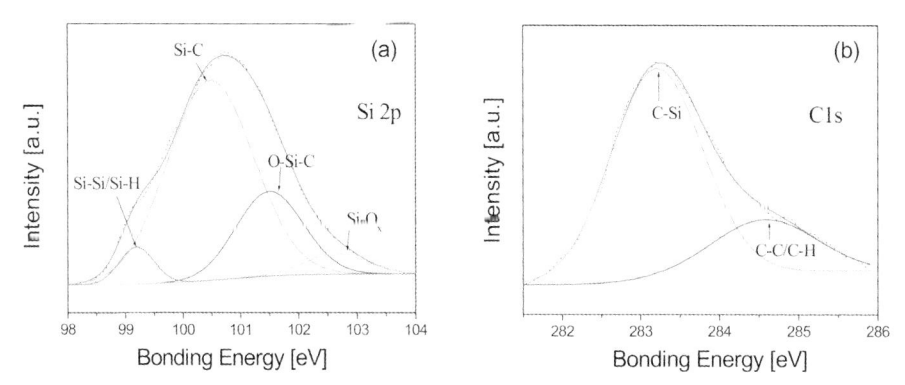

Fig. 5.13 XPS narrow scan spectra of (a) Si 2p and (b) C 1s of SiC QDs. [287].

From Fig. 5.13, one can easily notice that the SiC peaks are the strongest and the areas under them are also the largest compared to the peaks of other elements, which means that SiC is the main constituent in the film. Moreover, relative atomic concentrations of elements calculated by using transmission functions and sensitivity factors for each of the constituent elements are 49 % of Si and 47 % of C, with only minor traces of oxygen. Therefore, the SiC quantum dots are almost perfectly stoichiometric with almost equal percentages of silicon and carbon.

Glancing angle XRD measurements turned out extremely useful to substantiate the crystallinity of the SiC QDs [287]. Figure 5.14 shows the XRD spectra of samples deposited without the plasma assistance and at ICP power of 800 W at a low-substrate temperature of 400 °C. The two main diffraction peaks at $2\Theta = 35.7$ and $60.0°$ correspond to (111) and (220) of 3C–SiC, respectively [290, 300]. Comparison of the intensity and FWHM of these peaks suggests that the crystal growth is promoted with elevated ICP power.

Figure 5.15 shows FTIR spectra of SiC structures synthesized with and without the assistance of ICPs. The major band at 778 cm^{-1} (without ICP) is characteristic to the stretching mode vibrations of Si–C bonds, (Si–C)$_s$ [301, 302]. In the case of sputter-deposited films, it can be exclusively attributed to the (Si–C)$_s$ mode without any contribution from the Si–CH$_3$ vibration modes, which usually appear from samples grown by chemical vapor deposition in SiH$_4$+CH$_4$ gas mixtures.

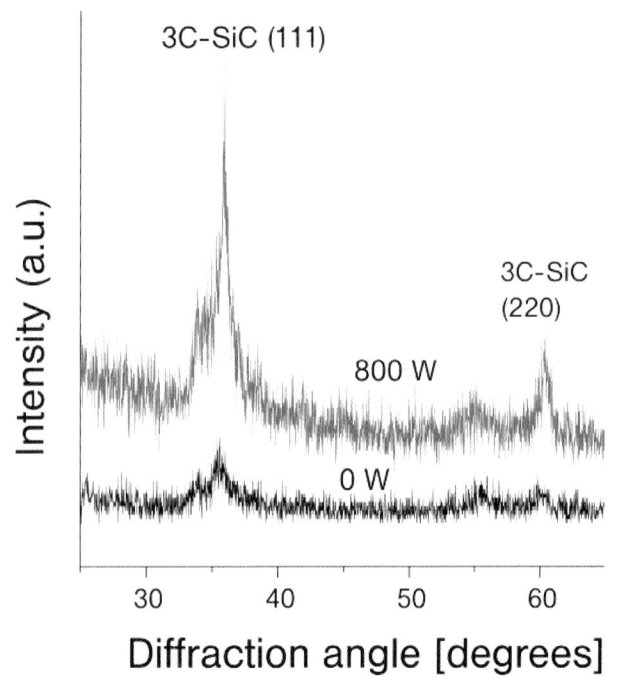

Fig. 5.14 XRD spectra of samples deposited at $T_s = 400\,°C$ and ICP power of 0 W and 800 W, respectively [287].

More importantly, after creating the inductively coupled plasma in the reactor chamber, the (Si–C)s peak position shifts into the shorter-wavelength domain (792 and 800 cm^{-1} at 400 and 1000 W, respectively). This is commonly attributed to the improvement of crystallinity [287]. Therefore, by using externally generated ICP, one can noticeably improve the crystallinity of SiC QD nanopatterns grown by reactive magnetron sputtering deposition.

The important salient features of the SiC quantum dots discussed in this section is a very low oxygen content, high-purity, uniform size distribution, and nonoverlapping (with other nanodots) in a very-high-surface-coverage (approaching to 1). More importantly, the nanodot sizes can be effectively controlled by varying the process parameters such as the external (ICP) power and substrate temperature (Figs. 5.11 and 5.12). Moreover, the process controllability and reproducibility are very high, which warrants its potential commercial applications.

We emphasize that some elements of the deterministic "cause and effect" approach [36] have been used to minimize the number of experimental trials. It is important to note that the nanofabrication method used by Cheng et al. [287] advances the existing magnetron sputtering techniques by separating the plasma production and sputtering of building units from the solid target.

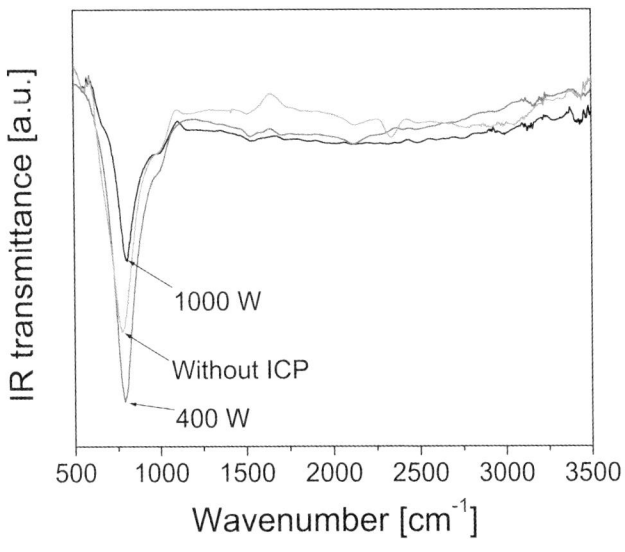

Fig. 5.15 Infrared transmittance spectra of SiC QD nanopatterns synthesized with the assistance of the inductively coupled plasmas (at P_{ind} = 400 and 1000 W) and without it. Other process parameters: T_s = 400 °C, P_m = 100 W, t_D = 60 min, p_0 = 0.25 Pa, p_{Ar}/p_{H_2} = 1 : 1 [287].

In this way, the external RF power which is used to sustain the plasma discharge becomes an efficient process control tool. In the above examples, by changing the ICP power P_{ind}, it became possible to control the deposition rates, and more importantly, nanodot size and crystallinity. We emphasize that the near-stoichiometric ([Si]/[C]=1) high-purity SiC sputtering target and the gas mixture with a low concentration of argon gas was chosen intentionally [287].

The choice of the near-stoichiometric target is actually dictated by the desire to sputter out a large fraction of SiC molecular building units that could directly incorporate into the growing patterns and thus improve the stoichiometry of the material. Moreover, RF magnetron sputtering at reasonably low powers is given preference to DC magnetron sputtering because at DC sputtering voltages exceeding 100 V (which is typical to most of DC magnetron sputtering systems) the rates of sputtering of silicon atoms (and also increase faster with the ion energy) much exceed those of carbon atoms, which could be detrimental for the synthesis of stoichiometric SiC QDs.

By choosing unconventionally low partial pressures of argon in the Ar+H$_2$ mixture (plasmas of hydrogen-dominated gas are more commonly used for the synthesis of polycrystalline, nanocrystalline, or polymorphous silicon), we have taken into consideration the results of our earlier computations, which suggest that when the input power increases, the fluxes of argon ions (which activate surface dangling bonds) usually increases faster than the fluxes of atomic hydrogen (which passivate the surface dangling bonds and etch amorphous material away) [36, 287].

Furthermore, higher plasma densities result in an increased sputtering yield, which in turn results in an increased delivery of building units to the growth surface. Not surprisingly, a combination of the above factors results in larger nanodot sizes seen in Fig. 5.12(b). It is important to note that the working pressure needs to be kept low ($p_0 = 0.25$ Pa) to enable the sputtered species travel to the deposition substrate unaffected by atomic collisions.

Finally, exciting and highly encouraging results on plasma-assisted synthesis of SiC nanodot patterns discussed in this section suggest that as yet elusive goal of deterministic synthesis of *highly uniform* arrays of *single-crystalline*, *pure* and *stoichiometric* SiC quantum dots is very likely to be achieved by using the innovative high-density inductively coupled plasma-assisted RF magnetron sputtering technique and a deterministic "cause and effect" approach [36]. Hopefully, the above results will encourage a wider nanoscience and plasma processing communities to more extensively use the low-temperature plasma-based nanofabrication tools.

5.4
Plasma-Aided Fabrication of Very Large-Aspect Ratio Si-Based Nanowires

Low-frequency inductively coupled plasma-assisted magnetron sputtering technique can also be used for the synthesis of amorphous silica nanowires on large area silicon substrates [303]. Such one-dimensional nanowire structures include nanohelicoids, nanosprings, nanohelices, and nanodrills and can be used to confine electrons in two dimensions in the plane perpendicular to the predominant growth direction.

Interestingly, the shape of all of the above nanostructures can be conveniently controlled by the arrangement of working gas flow and operation conditions of the plasma discharge. The growth process can be explained by the vapor–liquid–solid (VLS) mechanism [303]. More importantly, these nanowires could have potential applications in one-dimensional molecular devices, nanoscale biosensors and nanoresonators.

We now consider the issues related to the plasma-assisted growth of intricate one-dimensional structures in more detail. First, we note that one-dimensional (1D) nanomaterials such as carbon nanotubes, semiconductor [304–306], and oxide [307, 308] nanowires are being intensively studied because of their great potential applications in nanoelectronics and mesoscopic physics.

For example, wide-bandgap semiconductor light-emitting nanowires are ideal for fabrication of ultrahigh resolution visible display devices and optoelectronics [309]. On the other hand, amorphous silica is commonly used in silicon-based integrated devices as dielectric barriers and passivation layers. There are numerous recent reports on a successful fabrication of a variety of quasi-one-dimensional SiO_2 nanostructures, such as nanosprings [310], nanowires [311, 312] and some others.

In this section we show an example of successful use of reactive plasmas for the assembly of amorphous silica nanohelicoids, nanohelices and nanodrills. In this case, a plasma environment (high-density, low-frequency ($\sim$ 460 kHz) inductively coupled plasma-assisted RF magnetron sputtering in the IPANF reactor; frequency of RF signal applied to the RF magnetron is 13.56 MHz) quite similar to that used for nanofabrication of highly crystalline and ordered arrays of SiC/Si nanodots [287] discussed in the previous section, is used. We recall that the schematics of the IPANF facility is given in Chapter 3.

For the nanowire growth, Ni is usually chosen as a catalyst whereas lightly doped silicon or fused silica can be used as a substrate. In the experiments of Huang et al. [303] when the substrate temperature reaches a preset value, a silane and hydrogen gas mixture are introduced to the chamber and a low-frequency RF power is applied to ionize the working gas. Thereafter, a chemically pure Ni target magnetron electrode is then powered with an independent 13.56 MHz RF source. A typical process duration is $\sim$30 min. By adjusting the deposition parameters and substrate temperature, various one-dimensional nanostructures can be synthesized.

Figure 5.16 shows the field-emission scanning electron microscopy images (recorded using a JEOL JSM-6700F FE-SEM) of different 1D amorphous silica nanostructures. The overall lengths of the nanohelicoids (top left in Fig. 5.16) appears to be larger than 100 µm, with pitches varying from 1 to 1.5 µm. On the other hand, the nanosprings (top right in Fig. 5.16) are generally shorter than the nanohelicoids, with typical lengths not exceeding $\sim$10 µm; the pitches of the nanosprings usually change from 0.2 to 1 µm. The sizes of most of the nanosprings are fairly uniform along the preferential growth direction.

Interestingly, under certain growth conditions, a nanowire may split into two nanohelices (bottom left in Fig. 5.16). The diameters of the two helices are usually slightly smaller than that of the corresponding straight nanowire. The pitches of the two helices are not uniform, but they are quite symmetric about the central axis. Typical lengths of nanodrills (bottom left in Fig. 5.16) are slightly larger than 10 µm; their pitches usually exceed 200 nm [303].

Figure 5.17 shows a TEM and HRTEM (Philips CM 200) images of a typical nanohelix. Sample preparation involves a mechanical transfer of a nanohelix from the Si substrate to a holey carbon TEM copper grid. As one can see from Fig. 5.17(a), the diameter of the straight section is $\sim$126 nm, whereas the diameters of the two helical sections are 118 and 116 nm, respectively.

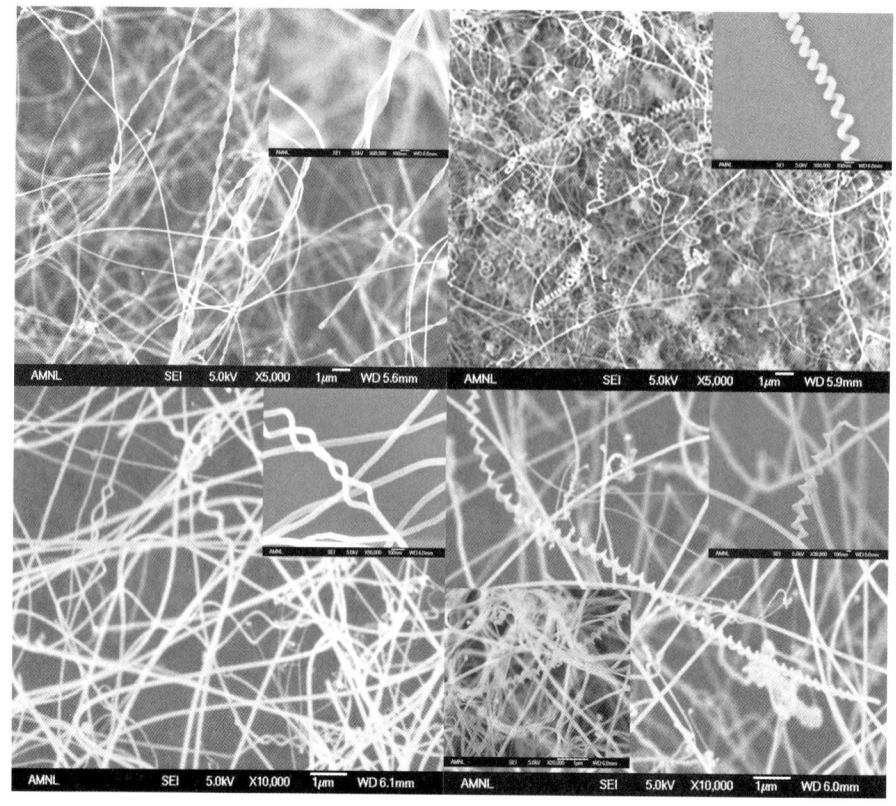

Fig. 5.16 FE-SEM images of one-dimensional nanowire-like amorphous silica nanostructures: nanohelicoids (top left), nanosprings (top right), nanohelices (bottom left), nanodrills (bottom right) [303].

Moreover, both directed and helical parts of the nanohelix have quite similar features in the HRTEM image, as shown in Fig. 5.17(b). The HRTEM image proves that the whole helix is purely amorphous.

The elemental composition of the amorphous helix is examined using an energy-dispersive X-ray (EDX) spectrometer (OXFORD INCA x-sight 7421). The EDX analysis confirms that the nanohelix is composed only of O and Si. Amazingly, the elemental ratio of oxygen and silicon is 55.2:26.4, which is very close to the stoichiometric [O]/[C] ratio of SiO_2 of 2:1. However, some EDX peaks attributed to Cu and C take their origin from the holey carbon TEM copper grid.

Furthermore, the results of the chemical mapping of silicon and oxygen in the nanowire structures suggest that both Si and O are uniformly distributed in the whole helix. The amorphous structure of the silica nanosprings is consistent with the established growth models of helical nanowires [313]. Huang et al. postulated that the changes of the diameter and pitch of the

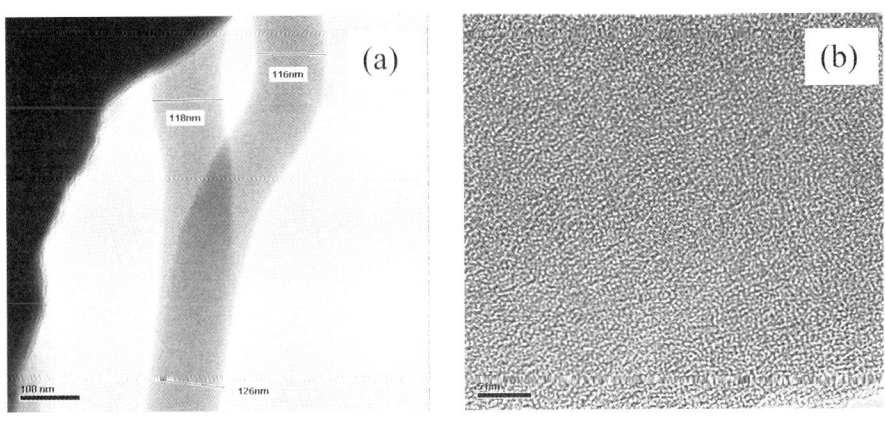

Fig. 5.17 TEM (a) and HR-TEM (b) images of a SiO_2 nanohelix [303].

nanosprings are caused by the composition, velocities, and flow directions of reactant gases during the growth process and put forward an inelastic collision theory [303]; however, this theory needs further justification before it can be widely adopted.

Therefore, low-frequency inductively coupled plasma assisted magnetron sputtering is a very efficient technique for nanofabrication of ultra-large-aspect-ratio one-dimensional nanostructures. To this end, highly reproducible synthesis and morphology control of the amorphous silica nanosprings, nanohelicoids, nanohelices, and nanodrills seem feasible. Finally, plasma-synthesized nanohelicoids, nanohelices, and nanodrills have thus joined the rapidly growing family of quasi-one-dimensional nanostructures, which have tantalizing prospects in nanotechnology.

5.5
Quasi-Two-Dimensional Semiconductor Superlattices Synthesized by Plasma-Assisted Sputtering Deposition

We now turn our attention to quasi-two-dimensional semiconductor nanostructures that can provide electron confinement in one direction. Such a fundamental possibility has already been briefly discussed in Section 5.3 in relation to AlN buffer interlayers in nanofabrication of SiC quantum dots.

Let us recall Fig. 5.6, which illustrates the basic idea of quantum confinement in quantum superlattices, which contain two alternating layers of two different materials. The thickness of these layers is usually in the range of a few to a few tens of nanometers. The contacting materials and the thickness of the layers in superlattices are chosen intentionally. Indeed, the electronic band structure of such a quasi-two-dimensional sandwich-like nanoassembly is engineered to meet a specific requirement, for instance, a room-temperature photoluminescence inefficient of impossible otherwise.

In this section, we discuss an even more complex example of such superlattice structures, namely SiCN/AlN nanoparticle superlattices (NPSLs), fabricated by using plasma-based nanotools. Specifically, alternating RF magnetron sputtering of SiC and Al targets in reactive plasmas of N_2+Ar+H_2 gas mixtures has been used [314]. Interestingly, the best periodic structures with the highest content of SiCN nanoparticles can be obtained when the thickness of alternating SiCN and AlN layers are 42 and 3 nm, respectively.

In this case, the NPSLs exhibit the strongest photoluminescence (PL) even without any postannealing. However, annealing of superlattice specimens at higher annealing temperatures T_a leads to an increase of the PL intensity. Moreover, the peak intensity experiences a strong redshift at annealing temperatures exceeding 650 °C; the maximum in the PL intensity is achieved at $T_a = 1100$ °C. Below, we will discuss the details of the plasma-assisted synthesis, structural properties, and room-temperature photoluminescence of SiCN/AlN nanoparticle superlattices [314].

As usual, we begin our discussion by mapping the main issues concerned and why the nanoparticle superlattices have recently been of so much interest. As has been mentioned several times elsewhere in this monograph, silicon-based nanostructures are indispensable in micro- and nanoelectronics, integrated photonics, and optoelectronics [315].

Since the early 1990s, porous Si structures have been in the spotlight of many research endeavors owing to their exceptional luminescence properties [316]. Unfortunately, it was soon revealed that relatively poor mechanical properties such as fragility and also unstable and uncontrollable luminescence of porous silicon are unlikely to keep the pace with continuously rising standards of modern nano- and optoelectronic technology.

To overcome such problems, research efforts in this direction had to focus on finding ways to synthesize Si nanocrystals embedded in oxide matrices [317–321]. In fact, a proper control of the size, passivation, and density of nanocrystalline Si (nc-Si) turns out to be a viable way to achieve a strong photoluminescence (PL) in the visible spectral range. For example, Zacharias et al. [322, 323] proposed a SiO/SiO_2 superlattice approach to control the nc-Si size and density and eventually improve the photoluminescence properties.

Besides the research of nc-Si, the photoluminescent porous SiC [324], a-SiN:H film [325, 326] and SiCN crystallites [327] have been paid a significant attention in the last decade. SiC-based nanomaterials discussed above in Section 5.3 is yet another alternative. However, as we have learned from Section 5.3, synthesis of crystalline nanostructured SiC still remains a major challenge. For example, to grow high-quality SiC film, i.e., cubic silicon carbide (β-SiC), very high growth temperatures are usually required.

Amazingly, as it often happens in nanofabrication (it is worth recalling how addition of In to AlN made it possible to achieve deterministic bandgap tunability in ternary $Al_xIn_{1-x}N$ quantum dot patterns, see Section 5.2), by adding one more element, one can synthesize a viable substitute capable of luminescing at room temperatures. By involving nitrogen gas in a high-temperature (~800 °C) synthesis process, Chen et al. [327] managed to synthesize nanoparticle SiCN films. It is also worth mentioning that controlled/guided self-assembly of nanoparticle films has recently become a matter of major importance because of its simplicity and cost-efficiency [314].

Due to the quantum confinement effects at nanoscales, the nanoparticle films also display promising characteristics for the fabrication of light-emitting diodes, nonlinear optical devices, and electronic devices. Recently, Xu et al. [314] reported an advanced plasma-based method of synthesizing SiCN/AlN superlattices with highly controlled thickness of alternating SiCN and AlN layers. In this way, it becomes possible to control the intensity of photoluminescence by varying the nanoparticle size and density. It is remarkable that when the SiCN sublayer is thin, there appears a notable size-dependent shift of the photoluminescence peak position in SiCN/AlN nanoparticle superlattices (NPSLs) [314].

In the experiments of Xu et al. [314], Si(100) substrates were chemically cleaned before being loaded into the sputtering chamber of the Integrated Plasma-Aided Nanofabrication Facility (see Chapter 3 for more detailed description of this facility). A typical base pressure of 3×10^{-3} Pa is routinely achieved by a turbomolecular pumping system. SiCN/AlN nanoparticle superlattices can be prepared by the following procedure: an AlN buffer layer is first grown on a Si(100) wafer by sputtering (for 30 min) a pure Al target under the gas flow rate N_2:Ar:H_2 = 32.1 : 21.4 : 3.2 (all flow rates in sccm) with a power of 300 W at the deposition temperature 350 °C.

Thereafter, SiCN nanoparticle films and amorphous AlN interlayers are alternatively deposited by a consecutive sputtering of a pure and stoichiometric SiC target ([Si]/[C] ~ 1:1) under the gas flow rates N_2:Ar:H_2 = 6.4 : 35.2 : 9.6 sccm with the RF power 500 W and Al target with the same growth condition as that of the buffer layer [314]. The substrate temperature is kept at 400 °C. After the deposition, the samples are annealed at 1100 °C for 20 min in a nitrogen flow. The details of the nanofilm characterization are standard and are similar to those used in previous sections. For more details of the plasma facility and analytical characterization the reader can be referred to the original work [314].

Figure 5.18 shows the transmission electron microscopy image of an as-grown SiCN/AlN nanoparticle superlattices sample (top panel), wherein the bright and dark layers are the SiCN and AlN layers, respectively. One can note regular and periodic structures with a large number of amorphous SiCN

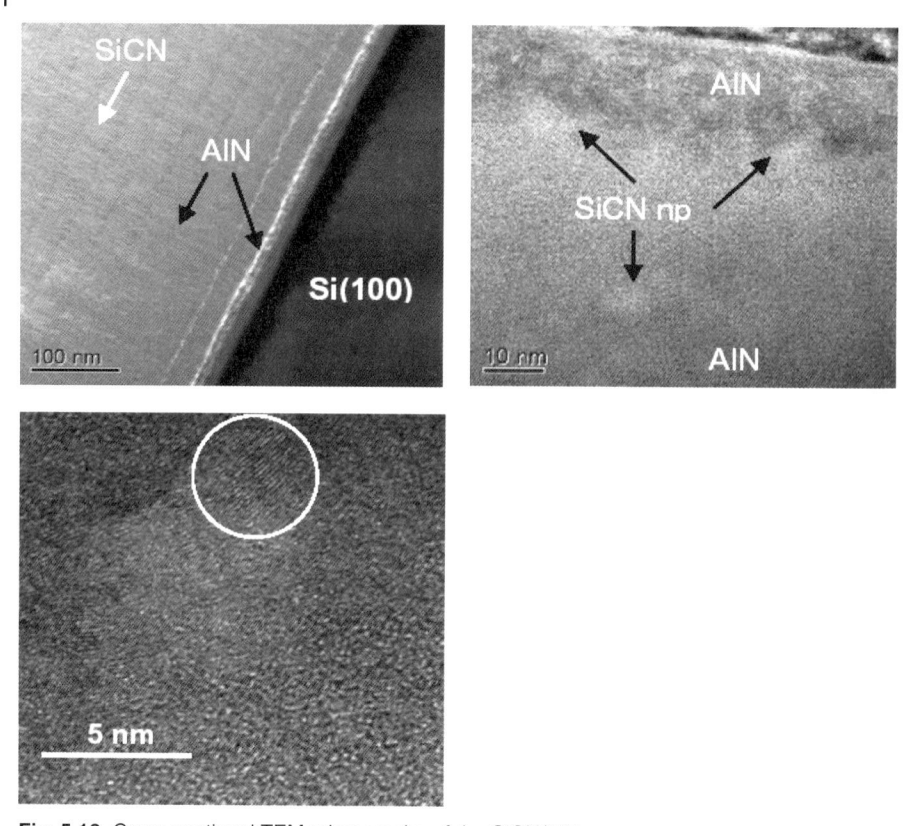

Fig. 5.18 Cross-sectional TEM micrographs of the SiCN/AlN nanoparticle superlattices sample: (top panel) as-grown SiCN/AlN nanoparticle superlattices; (panel in the middle) SiCN/AlN nanoparticle superlattices annealed at 1100 °C; and (bottom panel) a SiCN nanocrystal formed at the SiCN/AlN interface [314].

nanoparticles embedded in each SiCN sublayer. One can notice that the SiCN nanoparticles with a typical size of a few nanometers are located at the SiCN/AlN interfaces, as can be seen from Fig. 5.18 (panel on the right).

For pure SiCN nanoparticle films grown on an AlN buffer Xu et al. observed many amorphous cylindrical nanostructures with the size of a few tens of nanometers. Thus, SiCN nanoparticles in the superlattices can indeed be formed through the limitation of AlN sublayer. After annealing at 1100 °C, there are nanocrystals smaller than 5 nm at the SiCN/AlN interfaces (see Fig. 5.18, bottom panel); this appears to be consistent with the XRD analysis.

X-ray reflectivity (XRR) measurements have been used to investigate the effect of the sublayer thickness on the superlattice structures [314]. Figure 5.19 shows the XRR variation of the superlattices with the thickness of the SiCN

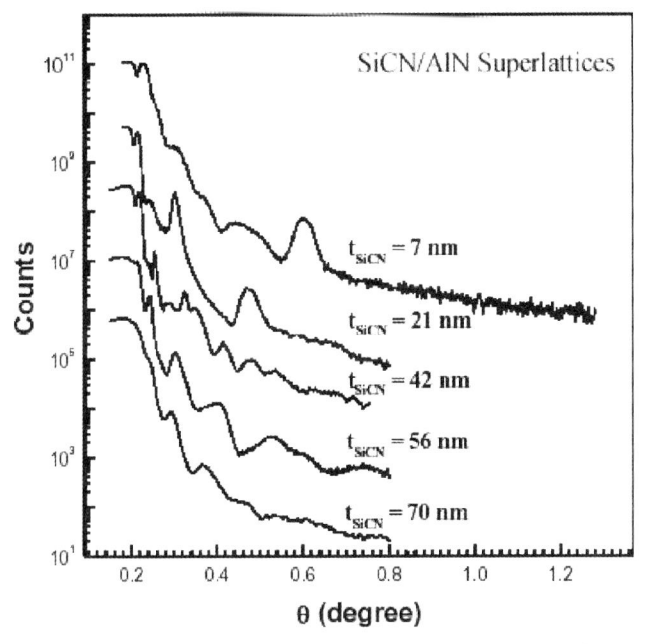

Fig. 5.19 X-ray reflectivity variation of the SiCN/AlN nanoparticle superlattices with the thickness of the SiCN alternating layer [314].

layer. When the thickness of SiCN and AlN alternating layers are 42 and 3 nm, respectively, one can observe the XRR peaks corresponding to up to three orders of diffraction; this is an indicator of the best periodic structure. The ratio of the intensities of the second-order and the first-order XRR peaks can be used to estimate the interface roughness [328].

Moreover, when the thickness of the SiCN and AlN alternating layers are 42 and 3 nm, respectively, the above intensity ratio reaches a maximum, which is indicative of the smoothest SiCN/AlN interface [314]. Meanwhile, the optimal value of the AlN sublayer thickness is approximately 3 nm.

Interestingly, the as-grown samples do not show any significant XRD peaks characteristic to SiCN nanocrystals. However, after annealing at 1100 °C, the XRD spectra (not shown here) of the SiCN/AlN nanoparticle superlattices exhibit broad diffraction peaks between 34 and 36°. Because the AlN buffer layer is too thin to give any diffraction information, the broad peak appearing at around 35° can be attributed to the SiCN crystalline structure [329].

More importantly, the intensity of the above broad peak reaches a maximum when the thickness of SiCN and AlN alternating layers is 42 and 3 nm, respectively. It is worth noting that the annealing temperature also shows a notable effect on the SiCN nanocrystal growth, which starts to form even at low T_a. The content of SiCN nanocrystallites in the film increases with T_a and peaks after annealing at 1100 °C. However, any further increase in T_a causes the content of SiCN nanocrystals to decrease quite significantly [314].

FTIR analysis of the SiCN nanofilms and SiCN/AlN nanoparticle superlattices reveals that under the same total thickness, the superlattices contain more SiCN nanoparticles compared to the case of a pure SiCN thin film. This is evidenced by strong C–N and Si–N wagging and Si≡O stretching bonds. Moreover, the FTIR results suggest that the broad peak at 35° in the XRD spectra can be attributed to the nanocrystalline structure of SiCN rather than to that of SiC. A comparison on the intensities of the photoluminescence of the SiCN films and the SiCN/AlN nanoparticle superlattices shows that the PL of the SiCN/AlN NPSLs appears to be much stronger than that of the SiCN film either before or after annealing.

Figure 5.20 displays the photoluminescence intensities of SiCN/AlN NPSLs with different SiCN sublayer thickness (t_{SiCN}). One can see that, when $t_{SiCN} < 70$ nm, the PL position is "blueshifted" as t_{SiCN} decreases. However, when $t_{SiCN} > 70$ nm, the PL position becomes independent of t_{SiCN} [314]. Moreover, the PL intensity increases with t_{SiCN} and reaches a maximum at $t_{SiCN} = 42$ nm. A further increase of the SiCN layer thickness causes a decrease of the intensity of photoluminescence.

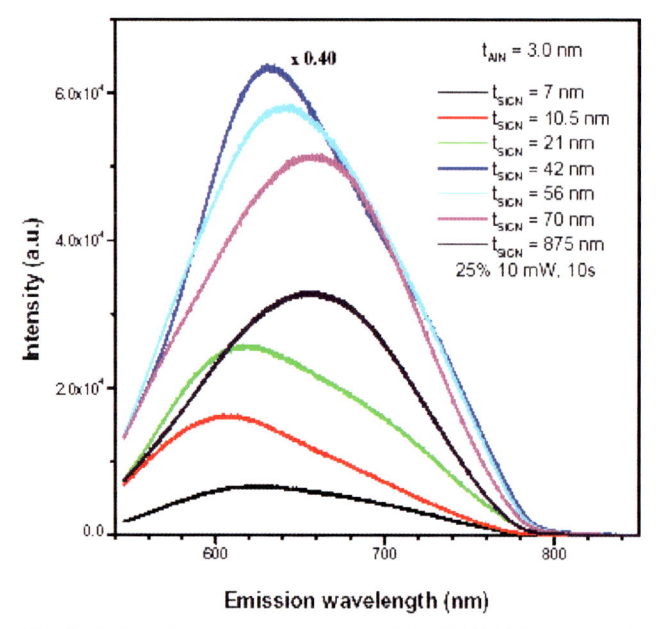

Fig. 5.20 Photoluminescence spectra of the SiCN/AlN nanoparticle superlattices with a different SiCN sublayer thickness [314].

Annealing the samples at 1100 °C can strongly enhance the intensity of the photoluminescence; however, the trend of the variation of the PL intensity with t_{SiCN} remains the same. The PL intensity also changes with the annealing temperature. In fact, at low temperatures the PL intensity first increases without any shift in the peak position; when $T_a > 650\,°C$, the PL increases again in intensity; this increase is accompanied by a redshift of the peak position. The PL intensity reaches a maximum when $T_a = 1100\,°C$; a further increase in T_a decreases the PL of SiCN/AlN nanoparticle superlattices [314].

It is worthwhile to mention that the superlattice method to limit the Si nanocrystal size was first explored by Zacharias et al. [322]. Using an empirical model that takes into account the different interfacial energies and materials, they successfully explained the exponential scaling of crystallization temperature with the layer thickness of amorphous Si sublayer and revealed a critical thickness for amorphous Si sublayer below which no crystallization can occur in the Si/SiO_2 system.

In this model, it was assumed that the crystallization nucleus is symmetrically embedded in the amorphous material between the oxide interfaces and is cylindrical in shape. In the SiCN/AlN nanoparticle superlattices of our interest here, the crystallization nucleus is most likely on the surface of amorphous AlN sublayer because of the close lattice constants of SiCN and AlN and thermal expansion coefficients; this certainly favors the formation of SiCN nanocrystals on the AlN surface at room temperatures. The TEM results in Fig. 5.18 clearly justify this assumption. On the other hand, any precise description of the nanocrystal nucleation kinetics would require an explicit assumption of the nanocrystal shape. Xu et al. assumed that the SiCN nanocrystals are also cylindrical in shape. For details of the SiCN nucleation kinetics, we refer the interested reader to the original work [314].

Here we pinpoint two interesting conclusions drawn from this elegant theory:

- there is a critical size of a SiCN structure for the amorphous-to-crystalline transition;

- smooth interfaces are more favorable for the crystallization of SiCN on the AlN surface.

As we have stressed above, the smoothest SiCN/AlN interface can be achieved when the thickness of SiCN and AlN interlayers are 42 and 3 nm, respectively. As such, it is not surprising that there exists more SiCN nuclei in the SiCN sublayers. As a result, the highest SiCN nanocrystal content was observed after postannealing at high temperatures.

The intensity of photoluminescence from SiCN/AlN superlattice can in general be described as [330]

$$I_{PL} \propto \sigma \phi N_{ec} \frac{\tau}{\tau_R}$$

where σ, ϕ, τ, τ_R, N_{ec} are the excitation cross section, photon flux, lifetime, radiative lifetime, and the total number of emitting centers, respectively. It is commonly believed that the passivation of nanoparticles can effectively enlarge the excitation cross section and increase the lifetime. Furthermore, in as-grown SiCN/AlN nanoparticle superlattices the emitting centers should be related to the SiCN nuclei or nanocrystals. It is also possible that after annealing, silicon oxides also affect the PL in SiCN/AlN nanoparticle superlattices. By considering so, the photoluminescence of SiCN/AlN nanoparticle superlattices can be understood as follows [314]:

First, in comparison with pure SiCN films, the SiCN/AlN NPSLs with the same total thickness of SiCN contain more SiCN nanoparticles (SiCN nuclei or nanocrystal), as revealed by the FTIR analysis. That is, the SiCN/AlN NPSLs have more emitting centers. Since the growth conditions of the SiCN interlayer are intermediate to those for the pure SiCN film and SiCN/AlN superlattices, σ, τ, and τ_R are quite similar. It is reasonable that the PL intensity mostly depends on the presence of SiCN nanoparticles in the superlattice. This is why the strongest PL is observed from the SiCN/AlN nanoparticle superlattices.

Second, the observed redshift of the photoluminescence position of SiCN/AlN nanoparticle superlattices at t_{SiCN} <70 nm can be attributed to quantum confinement effects of SiCN nanoparticles. In fact, the SiCN nanoparticle size is just a few nanometers at t_{SiCN} = 42 nm. The shorter the growth time of the SiCN sublayer, the smaller is the SiCN nanoparticle size, which naturally causes the blueshift of the PL peak position.

Third, the maximum of the photoluminescence intensity observed as t_{SiCN} = 42 nm can be also related to the abundance of SiCN nanoparticles in the superlattice. In this case, the annealed SiCN/AlN nanoparticle superlattices exhibit the highest content of SiCN nanocrystals [314]. Meanwhile, the as-grown SiCN/AlN NPSLs should have the largest number of SiCN nuclei. Hence, the maximum of the intensity of the photoluminescence is observed as t_{SiCN} = 42 nm and t_{AlN} = 3 nm, either with or without annealing.

Furthermore, the effect of postannealing on the PL of SiCN/AlN nanoparticle superlattices (which goes outside the scope of this monograph) can be satisfactorily explained in terms of the photoluminescence model suggested by Qin et al. [331]. This model takes into account two competitive processes, namely the quantum confinement and the luminescence center processes. When the most probable nanoparticle size is below the critical size, the quantum confinement processes determine the PL. Otherwise, the luminescence center processes control the photoluminescence.

Finally, the results of Xu et al. have convincingly demonstrated the possibility of using a plasma-based nanofabrication technique to synthesize quasi-two-dimensional semiconductor superlattices with ultrathin alternating SiCN and AlN layers. By adjusting the plasma process duration and other experimental parameters, it turns out possible to achieve a precise control and excellent reproducibility of the thickness and interface quality of the alternating layers. In turn, relative thicknesses of the SiCN and AlN layers significantly affect the electronic structure of the quasi-two-dimensional structure, and eventually its photoluminescence properties.

This is an exciting example of the successful use of the (plasma-based) process—structure—properties sequence in nanofabrication of light-emitting semiconductor nanostructures. In this way, by achieving a certain level of control of the structure and composition of the alternating layers, it is possible to ultimately achieve a high level of size-controlled tunability of the optical emission within a preselected spectral band.

5.6
Other Low-Dimensional Quantum Confinement Structures and Concluding Remarks

We now briefly discuss some other examples related to plasma-assisted nanofabrication of quantum confinement structures of various dimensionality. Owing to extreme flexibility of the Integrated Plasma-Aided Nanofabrication Facility in generating the broadest possible range of nanoassembly precursor species in solid (by sputtering), liquid (by *in situ* evaporation), and traditional gaseous (by PECVD) forms, the range of materials involved in the synthesis and postprocessing of functional nanostructures and their nanopatterns is virtually unlimited.

As we have seen from the previous sections, generation of multiple reactive species in a plasma environment makes it possible to synthesize a range of nanostructures made of binary and ternary compounds of elements of different groups (e.g., III–IV semiconductors). It is impossible to consider all possible options in a single monograph and this is why we will briefly discuss below some important cases that should be paid a particular attention.

In the first example, we show a microstructure and a clear relationship between the photoluminescence peak position and the sizes of silicon quantum dots grown on AlN interlayers. The second case is related to plasma-assisted self-assembly of ordered arrays of SiC-based nanodots. We will then show a striking example of the possibility to synthesize single-crystalline SiCAlN nanorods and by manipulating the plasma process parameters, to increase the nanorod's length and eventually "pull" ultra-high-aspect ratio nanowires. This section will conclude with a brief discussion of some of the important issues related to plasma-assisted nanofabrication of quantum confinement structures of various dimensionality.

Huang et al. observed photoluminescence from Si quantum dots embedded in amorphous AlN films and deposited (at low process temperatures on Si(111) substrates) in the IPANF plasma reactor in the plasma-assisted RF magnetron sputtering mode [332]. The main emission peak is located near 700 nm wavelength and has a full width at half maximum smaller than 80 nm. More importantly, the peak position in the PL spectrum shifts from 726 to 778 nm with the size of the Si quantum dots increasing from 9.1 to 14.5 nm.

It has also been deduced that the optical emission near 700 nm from Si quantum dots originates from the quantum confinement effect and not due to the surface state of the oxidized Si nanocrystals as is commonly believed [332]. Moreover, the mean diameters of the quantum dots can be effectively controlled by the plasma-based process parameters and properties of the amorphous AlN buffer interlayers.

The attempt to grow silicon quantum dots on AlN buffer layers by the plasma-assisted RF sputtering deposition is primarily motivated by the attractiveness of Si QDs for various applications in nanoelectronics. Their unique physical properties, size confinements effects, and Coulomb blockade phenomena make silicon quantum dots suitable for silicon-based nanodevices like single-electron transistors or quantum dot floating gate memories [333, 334].

To this end, a precise control of nanodot sizes and nucleation sites are of key importance in achieving stable operation and increasing the reliability of such devices [333, 334]. However, ultrasmall sizes of Si quantum dots (essentially in the nanometer range) are required for their efficient operation at room temperatures.

We emphasize that the AlN buffer interlayer and Si QDs can be synthesized by the RF magnetron sputtering technique within a single vacuum cycle [332]. The properties of the AlN buffer layer, which in fact significantly affect the assembly of Si QDs, are quite similar to what has been discussed in Section 5.1. After pretreatment of Si(111) substrates in a 2 % HF solution, AlN film is deposited on it; the sputtering parameters are as follows [332]: base pressure 1.1×10^{-3} Pa; ratio of mass flows of Ar and N_2 1:4; sputtering pressure 0.8 Pa; RF (sputtering) power 150 W; substrate temperature 350 °C; and deposition time 60 min.

Under the optimum sputtering conditions, the AlN films appear to be polycrystalline, with the mean optical transmittance in the ultraviolet and infrared ranges exceeding 96 %. In this case, the surface of the AlN buffer layer is very smooth, and the film thickness is approximately 54 nm. Interestingly, if the AlN film thickness is larger than 27 nm, it has a very weak effect on the formation of Si QDs.

Generally speaking, the size and pattern density of Si quantum dots are affected by a number of operation parameters that can be independently varied in the IPANF. However, it turns out that the mass flow ratios of Ar and H_2, substrate bias voltage, and RF power show the strongest influence on the Si

5.6 Other Low-Dimensional Quantum Confinement Structures and Concluding Remarks

QDs assembly. The deposition parameters of four samples in the experiments of Huang et al. are summarized in Table 5.5.

Tab. 5.5 Summary of deposition conditions of Si QDs [332].

Sample number	[Ar]/[H$_2$] ratio	RF power (W)	Bias (V)	Pressure (Pa)	Deposition time (min)	Substrate temp. (°C)
1	1:30	200	300	2.6	15	350
2	1:30	200	200	2.6	20	350
3	1:37	200	100	2.3	25	350
4	1:10	300	0	3.3	30	400

Figure 5.21 shows the TEM images of Sample 1. From Fig. 5.21 and also from relevant SEM images (not shown here), one can infer that the mean diameter of Si quantum dots is ~7.5 nm, and the deposited silicon material is in fact polycrystalline. Photoluminescence spectra of Sample 1 measured at room temperatures by using an Ar-ion laser (514 nm wavelength) suggest that the dominant emission peak is located at 738.41 nm [332]. The full width at half maximum (FWHM) of this peak is ~121 nm. The mean diameters of Samples 2, 3, and 4 are 9.1, 14.1, and 29.5 nm, respectively. The photoluminescence peaks of a higher energy of Samples 2 and 3 are located at 749.83 and 778.34 nm, respectively, and their FWHMs are 111 and 97 nm, respectively.

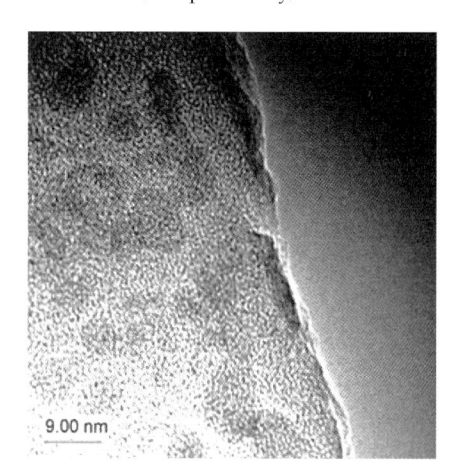

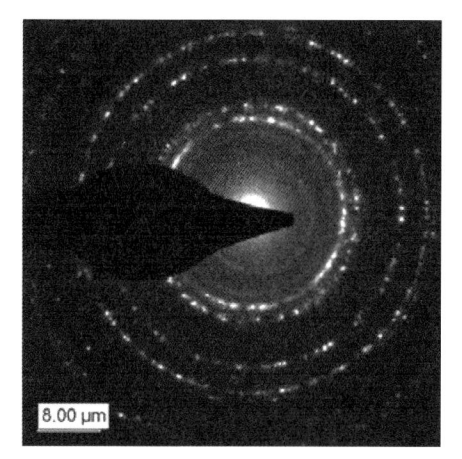

Fig. 5.21 Transmission Electron Microscopy of Si quantum dots embedded in an amorphous AlN matrix [332].

One can notice that the peak positions shift to the longer wavelength domain, and the FWHM becomes larger as the average size of the quantum dots decreases. Moreover, it turns out that the PL spectra of Si nanodots originate due to the quantum confinement effect rather than the surface states of oxidized Si nanocrystals [332]. Interestingly, a similar conclusion also applies to other luminescent nanoassemblies, such as nanoparticle superlattices considered in this Chapter.

More importantly, when the silicon islands are small, their shapes are almost hemispherical. As the QDs grow, their shapes became cylindrical, with their tops remaining hemispherical. Huang et al. [332] assumed that such features are caused by the interface binding energy of the constituents, and the lattice mismatch and binding energy of the Si QDs on amorphous AlN [337]. We emphasize that one can easily control the AlN surface coverage by silicon at low deposition rates.

Another very important issue is related to nanofabrication of highly ordered quantum dot arrays in plasma environments. Huang et al. [338] reported on a successful fabrication of highly ordered amorphous SiC quantum dots on a polycrystalline AlN buffer layer by directed nanodot self-assembly in the IPANF reactor operated in the low-frequency inductively coupled plasma-assisted RF magnetron sputtering mode. Amazingly, the ordered arrays of SiC nanodots (Fig. 5.22) usually align along the parallel lines, which are perpendicular to the surface of the SiC sputtering target [338].

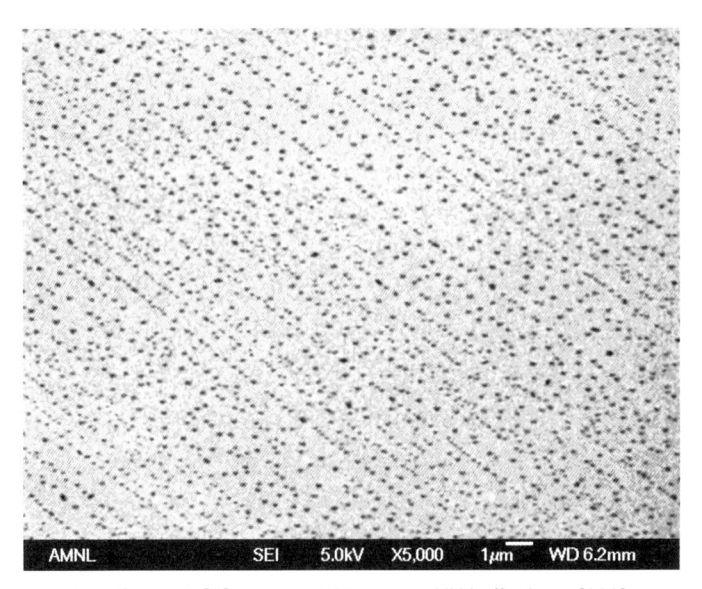

Fig. 5.22 Ordered SiC quantum dots on an AlN buffer layer [338].

Generally speaking, in most cases the nanodot growth is very difficult to control, and the size uniformity and spatial order of QDs required for many technological applications are still not readily achievable [339]. To this end, it would be prudent to mention that various approaches that in some manner direct the growth and ordering of QDs have been explored to improve their spatial ordering and size uniformity [343–349]. One of such approaches is to use nanopatterned substrates as growth templates. The SiC quantum dots have been grown with very good spatial order and size uniformity on nanopatterned and terraced surfaces [343, 344]. However, major practical and fundamental difficulties limit a widespread commercial use of such techniques. In fact, nanopatterning is quite difficult and time consuming, and thus negates many of the advantages of self-assembly. Another possible approach is to exploit modified growth kinetics that occurs on high-index vicinal surfaces with regularly ordered atomic steps [350–352].

As has already been mentioned above, highly ordered SiC nanodot arrays have been synthesized by the ICP-assisted magnetron technique on Si(100) substrates precoated by a thin AlN buffer interlayer [338]. Notably, the size, density, and ordering degree of SiC quantum dots can be effectively controlled by the process parameters. Specifically, the ordering direction appears to be orthogonal to the surface of the SiC sputtering target. We emphasize here that compared to molecular beam epitaxy, LFICP-assisted RF magnetron sputtering has the advantage of simple and flexible control of elemental composition of the nanofilms over large scales at a relatively low cost [338].

The most important process parameters in the deposition of the AlN interlayer are as follows [338]: base pressure of 8.64×10^{-4} Pa; mass flow ratios of Ar, N_2, and H_2 2:4:1; total pressure in the IPANF chamber 1.5 Pa; RF powers of the external ICP antenna and RF magnetron 1.2 and 0.3 kW, respectively; and substrate temperature 450 °C. Under the optimized sputtering conditions, the AlN appears to be polycrystalline, with a smooth surface, quite similar to the case of nanofabrication of pure Si nanodots on AlN buffers. It is important to note that the deposition rate in this case is quite low, ~ 0.15 Å/s. Thus, the thickness of the AlN interlayers can be effectively controlled by varying the deposition time.

It is worth emphasizing that the use of independently generated inductively coupled plasmas leads to much better quality AlN films (interface quality, surface morphology, crystallinity, oxygen content, etc.) as compared to the purely RF magnetron-based deposition under the same process parameters. During the deposition of the SiC quantum dots, the following parameters are kept constant: ICP power (1.0 kW), SiC magnetron power (0.2 kW), substrate temperature (650 °C), total pressure in the IPANF chamber (2.6 Pa), argon mass flow rate (19.5 sccm), and the substrate position. The hydrogen mass flow ratio, deposition time, and AlN buffer layer thickness can be varied. For example, for the sample shown in Fig. 5.22 the above three variable parameters are 3.6 sccm, 10 min, and 18 nm, respectively. We stress that the density, ordering degree, and average size of SiC QDs show a direct dependence on the plasma process parameters and also the thickness of the AlN buffer layers.

Generally speaking, the density increases at higher hydrogen mass flow rates, shorter deposition time, or thinner AlN interlayers. On the other hand,

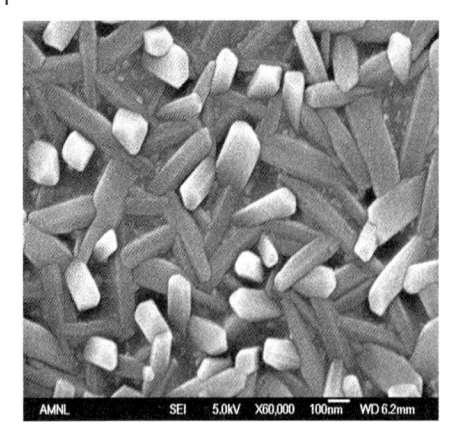

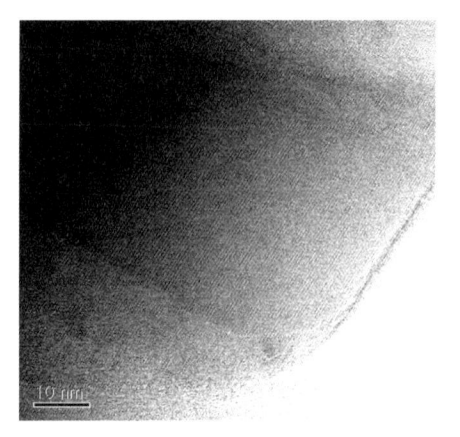

Fig. 5.23 SEM and TEM of SiCAlN nanorods [356].

the average size usually increases at longer process durations, lower hydrogen mass flow rates, or thinner AlN films. Furthermore, the degree of ordering can be reduced by increasing hydrogen mass flow rate, reducing deposition time, or depositing thicker buffer interlayers. It is also important to mention that Huang et al. [338] established lower thresholds for the successful synthesis of ordered SiC nanodot arrays. Specifically, SiC nanodot assembly is impossible at the substrate temperatures below 600 °C, and AlN buffer layer thickness less than 7 nm.

A striking observation [338] is that rotation of samples by 90° in either direction does not affect the alignment of the SiC nanodots. It is thus likely that the ordering/alignment direction in this particular experiment is neither related to preferential crystallographic directions of the substrate nor to the surface morphology of the AlN buffer layer. However, this assumption requires further experimental verification.

Cross-sectional high-resolution TEM and high-magnification plane-view FESEM microscopy reveals that Sample 1 shown in Fig 5.22 is, in fact, polycrystalline. Moreover, there is a thin (∼30 nm) amorphous SiC wetting layer between the SiC QDs and AlN buffer layer, which is quite typical to self-assembly via the Stranski–Krashtanov (SK) growth mode in heteroepitaxial systems [353–355]. Another argument that supports the likelihood of the SK growth mode is that no SiC nanodot patterns can be formed in short deposition processes that do not last more than 5 min. Another interesting observation is that the degree of crystallinity evidenced by X-ray diffractometry appears to be higher for highly ordered nanodot arrays (e.g., similar to the pattern shown in Fig. 5.22).

The mechanism of long range, large area (over surface areas up to 2×2 cm^2) alignment of the SiC quantum dots remains essentially unclear. Huang et

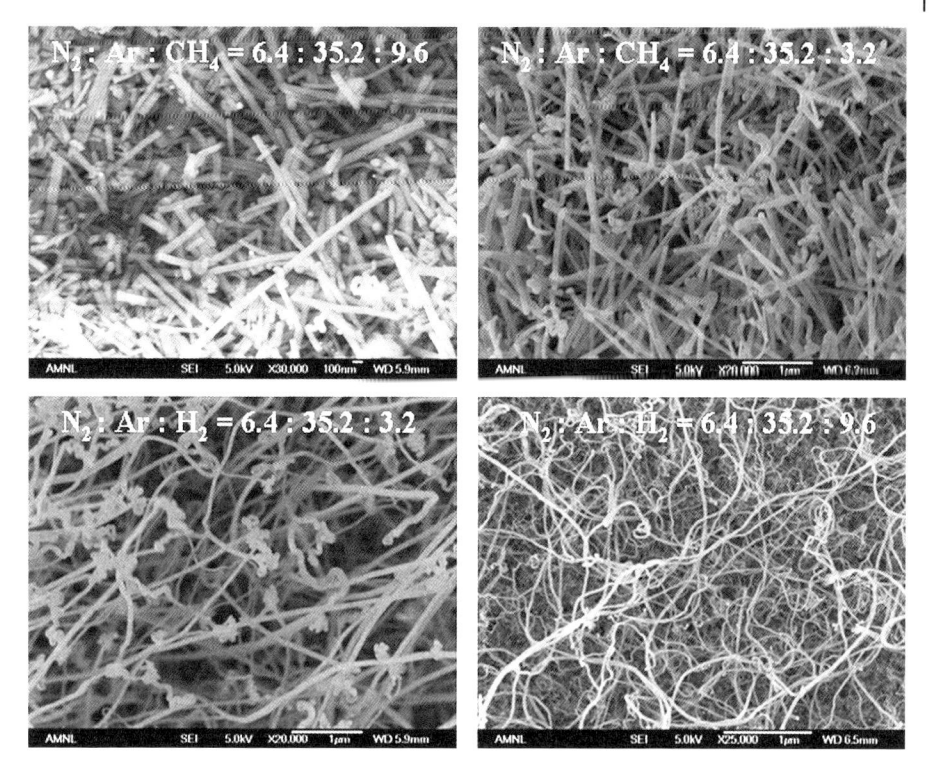

5.6 Other Low-Dimensional Quantum Confinement Structures and Concluding Remarks

Fig. 5.24 SEM of transition of SiCN nanorods to nanowires with changing process parameters [357].

al. suggested that parallel and aligned undulations (these might be some groove- or tongue-like nanofeatures) somehow form on smooth AlN surfaces and serve as preferential sites for the development of aligned quantum dot arrays. Quite similar cases of nanodot growth on concave features have been reported elsewhere [339, 347]. This undoubtedly very interesting phenomenon still requires the development of the adequate growth model and would certainly benefit from numerical simulations.

Let us now briefly examine an example of three-dimensional nanostructures that can also be synthesized by the plasma-assisted RF magnetron sputtering deposition in the IPANF plasma facility. One such structure is depicted in Fig. 5.23, which shows scanning electron and high-resolution transmission electron micrographs of single-crystalline SiCAlN nanorods [356]. From the left panel in Fig. 5.23 one can notice that typical widths of the nanorods belong to the range of 100–200 nm, whereas their lengths to 0.6–1.5 µm. Amazingly, the surface morphology resembles a stack of potato chips randomly oriented and interconnected in a small pocket-like paper bag from a fast food outlet!

The preferential growth directions (along the longest dimension in the case considered) of individual nanorods appear to be randomly oriented, with some of them laying parallel to the surface and only a few of them aligned along the normal to the surface. Axes of most of the nanorods form various angles with the substrate, with the most popular tilt belonging to the range 30–60°. This random orientation, on one hand, reveals clear nanosized features on the surface, and on the other hand, evidences a relatively poor controllability of the growth direction.

On a positive note, the nanorods are single crystalline as is clearly seen from the HRTEM image on the right panel in Fig. 5.23. Quite similar to single-crystalline conical carbon nanotip microemitter structures discussed in the previous chapter of this monograph, the single-crystalline SiCAlN nanorods are made of parallel crystalline planes stacked along the preferential growth direction, which, in fact, varies from one individual nanorod to another. Future research should be focused on finding ways to improve the size uniformity, alignment and orientation of the nanorods over large substrate areas.

The final example of this chapter shows the possibility to effectively control the aspect ratio of SiCN nanorod-like structures by varying the parameters of the plasma-based process [357]. Four SEM images in Fig. 5.24 correspond to the different compositions (partial pressures) of the reactive gas feedstock in the Integrated Plasma-Aided Nanofabrication Facility. The two electron micrographs in the top two panels in Fig. 5.24 show SiCN nanorod-like structures synthesized under the same partial pressures of nitrogen and argon and different (in three times) partial pressures of methane.

It is seen that the density of nanorod-like structures increases when the CH_4 inlet is reduced. In the latter case, the nanostructures become slightly thinner and also feature larger aspect ratios compared to the case shown in the top left panel in Fig. 5.24. However, replacing methane gas with hydrogen makes a dramatic change which is evidenced by the SEM photographs in the two bottom panels in the same figure.

First, comparison of the two images on the left shows that such a change in the gas feedstock composition leads to the growth of substantially longer, wire-like, quasi-one-dimensional nanostructures. When the hydrogen partial pressure is tripled, the aspect ratio of the SiCN nanowires increases even further, reaching in some cases ∼1000 and even more [357]. This amazing transition is most likely attributed to passivation of the growth surface by hydrocarbon radicals and atomic hydrogen in a reactive plasma environment. However, our assertion is purely speculative and more detailed experimental and theoretical investigations into this effect are warranted.

Finally, the examples of semiconductor quantum confinement nanostructures of different dimensionality (changing from 0 in the case of ultrasmall quantum dots to 3 in the case of relatively short nanorod-like structures) synthesized in the same versatile plasma-aided nanofabrication facility, evidence an outstanding degree of flexibility of plasma-based nanofabrication environments in providing control over main properties (e.g., composition, stoichiometry, crystallinity, size, shape) of the desired nanostructures and their growth patterns (e.g., alignment, ordering, spacing, etc.). At this stage plasma-aided nanofabrication of low-dimensional semiconductor nanostructures has not yet reached the deterministic level. However, exciting results presented in this chapter suggest that such a challenging goal can indeed be achieved in the near future.

6
Hydroxyapatite Bioceramics

In this chapter, we describe the unique features of hydroxyapatite (HA) bioceramic that make it particularly attractive for numerous applications in orthopedics and dentistry, discuss the most common specific applications and fabrication methods and identify the main problems that still need to be solved in view of the existing and envisaged applications and most recent progress in the field. A particular attention is paid to the fact that intentional disintegration of commercially available HA powders into smaller nanometer-sized building units in plasma environments proved a successful strategy for the improvement of structural, mechanical, and biocompatible properties of this unique bioceramic material.

Logically, this chapter is structured as follows. We begin from introducing the reader to the present-day applications of hydroxyapatite in orthopedics and dentistry and map the key fabrication requirements for medical device-grade bioceramics (Section 6.1). In Section 6.2, referring to the existing fabrication methods and their shortcomings, we introduce the hydroxyapatite synthesis concept that builds up on generation of subnanometer- and nanometer-sized building units in plasma environment of RF concurrent sputtering deposition system. The following section 6.3 details the means of optimizing the plasma-based process parameters to achieve the desired parameters of the coating. The details of mechanical assessment of the HA films are presented in Section 6.4. In Section 6.5 we discuss the results of *in vitro* assessment of hydroxyapatite coatings of the T6Al4V orthopedic alloy. The studies include the simulated body fluid (SBF) and cell culture tests. This chapter ends with concluding remarks and a brief discussion of the adopted approach in Section 6.6. The material of this chapter is primarily based on the original works of the authors and their colleagues ([358] and references therein).

6.1
Basic Requirements for the Synthesis of HA Bioceramics

For many decades, titanium, stainless steel, and cobalt–chromium alloys have been widely used as load-bearing biomedical or biodental implants. Several applications, such as fracture fixation and joint replacements require outstanding mechanical properties, such as high tensile strength, stiffness, and resistance to fatigue [359]. Titanium–aluminum–vanadium alloy Ti6Al4V satisfies all the above requirements and, in addition, features exceptional specific strength, chemical stability, and close values of the elasticity modulus to that of natural bone apatites.

However, due to its metallic nature, Ti6Al4V exhibits poor bioactivity, which implies the ability of a material to sustain the appropriate host response in a specific application [360]. Consequently, inadequate compatibility of the implant surface and bone tissues can result in longer healing time, fixation failures, and undesired inflammatory responses.

The biocompatibility of Ti6Al4V alloy can be improved by coating it with a bioceramic film. Hydroxyapatite (HA, $Ca_{10}(PO_4)_6(OH)_2$) has been widely used to promote biological functions of various implant materials [361]. This bioceramic material is the main mineral constituent of bone and tooth tissues and features excellent bioactivity, biocompatibility, and chemical and mechanical properties. HA coatings have revealed inspiring clinical advantages in promoting efficient implant fixation and implant-to-bone adhesion shortly after the implantation, as well as faster bone remodeling due to enhanced bidirectional growth and formation of a bonding interlayer between bone and implant [362].

Recently, nanocrystalline HA dental enamel paste has been used for rapid repair of early caries lesions without the need of removing healthy tooth material, commonly practiced in dentistry to ensure the filling stick [363]. The outstanding performance of the HA in orthopedic and dental applications is, in particular, due to its chemical composition similar to the mineral phase of natural bone tissues and extraordinary bioactivity, which is critical in promoting the interactions of biomolecules and proteins with the functionalized implant surface that ultimately result in sustainable bone or dental enamel remodeling (for example, via nanocrystalline growth [363]) and excellent implant fixation [362]. Furthermore, excellent osteoconductivity of hydroxyapatite makes it an ideal scaffold for the formation of new bone or dental tissues.

In spite of the many outstanding features, functional hydroxyapatite implant coatings often fail to meet numerous specific requirements of some clinical applications. For example, intrinsic brittleness of the bioceramic limits its use in heavy-load-bearing body areas, such as knee joints. On the other hand, poor molding ability restricts the HA's kneading into a desired consistency or shape, which is essential in bone fracture repairs and dental fillings applications.

A typical implant-bone fixation has a three-layered structure, where the hydroxyapatite coating acts as an interlayer between the metal implant and bone tissue. To this end, each of the two interfaces bears their unique functions. Indeed, the interface between the metal alloy and the bioceramic must be stable enough to withstand heavy mechanical loads. The other interface will then be primarily responsible for the interactions of the bioceramic with body fluids. However, several *in vitro* and *in vivo* studies have shown that dissolution rates of amorphous calcium phosphate films exceed reasonable limits to warrant their commercial use [364]. Therefore, a high degree of crystallinity appears to be another crucial requirement in the HA synthesis process. Additional requirements include well-defined elemental composition, lattice structure, and morphology of the interfacial matrix to match those of natural bone apatites to enhance epitaxial bone material ingrowth and recrystallization on the HA surface, and hence, faster implant fixation and bone remodeling.

Table 6.1 summarizes the key requirements the HA bioceramic coatings must meet before they can be successfully used in biomedical implants. More details about standard orthopedic requirements for devising hydroxyapatite bioceramics for surgical implants can be found in technical specifications of the American Society for Testing and Materials [365]. In view of the most recent achievements [363], one can formulate an additional key requirement, not yet reflected in the literature and technical specifications, namely, wherever possible, achieve ultrafine nanocrystalline structure of the hydroxyapatite material to achieve perfect incorporation and adhesion of the bioceramic in tiny pores of natural bone and dental materials.

Tab. 6.1 Key requirements of hydroxyapatite coatings of Ti6Al4V implants and specific functions [358].

Requirement	Definition/Function
Biocompatibility	Ability of a material to perform with an appropriate host response in a specific application
Bioactivity	Ability to interact with the surrounding bone and soft tissues
Osteoconductivity	Ability to provide scaffold for the formation of new bone
Dominant crystalline phase	Prevent resorption (dissolution) of coating in body fluids
Defined elemental composition	Match elemental composition of bone mineral phase
Specific lattice structure and morphology of interfacial matrix	Enable bone material ingrowth and epitaxial recrystallization on the HA surface
Interfacial stability and strong adhesion to Ti6Al4V implant alloy	Prevent mechanical failures under load-bearing conditions

The ongoing international research efforts attempt to meet the basic regulatory clinical requirements of the HA bioceramic, which include *in vitro* biocompatibility assessments in SBF and bone cell culture. In general, these techniques can be used to evaluate both the biocompatibility and cytocompatibility of the devised HA-coated implant to the physiological environment and bone cells to sustain acceptable bioactive and osteoconductive responses required to support growth and remodeling of new bone or dental material. However, the issues of devising the most suitable fabrication process and optimization of the process parameters to meet the key biomimetic material requirements altogether still remain open and demand an adequate solution, which is one of the main aims of the research efforts described in this chapter.

6.2
Plasma-Assisted RF Magnetron Sputtering Deposition Approach

6.2.1
Comparative Advantage

The requirements of Table 6.1 pose numerous challenges to the coating fabrication processes. To date, several techniques including thermal plasma spraying [362, 366–371], laser ablation deposition [372, 373], ion-beam assisted deposition [374], electrophoretic deposition [375, 376], and dip coating [377, 378] have extensively been explored to fabricate HA-coated implants.

The most widely used method of depositing hydroxyapatite coatings on titanium-based implant alloys is the thermal plasma spraying [362, 366–371]. This method is usually associated with atmospheric-pressure thermal plasma jets (plasmatrons) carrying micro-dispersed HA powder toward deposition surfaces. However, despite its apparent simplicity, reasonably high deposition rates, and relatively low cost, the thermal plasma spraying method is unlikely to meet all the requirements specified in Table 6.1.

The key reason behind this is that under typical operating conditions micron-sized hydroxyapatite powder particles usually deposit (with typical flows of hot ionized gas in the plasma jets in the 1–10 m/s range) onto the metal surface without any significant disintegration. As a result, HA coatings prepared by the plasma spray method often feature highly porous structure and predominant amorphous content [366, 379], which ultimately leads to disastrous coating performance in the envisaged application.

It is amazing that this conclusion is also valid even when the fine powder particles are perfectly crystalline, which is rarely the case for commercially available or laboratory-synthesized HA powders. Another critical drawback of many existing methods of coating of metal implants by hydroxyapatite is a weak adhesive strength of the bioceramic-metal implant interface [380]. Numerous *in vivo* studies reveal that violent mechanical failures and significant implant degradation often occur at metal-bioceramic interfaces [366, 379].

Several years ago, an alternative approach, based on intentional disintegration of compressed commercial hydroxyapatite powder into smaller sub-nano (atomic, molecular) and nano- (macromolecular, nanocluster, and nanoparticle) fragments in low-pressure reactive plasma environments, has been initiated [358, 381]. Since the HA is a dielectric material, plasma-assisted RF magnetron sputtering was chosen as an ideal practical framework of the above bioceramic synthesis concept [382–387].

Comparatively, RF magnetron sputtering deposition offers a great deal of film thickness uniformity over large substrate areas, straightforward control of elemental composition, sputtered species, and their energies, etc. A reasonable combination of the sputtering yield with the substrate bias and surface temperature is a viable pathway to achieve excellent bonding strength of various metal-dielectric/calcium phosphate bioceramic interfaces.

This common fact laid a foundation for the concurrent sputtering (hereinafter referred to as co-sputtering) of hydroxyapatite and titanium targets as a means of enhancement of adhesive strength of the bioceramic–implant alloy interface. The interfacial stability exceeded the expectations as was confirmed by numerous micro-scratch tests [388].

Physically, the co-sputtering technique involves concurrent plasma-assisted sputtering of titanium and HA targets, forming a bioactive hydroxyapatite thin film that contains a small, but crucial amount of calcium titanate ($CaTiO_3$). This compound serves as a bridging system between the Ti6Al4V orthopedic alloy and the bioceramic film, leading to an enhanced interfacial adhesion and ultimately a higher quality HA-coated implant [388].

6.2.2
Experimental Details

In this section, we describe the RF magnetron concurrent sputtering system and essential details of process control and materials characterization [358]. A 13.56 MHz RF magnetron sputtering system shown in Fig. 6.1 is used in the experiments. The setup comprises four main components, namely the deposition, RF power, substrate heating, and vacuum pumping systems. The major functions and special features of each of the setup components are described below.

The *deposition system* consists of the temperature- and DC bias-controlled substrate stage, a combined HA+Ti sputtering target (also referred to below as a bi-target), and Ti6Al4V substrates in a stainless steel vacuum chamber as can be seen in Fig. 6.1. The chamber has an inner diameter of 32 cm and a height of 30 cm and features four functional ports (with the diameter of 10 cm each) that are symmetrically arranged around the chamber. One of the ports is connected to a turbomolecular pump and the other three are used for sample loading and diagnostic purposes during the deposition.

The target comprising the bioceramic HA and chemically pure titanium (cp-Ti) is located on the RF powered electrode. The surface of the substrate stage,

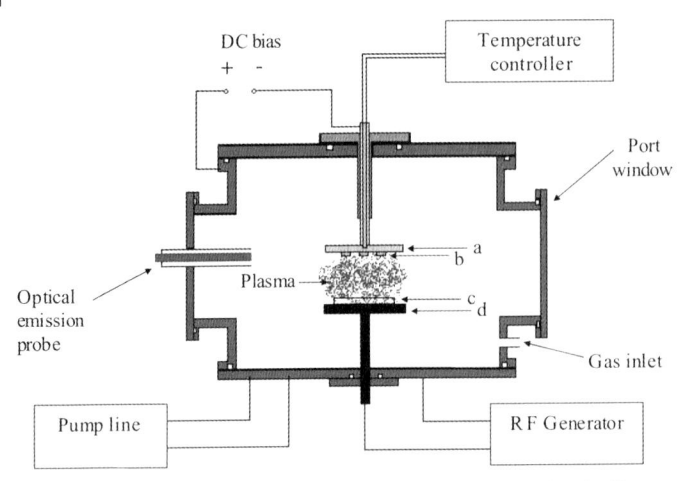

Fig. 6.1 Schematic diagram of the RF magnetron sputtering facility: (a) temperature-controlled substrate stage; (b) Ti6AlV substrates; (c) HA + Ti targets; (d) RF powered electrode [358].

with firmly clipped Ti6Al4V substrates, is placed ~6 cm above the target and faces downward. This substrate orientation is chosen to minimize film contamination by unwelcome micron-sized powder particles that are strongly affected by the pull of gravity. The substrate holder and the target electrode are electrically insulated from the chamber and are mounted through the top and bottom flanges as shown in Fig. 6.1. For better flexibility, the spacing between the substrate holder and target electrode has been made adjustable.

The *RF power system* consists of three units—a RF power generator, an impedance matching network, and a power meter. The generator produces a 13.56 MHz RF signal and has a total output power of 1 kW. The output impedance of the generator is 50 Ω and is self-protected by an internal feedback network. As such, any excess RF power above the preset threshold is reflected. The impedance of the plasma load varies with the geometry of the target and the discharge parameters. To reduce the reflected power from the load to the generator, an impedance matching network is employed to convert the external impedance to 50 Ω between the generator and the load system. Almost perfect matching can be achieved by a repeated adjustment of the two variable capacitors in the matching box.

An internal heater and a temperature controller are the two main components of the *substrate heating system*. This system enables one to measure and control the substrate temperature T_s. A nickel–chromium coil heating element is placed in contact with the internal surface of the substrate stage. To enhance the thermal efficiency, the heating element is thermally insulated by ceramic wool. This ensures that the heat is transferred to the substrate only. To measure and control the substrate temperature, a K-type thermocouple connected to the Athena Series 32C temperature/process controller is brought in contact with the internal surface of the substrate stage.

A two-stage *vacuum system* consisting of a rotary (2XZ-8, 8 l/s) and a turbomolecular (TP450, 450 l/s) pump is used to evacuate the vacuum chamber. The pressure in the vacuum chamber is measured by a Varian vacuum multi-gauge system. A typical base pressure of the order of $\sim 10^{-5}$ Torr is routinely achieved. The gas feedstock inlet is controlled by MKS mass-flow controllers. The working pressure is monitored by a MKS 146 measurement and control unit. The gas pressure is normally maintained in the range of a few tens of mTorr.

The *magnetron electrode* serves both as a powered RF electrode and a sputtering target stage. The electrodes consist of a copper casing and a magnetron insert, cooled by circulating water from the chiller. The magnetron insert consists of two concentric circular arrays of cylindrical magnets, each having a diameter of 13 mm, embedded in an aluminum holder. Any two adjacent magnets in the same circular array have opposite polarities. Sputtering targets are made of circular HA tablets and chemically pure titanium (cp-Ti). The tablets are molded into a cylindrical shape and are 13 mm in diameter and 5 mm thick each. The HA tablets and a 10 cm in diameter cp-Ti disc are placed atop of the magnetron electrode.

Prior to the surface cleansing process, the Ti6Al4V substrate material is cut into square pieces of approximately 1 cm × 1 cm. The Ti6Al4V substrates are immersed into HF+HNO$_3$ for about 1 min to remove the layer of oxides from the surface, followed by supersonic decreasing for 5 min in de-watered acetone (CH$_3$·CO·CH$_3$) in an LC20H ultrasonic cleaner. Thereafter, the substrates are blown dry with purified nitrogen. The cleaned substrates are clipped onto a stainless steel substrate holder and loaded into the vacuum chamber.

For the HA *target preparation*, a mixture of commercial microdispersed hydroxyapatite powder and distilled water is used. The mixture is molded into a 13 mm diameter GRASEBY SPECAC mold by a hydraulic press with a load of approximately 4 tons. The resulting HA target is 13 mm in diameter and 5 mm in height. The tablets are then sintered in the CARBOLITE chamber furnace, where they are heated up to 950 °C. Finally, after soaking in the furnace for three hours, the tablets are ready to use as sputtering targets.

After the cleaning steps, all the substrates are mounted onto the substrate holder before being loaded into the chamber via one of the chamber ports. A combination of the rotary and turbomolecular pumps allows one to evacuate the chamber down to $\sim 10^{-5}$ Torr. Shortly after inlet of gas feedstock into the chamber, the substrate heater and the RF generator are turned on.

Low-pressure discharge glows (Fig. 6.2) are sustained with approximately 700 W RF powers. The reflected power is minimized to less than 4% by adjusting the impedance matching network. During each deposition process,

6 Hydroxyapatite Bioceramics

Tab. 6.2 Summary of process conditions [358]

Parameter	Notation	Conditions
Base pressure	p_{base}	$\sim 10^{-5}$ Torr
RF input power	P_{in}	700 W
DC substrate bias	V_s	0–100 V
Substrate temperature	T_s	130–550 °C
Working gas pressure	p_0	9–78 mTorr
Deposition time	t_D	15–120 min

the gas pressure and temperature are kept constant. In most of the experiments described here, high-purity argon has been used as a working gas. A summary of deposition conditions in the original experiments of Long et al. is given in Table 6.2.

Fig. 6.2 A typical plasma glow in the process of the plasma-assisted co-sputtering deposition of bioactive hydroxyapatite coatings.

In situ optical emission spectroscopy is used to study the role of various gaseous species in the coating synthesis process. The details of the OES and auxiliary equipment have been discussed in previous chapters and original articles [93, 94].

Characterization of the deposited films involves five main *analytical tools*. Fourier transform infrared (FTIR) and Raman spectroscopy are used to investigate the bonding states of the elements in the coating. Analysis of the HA crystalline structure is performed by using X-ray diffractometry (XRD), and information on elemental composition is gathered from the energy dispersive X-ray (EDX) spectroscopy. Finally, scanning electron microscopy (SEM) is used to examine the surface morphology of the films. Infrared spectra are studied by a Perkin Elmer Spectrum One FTIR spectrometer. Furthermore,

Raman spectra are studied at room temperatures by using a RENISHAW micro-Raman system 1000 spectrometer coupled with a CW argon–ion laser excitation source operating at 514.5 nm. A laser beam is focused onto the area of ∼1–2 μm in diameter, giving a spectral resolution of ∼1.5 cm^{-1}. The XRD measurements are performed by SIEMENS D5005 X-ray diffractometer equipped with a vertical Bragg–Brentano focusing system. An X-ray wavelength of 0.15406 nm (CuK$_\alpha$ line) with a fixed power of 1600 W (40 kV × 40 mA) is used. Finally, surface morphology of the HA films is studied with a JEOL JSM-6700F field emission scanning electron microscope coupled with an Oxford Instruments EDX spectrometer.

6.3 Synthesis and Growth Kinetics

6.3.1 Optimization of the Plasma-Aided Coating Fabrication Process

In this section, it will be shown how the process conditions can be optimized to enable the required film stoichiometry, elemental bonding states, crystallinity, surface morphology, and interfacial matrix structure [358]. Logically, at a guessed substrate temperature, one can optimize the substrate temperature and working pressure to achieve the desired [Ca]/[P] stoichiometric value, which is 1.67 for the hydroxyapatite and can be measured by the EDX technique.

Next, under optimized DC bias and gas pressure, the required HA content in the films is confirmed via studying the elemental bonding states by using the FTIR and Raman spectroscopy. Thereafter, the substrate temperature is optimized to enable the highest crystalline phase content (analyzed by the XRD) in the coating. Finally, the surface morphology and porosity of the interfacial matrix is studied by the SEM.

6.3.1.1 Optimization of the DC bias voltage

For the purpose of optimization of the DC bias voltage, the deposition parameters have been set to $P_{in} = 700$ W, $T_s = 400\,°C$, $p_0 = 21$ mTorr, and $t_D = 120$ min, and we recall that P_{in}, p_0, and t_D are the RF input power, argon gas pressure, and deposition time, respectively. The relative elemental composition of calcium and phosphorus ([Ca]/[P] ratio) has been studied by the EDX for four different values of the negative DC bias V_s: 0, −50, −70, and −100 V.

The [Ca]/[P] ratio shows a general trend to increase with negative DC bias voltage. In particular, [Ca]/[P] soar to a high value ranging from about 4 at −50 V to 4.8 at −100 V. These values are far from the stoichiometric ratio of hydroxyapatite of 1.67. Comparatively, a closer ratio of about 1.7 is obtained without any bias voltage. One can thus conclude that a bias voltage should not be applied for the synthesis of the HA film.

Interestingly, the study of optical emission spectra of RF sputtering discharges combined with the EDX analysis of the film elemental composition suggest that the content of calcium in the films is determined by by the abundance of CaO^+ cations in the ionized gas phase, whereas the phosphorus content critically depends on the number densities of PO_4^{3-} anions [358, 388, 389]. The OES reveals that the emission intensities from most of positive ions in the discharge, including CaO^+, increase with the DC bias. A representative spectrum of the optical emission from the RF magnetron sputtering plasma discharge is shown in Fig. 6.3.

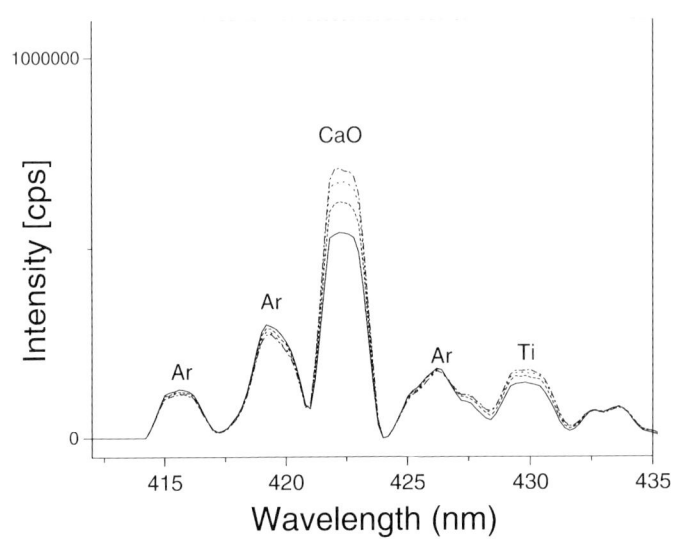

Fig. 6.3 Representative optical emission spectra in the process of the plasma-assisted co-sputtering deposition of bioactive hydroxyapatite coatings. RF power and working gas inlet are 700 W and 120 sccm, respectively. Solid, dashed, dotted, and dash-dotted lines correspond to the following values of DC bias voltage: 0, −75, −150, and −200 V, respectively [358].

It is also remarkable that the relative abundance of CaO^+ anions is the highest among other calcium and phosphorus-containing charged species. Thus, higher negative substrate biasing results in higher fluxes of CaO^+ cations onto the surface. On the other hand, at higher negative DC biases, it becomes more difficult for PO_4^{3-} anions to reach the surface, which explains higher [Ca]/[P] ratios at higher negative DC biases.

We also note that the [Ca]/[P] ratio can also be measured by using the X-ray photoelectron spectroscopy (XPS). In particular, peak fitting of the P2p narrow scan spectrum yields two binding energies of 132.8 and 133.7 eV, which confirm the presence of the phosphate phase in the films. On the other hand, deconvolution of Ti2p narrow scan spectra gives two binding energies of 458.5 and 459.2 eV, characteristic to titanium bonding states in TiO_2 and $CaTiO_3$, respectively.

Thus, the XPS analysis confirms the presence of $CaTiO_3$ phase, which is essential for better interfacial stability of the implant–bioceramic interface. It is noteworthy that the [Ca]/[P] ratio can range between 1 and 7.5 for Ca–P–Ti bioceramic films deposited by RF concurrent sputtering technique under various process conditions [381, 388, 390, 391]. Summarizing the results of the DC bias optimization, we can state that the best fit to the HA stoichiometric [Ca]/[P] ratio can be achieved when the deposition substrates are unbiased.

6.3.1.2 Optimization of working gas pressure

In this set of experiments, the following set of invariable parameters $P_{in} = 700$ W, $T_s = 550\,°C$, $V_s = 0$, $t_D = 120$ min and six values of argon gas pressure, namely, $p_0 = 9, 11, 14, 21, 38$, and 78 mTorr, are used [358]. The measured dependence of the [Ca]/[P] ratio on the working gas pressure suggests that the optimal pressure range allowing the near-stoichiometric values of the [Ca]/[P] is 20–30 mTorr. It is interesting to note that the effect of an increasing pressure on the Ca- and P-bearing species appears to be different. As the pressure increases, the amount of sputtered atomic and molecular species also increases, as has been noted from optical emission spectra. When number densities of the plasma species increase, collisional effects become more important. In addition, at higher pressures, lighter phosphorus-bearing species can reach the film faster than heavier calcium-bearing species.

Consequently, the [Ca]/[P] ratio decreases as observed in the experiments. Therefore, the negative bias voltage and working gas pressure strongly affect the ratio of calcium and phosphorous atoms in the films. It is imperative that in order to deposit an HA coating with the desired elemental composition, no negative DC bias should be applied. Likewise, the working gas pressure in the range between 20 and 30 mTorr is favorable to obtain the [Ca]/[P] ratios close to the HA stoichiometric ratio of 1.67. Hence, further film synthesis and characterization is performed for unbiased substrates and a fixed value of the gas pressure ($p_0 = 21$ mTorr).

6.3.1.3 Optimization of substrate temperature

We recall that a high degree of crystallinity is one of the crucial components in the development of HA-coated bioimplants. It is a common knowledge that relative contents of crystalline and amorphous phases in the film can be efficiently controlled by the substrate temperature. For this purpose, the deposition has been performed at $T_s = 130\,°C$, $300\,°C$, $400\,°C$ and $550\,°C$, respectively. Other deposition conditions are the same as in the previous subsection.

Figure 6.4 displays the XRD spectra of the HA samples synthesized at different substrate temperatures. One can notice that the films feature polycrystalline structure as suggested by the appearance of multiple diffraction peaks in the spectrum. Moreover, several new growth directions such as (222), (213), (402), (102), (210), (111), and (200) also show signs of growth at higher temperatures, which indicates the development of the polycrystalline structure, which can enhance the performance of the HA coating in applications. Likewise, one can observe that the crystalline content of HA increases with substrate temperature, as suggested by higher amplitudes of the main diffraction peaks, such as (112) and (211) at higher T_s.

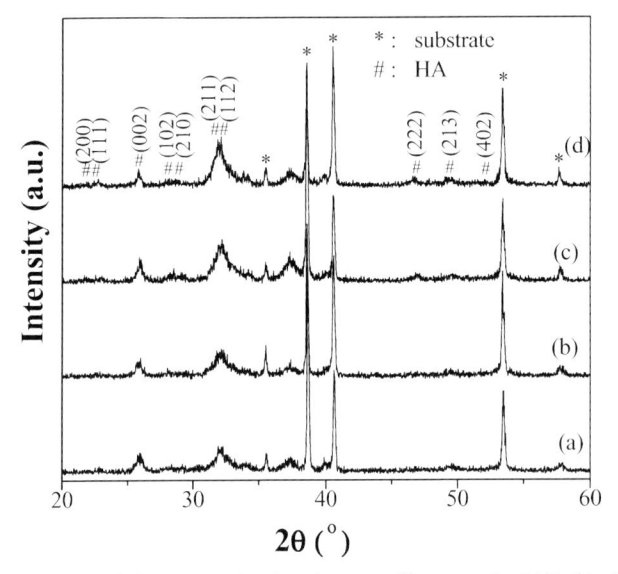

Fig. 6.4 XRD spectra showing the crystalline growth of HA thin film at (a) $130\,°C$; (b) $300\,°C$; (c) $400\,°C$; and (d) $550\,°C$ [358].

The effect of the substrate temperature on the surface morphology of the films has been studied by the SEM. Figure 6.5 shows the SEM micrographs of the HA films deposited at two different temperatures. It is seen that more uniform and lower porosity films can be deposited at elevated substrate temperatures. For the sample shown in Fig. 6.5(a), quite many pores are clearly visible. On the other hand, the entire surface of the sample synthesized at a higher temperature is covered quite homogeneously with hydroxyapatite crystallite grains (Fig. 6.5(b)).

Thus, crystallinity of the HA film indeed becomes better at higher substrate temperatures. Therefore, in the following studies, the temperature is fixed at $550\,°C$.

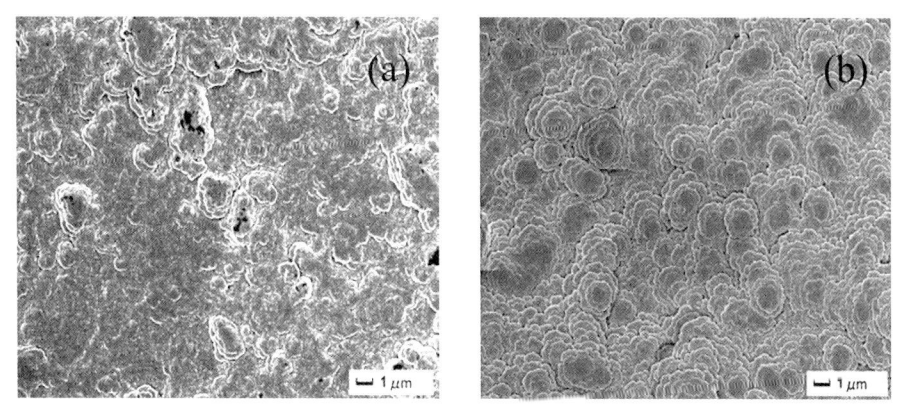

Fig. 6.5 SEM images showing that uniformity and low porosity can be achieved with increasing temperature: (a) 130 °C and (b) 300 °C [358].

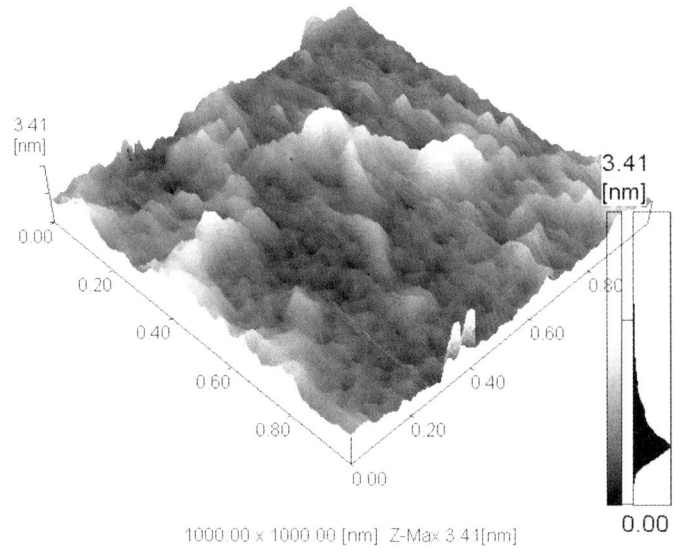

Fig. 6.6 A typical AFM image of the nanostructured hydroxyapatite surface [358].

It is noteworthy that the above SEM results are consistent with our earlier studies of the HA surface morphology by means of atomic force microscopy (AFM) [392]. A typical AFM micrograph is shown in Fig. 6.6. The AFM results also suggest that the surface roughness and morphology of the hydroxyapatite films change remarkably with the process parameters, such as the substrate temperature, gas pressure, and substrate bias.

For example, variations of the DC substrate bias and/or T_s affect the homogeneity of distribution and architecture (e.g., sharpness) of the elements of morphology as well as inter-element spacing and surface areas in the "humps" and "dips." To this end, homogeneous distribution of similar by shape and size nanosized morphology elements over larger (more than a few square microns) surface areas is ideal for promoting biomolecule/protein attachment and growth [392].

The AFM results thus confirm a challenging opportunity of controlling the nanoscale surface morphology of the HA coating by varying the process parameters. We emphasize that the surface roughness of the HA coatings has been reported as one of the key factors that controls the bone–implant interface shear strength [393]. On the other hand, *in vivo* toxicity of the hydroxyapatite coating increases with the size of crystallites (morphology elements) on the surface [394].

The effect of the substrate temperature on the growth of crystalline hydroxyapatite films can be explained from the perspective of surface structure of HA. When the temperature is increased, the atoms acquire more thermal energy and hence vibrate more vigorously. As the atoms on the surface have fewer neighbors and are more loosely bound than in the bulk, the amplitude of their vibrations becomes larger. This also implies that the energy of the atoms on the surface is higher than in the bulk.

When the temperature is high enough, the surface atoms gain sufficient energy to leave their sites, migrate over the surface, and eventually rearrange to reduce the overall internal energy. This rearrangement is most efficient at higher surface temperatures promoting higher atom mobility and more efficient surface diffusion, which eventually results in the long-range ordering and growth of crystalline structures.

However, at lower temperatures, the surface atoms have a lower mobility, which results in the build-up of the amorphous phase. This analysis further elucidates the need of higher substrate temperatures to synthesize highly crystalline hydroxyapatite films.

6.3.2
Film Growth Kinetics

To allow a more in-depth understanding on the evolution the HA crystalline structure during the deposition process, the effect of the duration of the deposition process on the hydroxyapatite film properties has been investigated. The film deposition time t_D is fixed at 15, 30, 60, 90, and 120 min. A systematic study into the growth dynamics of the HA grains includes the XRD (wide and narrow scans) and SEM analyses. The XRD wide-scan spectra provide the information on the overall development of the hydroxyapatite crystallites, whereas the narrow-scan spectra reflect the evolution of the HA grain size.

Figure 6.7 shows the XRD spectra from the HA films deposited under different process durations. It is apparent from the wide scan spectra in Fig. 6.7 that the HA crystalline content augments with t_D, as evidenced by a substantial in-

6.3 Synthesis and Growth Kinetics

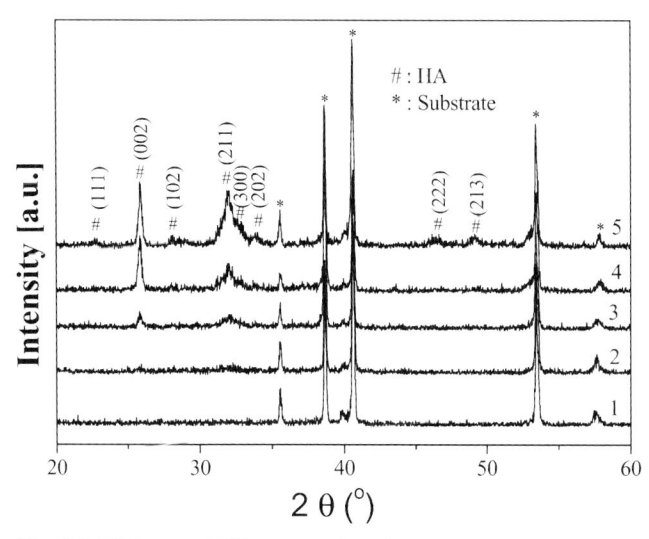

Fig. 6.7 Wide scan XRD spectra depicting the HA crystallization process. Curves 1–5 are plotted for the film deposition time of 15, 30, 60, 90, and 120 min, respectively [358].

crease of the amplitudes of the (002), (211), (112), and (300) crystalline peaks. Furthermore, from the narrow scan of the (002) peak, one can conclude that crystalline growth in this direction becomes notable after 30 min of deposition and even more pronounced after 90–120 min into the process.

Apart from that, at longer deposition times, several new diffraction planes emerge. For example, one can observe the origin and further development of new planes (102) and (222) in the XRD spectra recorded for $t_D = 90$ and 120 min, respectively. Generally, from the XRD wide scan spectra one can conclude that longer deposition times result in higher crystalline phase contents and a wider variety of growth directions.

To analyze the dynamics of the grain size growth, narrow-scan (from 25 to 27°) spectra of the (002) peak have been recorded. The full width at half maximum (FWHM) of the (002) peak decreases when t_D is increased from 60 to 120 min. By using Scherrer's relation [395]

$$a_g = 0.9\lambda / B \cos \Theta \tag{6.1}$$

where a_g is the grain size, B is the FWHM of the peak, and Θ is the diffraction angle, one can conclude that the crystalline size increases with the deposition time. Noting that $2\Theta = 25.87°$ and $\lambda = 0.15406$ nm, one can estimate that the grain sizes are 46.58, 54.34, and 62.70 nm for $t_D = 60$, 90, and 120 nm, respectively. The above estimates reveal that the hydroxyapatite crystallites grow steadily in size with the deposition time.

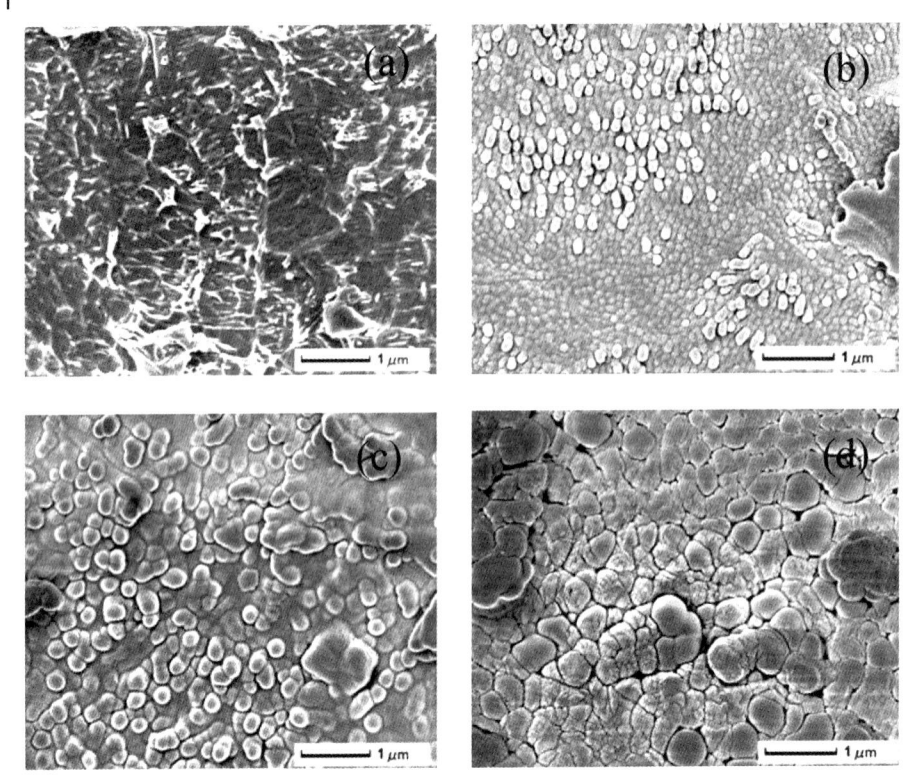

Fig. 6.8 SEM images of (a) Ti6Al4V alloy and as-deposited films after (b) 5; (c)15; and (d) 60 min of deposition.

The XRD results are consistent with the SEM analysis (Fig. 6.8) showing the surface morphology of the uncoated Ti6Al4V alloy (Fig. 6.8(a)) and HA films deposited at different process durations (Figs. 6.8(b)–(d)). From Fig. 6.8(b), one can see that the nucleation and crystallization process can be observed after 5 min of deposition. However, only a small amount of HA crystallites is formed and an uncoated Ti6Al4V alloy is still visible. One can notice larger grains after 15 min of deposition, as can be seen in Fig. 6.8(c).

The HA grains continue to grow in size and after 30 min of deposition cover the entire substrate surface. This tendency continues, and larger agglomerates can be seen after 60 min into the deposition process (Fig. 6.8(d)) A representative thickness of the HA coating after 90–120 min deposition is 1.5–2 µm, which gives a few tens of nanometers per minute as an estimate for the deposition rates. This is less than the rates achievable by the plasma spray technique but higher than typical rates of the RF sputtering deposition [396].

Thus, the SEM observations suggest that longer deposition times are favorable to synthesize higher quality, lower porosity HA films. Interestingly, crystalline grain size also increases, as has been derived from the XRD spectra in

Fig. 6.7. However, the sizes of the grains seen in Fig. 6.8 are typically several times larger than those obtained by using Sherrer's relation (6.1).

This can be best understood by noting that the coatings of our interest here are polycrystalline and contain many small (~40–60 nm in size) nanocrystals with different orientations of crystallographic planes, as suggested by the XRD spectra. Presumably, a large number of nanocrystals populate larger (typically, ~300–400 nm in size) grains seen in the SEM of Fig. 6.8. In addition to a large number of nanocrystals, the latter grains also contain the amorphous phase.

We emphasize that an increase of the nanocrystals and "larger grains" in size, smaller intergrain pores, densification of the material, and more clear brittle fractures in Fig. 6.8(d) (after 60 min growth) as compared with the films grown for 30 min, are clear indicators [396] of an increasing content of the polycrystalline phase in the coating. Interestingly, the as-sputtered CaP coatings previously synthesized by other authors were confirmed amorphous by the X-ray diffraction analysis [396–400]. It is worth emphasizing that the amorphous structure of CaP films is one of major disadvantages of the existing plasma-assisted sputtering deposition techniques [396].

We now briefly discuss a possible hydroxyapatite crystalline growth scenario [358]. It is well known that relatively loose and bulk electron clouds can easily be separated from the lattice by en external action. Therefore, chemical bonds of the surface atoms "dangle" into the near-substrate plasma sheath area. This implies that the surface atoms have a higher energy than the atoms in the bulk.

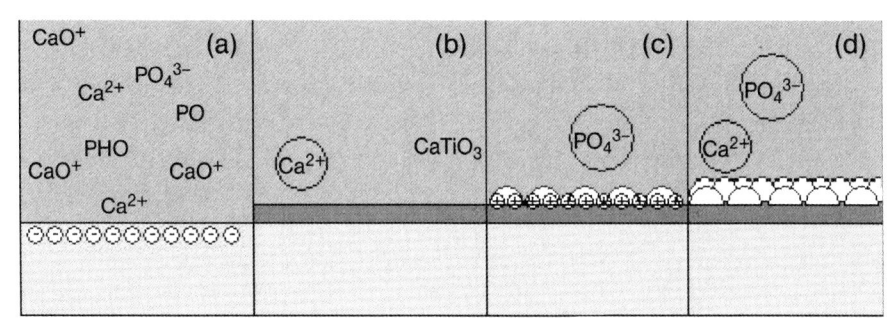

Fig. 6.9 Kinetics of the HA film formation: (a) chemical bonds of the substrate "dangle" into the space quickly attracted the sputtered positive radicals; (b) with large amount of Ca species deposited, the surface becomes increasingly positive (c); (d) negatively charged PO_4^{3-} are attracted to surface, combined to form initially a layer of amorphous calcium phosphate; with time, the crystalline HA film is formed [358].

To reduce the free energy, the surface atoms thus attempt to rearrange and/or bond to favorable reactive species impinging onto the growth surface from the gas phase. In this case, positive CaO^+, Ca^{2+}, PHO^+, PO^+, P^+, TiO^+, Ti^+, Ti^{2+}, (and several other) species sputtered out from the target are quickly attracted by the negatively charged (due to a much higher mobility of the plasma electrons) metal surface, as shown in Fig. 6.9(a).

As a result of interactions of the impinging positively charged and neutral species with activated surface atoms of the Ti6Al4V alloy, amorphous calcium titanate ($CaTiO_3$) and hydroxyapatite prenucleation layer develop (Fig. 6.9(b)). Indeed, the amorphous phase that usually develops at early film growth stages has previously been confirmed as $CaTiO_3$ [401].

We emphasize that in our case this intermediate layer serves as a bridging system between the Ti-based alloy substrate and the ceramic HA film. As there is a large amount of positively charged Ca-bearing species in the plasma, they accumulate on and cover the entire surface. The outer layer of the film becomes insulating and a continuous build-up of positive charge on the surface make the overall surface charge increasingly positive (Fig. 6.9(c)).

This positively charged surface attracts the negatively charged phosphate ions to form amorphous calcium phosphate (Fig. 6.9(d)) When the substrate surface temperature is high enough, surface migration and diffusion processes transform the amorphous calcium phosphate film into crystalline HA.

6.4
Mechanical Testing of HA Films

As we have mentioned in Section 6.1, adequate mechanical strength of the bioceramic–implant alloy interface is one of the crucial requirements in the development and fabrication of HA-coated Ti6Al4V bioimplants. In this section, the results of testing of adhesive strength, an important characteristic of mechanical performance of the HA films on the Ti6Al4V alloy, are critically examined [358]. We will also briefly discuss the results of the film failure analysis.

6.4.0.1 Adhesive strength assessment

A microscratch test method has been adopted for the assessment of adhesive strength of bioceramic–implant alloy interface. The tests have been performed by using a CSEM MST MICROSCRATCH tester. The applied load is varied from 0.3 to 15 N at a rate of 3 N/min over the length of 10 mm. For better clarity, it is worth mentioning that for the description of the integrity and adhesion of the relatively brittle coatings during the scratch test, two characteristic loads are usually used. Application of the first characteristic load produces the first crack on the coating. It is called the cohesive load and is a measure of the cohesive (intrinsic) strength of the coating [402].

Fig. 6.10 Delamination of the HA coating from a Ti6Al4V implant alloy [358].

The second characteristic load is the minimum load required to delaminate the coating from the substrate, which is a measure of the adhesive strength of the coating. Here, our focus is on the second characteristic load, which can be determined by a combination of optical microscopy and frictional force measurements. The adhesive strength assessment is carried out by analyzing the critical load of the samples synthesized under various experimental parameters, such as working gas pressure, bias voltage, and substrate temperature.

During the test, the frictional force changes smoothly until reaching a critical point when it changes abruptly. This indicates that the ceramic film has been detached from the metal surface. A typical optical image of the delaminated HA coating from the Ti6Al4V substrate is shown in Fig. 6.10.

Figure 6.11 shows the dependence of the critical load on working pressure, DC bias, and substrate temperature. It is remarkable that under lower pressures, DC biases, and higher substrate temperatures, higher critical loads are required to scrape the HA coating off the alloy surface. Indeed, a no-DC bias condition (when the near-stoichiometric [Ca]/[P] ratios are achieved) is ideal to maintain the critical load of the order of 10 N (Fig. 6.11(b)).

On the other hand, films synthesized under the optimized (from the crystallization viewpoint) substrate temperature of 550 °C show a better interfacial stability than those deposited under lower substrate temperatures (Fig. 6.11(c)). Furthermore, as can be seen from Fig. 6.11(a), the optimized (from the best [Ca]/[P] values point of view) working gas pressure 21 mTorr, is also within the range of reasonably high critical loads. However, since the

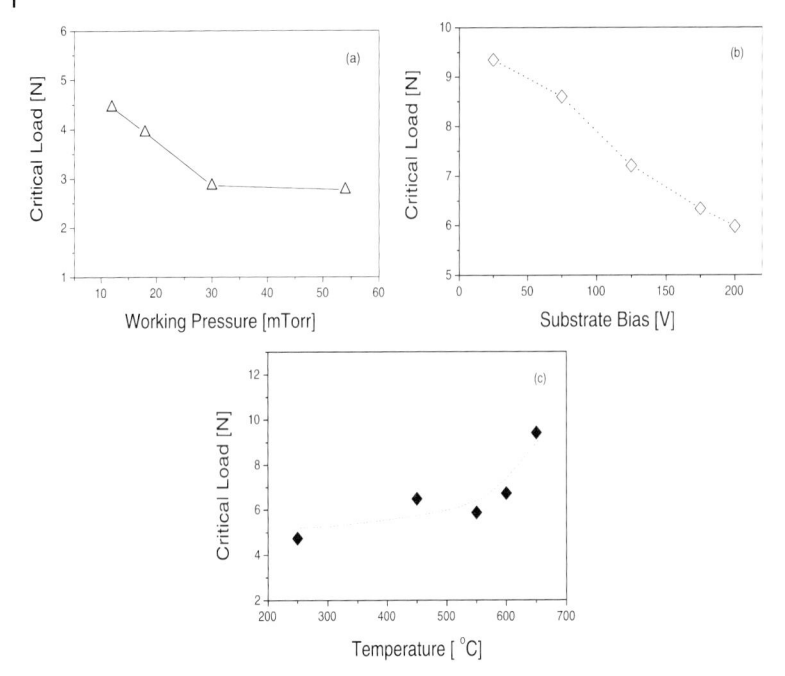

Fig. 6.11 Critical load as a function of (a) working gas pressure, (b) substrate bias, and (c) substrate temperature. Deposition parameters: input power 700 W; DC bias −70 V ((a) and (b)); working gas pressure 18 mTorr ((b) and (c)) [358].

assessments have been done for the films synthesized under slightly different, than those of Section 6.3, process conditions, one should consider these results as indicative. Nonetheless, the qualitative trends inferred from Fig. 6.11, further support the choice of the process conditions in Section 6.3.

It is important to note that the films synthesized at lower DC bias, working gas pressure, and higher substrate temperature, feature a high content of the crystalline phase (see Section 6.3). Therefore, excellent crystallization is also beneficial to achieve higher interfacial bonding strength. Physically, atoms in the HA and/or calcium titanate ($CaTiO_3$) compounds follow long-range ordering patterns in a crystalline film. When crystalline grains of the coating are located at the interface between the thin film and the titanium alloy, the probability of cross-interfacial atom migration is high.

Thus, Ca and O atoms from the bioceramic film can combine with titanium atoms in the substrate to form an interfacial compound $CaTiO_3$. More importantly, atoms in boundary compounds are shared by the bioceramic and metal alloy crystalline grains. The probability of the cross-interface migration is higher when the crystalline phase content in the film is higher. As a result, a perfect contact between the thin film and the substrate is formed, which is reflected by the improved adhesive properties of the coating.

On the other hand, if the films are predominantly amorphous, atoms mainly follow short-range ordering patterns, which results in lower cross-interfacial migration rates. This is one of the reasons why adhesion of amorphous HA films to titanium-based alloys is usually quite poor.

6.4.0.2 Film failure analysis

The failure analysis of the CaPTi films is performed by observation of the scratch tracks (similar to Fig. 6.10) and failure mode. This analysis shows that before the thin film is completely scraped off, there appears through-thickness cracks that result from the tensile cracking in the track, which implies the ductile failure. By means of scanning electron microscopy the details of the through-thickness cracks can be observed.

In particular, a slight deformation appears in the area near the cracks, where the bioceramic film is polycrystalline. In this case, the deformation probably involves relative shear between the crystal grain boundaries. When the deformation builds up to a critical extent, the microcracks are initiated and spread to form the through-thickness cracks.

In the case of ductile failure, the stress and strain generated in the coating by the indenter can be released via the deformation and through-thickness cracking so that the thin film can be prevented from detaching from the substrate. Thus, the film failure analysis also evidences an excellent adhesive strength of the HA coating. Further details can be found elsewhere [403].

6.5
In vitro Assessment of Performance of Biocompatible HA Coatings

In this section, we examine the results of *in vitro* biocompatibility assessments of the HA films synthesized under the optimum conditions as described in previous sections [358]. This study is particularly important to predict the actual biological and implant material responses to the physiological environment during the application. It is expected that the devised HA-coated Ti6Al4V alloy will exhibit the desired biocompatibility properties outlined in Section 6.1. More specifically, the HA film should ensure a higher possibility of enhanced bone reconstruction at the interface with the implant to result in a faster and better osteointegration [379].

In vitro SBF and bone cell culture assessments have been administered to investigate the associated biocompatibility responses from the synthesized HA films [358]. For the SBF assessment, the HA-coated implant alloy substrates are immersed in a solution with ion concentrations similar to those of human blood plasma. After immersion for different periods of time, the HA-induced responses of the coating to the SBF environment are studied by using the EDX, Raman, XRD, and SEM.

In addition to the SBF evaluation, a bone cell culture assessment has been carried out [358]. Basically, the mouse osteoblast cells MC3T3-E1 are seeded and allowed to adhere to and growth on the HA film surface. Scanning electron and optical microscopy are used to observe the cell growth and proliferation.

6.5.1
Simulated Body Fluid assessment

An SBF solution buffered at pH 7.4 with 50 mM trihydroxymethylaminomethane and 45 mM hydrochloric acid at 36.5 °C is used. The details of preparation of the SBF solution are presented elsewhere [404]. Before using the solution, the composition of the SBF has been tested to ensure that the required ion concentrations were achieved. The results of comparison of the ionic concentrations in the SBF of our experiments and in the real blood plasma are shown in Table 6.3.

Tab. 6.3 Ion concentrations in the SBF and human blood plasma [358].

Ion	SBF (mM)	Blood plasma (mM)
Na^+	142.0	142.0
K^+	5.0	5.0
Mg^{2+}	1.5	1.5
Ca^{2+}	2.5	2.5
Cl^-	147.8	103
HCO_3^-	4.2	27
HPO_4^{2-}	1.0	1.0
SO_4^{2-}	0.5	0.5

We recall that the HA-coated samples are prepared under the optimized process conditions, enabling one to achieve the required film composition and crystallinity (for details, see Section 6.3). The samples are immersed in the SBF solutions for 6, 12, 24, 48, and 168 h in a water bath maintained at 37 °C. After the immersion, the SEM and XRD measurements have been carried out to investigate the apatite growth.

Figure 6.12 depicts the evolution of the apatite growth from the SEM analysis. From Fig. 6.12(b), one can note that the film has been partially dissolved in the SBF and features a less compact structure as compared to that of the as-deposited film (Fig. 6.12(a)). However, after 12 h into the immersion, one can observe small nucleates on the surface (Fig. 6.12(c)). After 24 h of immersion, the nucleates form larger agglomerates as can be seen from (Fig. 6.12(d)) Thereafter, elongated nanosized apatite granules are formed after 48 h of immersion in the SBF, as (Fig. 6.12(e)) suggests. This is probably the final stage of

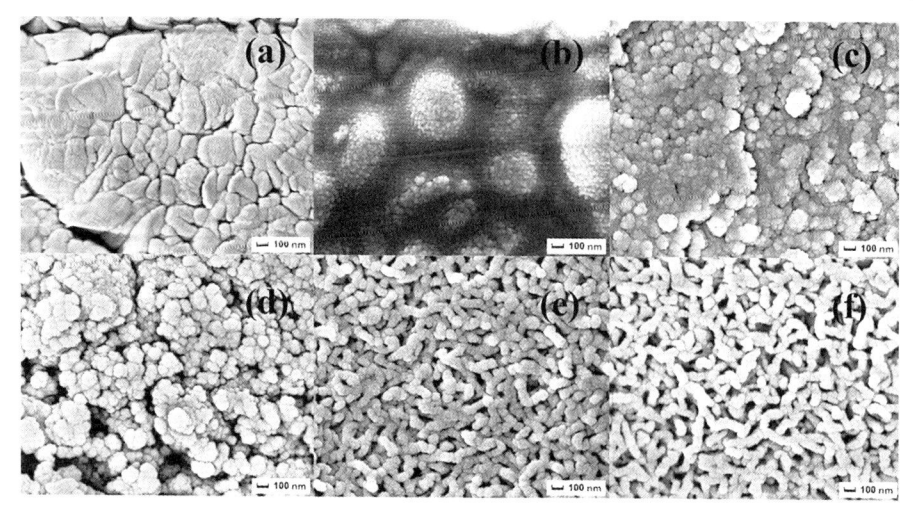

Fig. 6.12 SEM images of (a) As-deposited film; (b) 6 h—dissolution observed; (c) nucleation process is initiated after 12 h soaking; (d) 24 h; (e) and (f) apatite growth is observed after 48 h to 168 h of immersion [358].

the surface morphology evolution, which does not change significantly even after 168 h of immersion (Fig. 6.12(f)) when the apatite layer covers the entire surface. The XRD measurements have been performed to investigate the development of the hydroxyapatite crystalline phase on the HA-coated samples immersed in the SBF.

From the XRD spectra shown in Fig. 6.13, one can notice that during the first 6 h of the immersion, the hydroxyapatite phase is resorbed, which is attested by diminishing (002) and (112) peaks as compared to as-deposited films. At this stage, we have recorded several diffraction peaks corresponding to dicalcium phosphate (DCP, $Ca_2P_2O_7$). As we have seen from the results of the SEM analysis, precipitation of the bone material becomes pronounced after 12 h of immersion.

The hydroxyapatite phase begins to restructure and build up as evidenced by the stronger (002) peak. In the meantime, the amount of the DCP increases, which is confirmed by the (213) diffraction line, which also remains strong for the samples immersed for 24 h. It is interesting to note that the DCP is one of the intermediate products of biological precipitation processes, which involve dynamically counteracting resorption and re-deposition of different calcium phosphate phases [362]. Subject to longer immersion times, the growth of crystalline HA content proceeds along the (002), (112), (202), (301), and (213) crystallographic directions, as can be seen in Fig. 6.13.

It is also remarkable that the restructured calcium phosphate coatings shown in Figs. 6.12(e)–(f) also feature notable amounts of tricalcium phos-

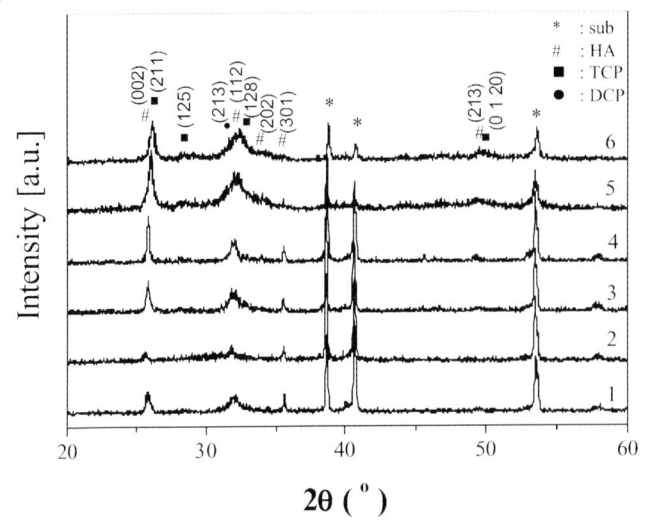

Fig. 6.13 Variation of XRD spectra with different SBF immersion times. Curves 1–6 correspond to as deposited films, and films assessed after 6, 12, 24, 48, and 168 h of immersion in the SBF [358].

phate [TCP, $Ca_3(PO_4)_2$]. Similar to the HA, the TCP is another constituent of natural bones. It is highly reactive and resorbable, and is considered as a natural enhancer of the hydroxyapatite biocompatibility function [405]. However, in assigning specific diffraction reflections in Fig. 6.13 to DCP or TCP carries a certain degree of ambiguity due to relatively poor resolution of the apatitic diffraction pattern.

The changes in the HA film composition with the SBF immersion time have been studied by the EDX technique. It has been found that the calcium content, and hence, the [Ca]/[P] ratio are lower in the films soaked for shorter periods of time, as compared to the as-grown samples. At this stage, the measured [Ca]/[P] ratio is closer to the stoichiometric [Ca]/[P] ratio of the dicalcium phosphate rather than to that of the HA phase. Nonetheless, the [Ca]/[P] ratio increases afterward with the immersion time and approaches to the HA stoichiometric value in the films soaked for 48 and 168 h. The EDX results are thus consistent with the results of the SEM and XRD analyses.

An increasing, with the immersion time, content of the HA phase in the films is further confirmed by the results of Raman spectroscopy. The intensities of Raman signals corresponding to the HA "fingerprint" Ca–O–P and P–O bonds at 430, 585, and 960 cm^{-1} are much higher in the films soaked for 48 and 168 h. Thus, the apatite content is indeed higher in the films shown in Figs. 6.12(e)–(f).

It is remarkable that the film contains not only the HA, but also the TCP phase. Therefore, the results of the SBF immersion analysis confirm that the devised HA film can indeed induce the bone apatite growth. As such, this is an indication that the films are remarkably biocompatible and can offer a great deal of bone remodeling in orthopedic and dental applications.

6.5.2
Cell Culture Assessment

In this subsection, we discuss the results of the cytocompatibility assessment of the HA-coated Ti6Al4V implant alloy [358]. In clinical applications it is essential that an HA-coated implant is able to provide an adequate scaffold to support, stimulate, and direct the growth of osteoblast cells. The cytocompatibility test is based on the bone cell culture method and aims at studying the bioscaffolding properties of the devised HA films. The culturing of cells has been carried out according to standard protocols [406].

Following the original work [358], we now recall the most essential details of the hydroxyapatite cytocompatibility assessment. The initially frozen in liquid nitrogen storage cells are thawed, put into a 150 mm Petri dish and examined by an optical microscope. The cells are then incubated at 37 °C (mammalian cell lines growth temperature) in a CO_2 environment to allow their growth. In the meantime, the HA-coated substrates are prepared and sterilized. This is done by soaking the samples in a dish with 70 % of ethanol for 10 min, drained off and repeating the same procedure again, this time for one hour. The samples are then UV sterilized for 2 h and neutralized in a Petri dish for 3 days.

After three days of incubation, the cell cultures have been examined by an optical microscope for their level of confluence, morphological appearance, and any signs of microbial contamination. The cells then have been moved into another dish containing the HA-coated substrates and put back into the incubator where they attach themselves to the substrates. This subculture procedure should be repeated every three days. We recall that the substrates have been prepared under the optimized deposition conditions detailed in Section 6.3.

Optical and scanning electron microscopy has been used for the observation of the cell growth patterns [358]. The results of the optical microscopy in Fig. 6.14 show the osteoblast cell growth pattern in the cell culture experiment. After one day of culture, cellular proliferation around the substrate can be observed. It is noteworthy that a high-resolution SEM imaging has confirmed efficient cell attachment to the HA-coated samples only after 24 h into the cell culture (Fig. 6.15(a)).

More encouragingly, the cells show signs of attachment to the HA crystals and cellular differentiation after 2 days of incubation. Cell growth and replication can also be seen on SEM micrographs shown in Figs. 6.15(b) and (c).

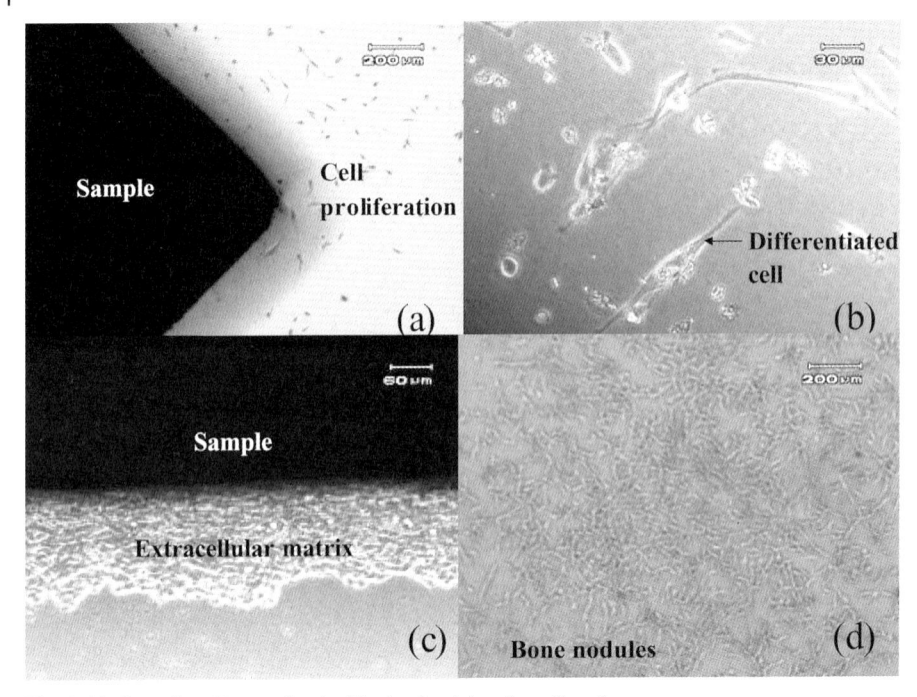

Fig. 6.14 Growth patterns of osteoblast cells; (a) cell proliferation; (b) cellular attachment and differentiation after two days. After 8-day culture, extracellular matrix (c) and bone nodules (d) are formed [358].

It is remarkable that the cells consistently grow in size throughout the entire duration of the cell culture experiment. After eight days, a thick layer of extracellular matrix and bone nodules are formed, as depicted in Fig. 6.14.

It is also notable that the hydroxyapatite coatings can support growth and proliferation of different cell types [381, 388]. Typical cytoskeletal structure of the COS7 monkey kidney cell line after 48 h incubation, visualized by using a Leica DMRXA fluorescent microscope is shown in Fig. 6.16.

Therefore, proliferation, attachment, growth, and differentiation of the osteoblast cells and also the formation of bone nodules and mineralized extracellular matrix, have been confirmed by the cell culture experiments, which suggest that the hydroxyapatite coatings are indeed cytocompatible [358]. As such, the HA-coated Ti6Al4V implant alloys have an outstanding potential to support the bone material ingrowth and remodeling in orthopedic and dental applications.

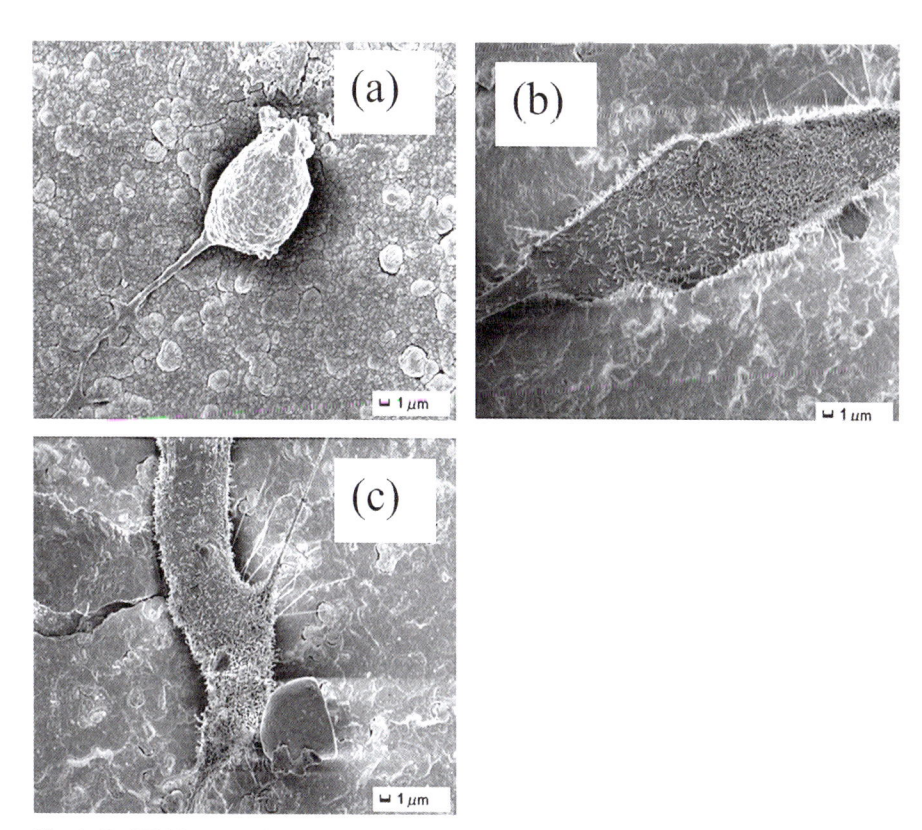

Fig. 6.15 SEM images of osteoblast cells: (a) after 1-day incubation; (b) and (c) differentiated cells after 2-day incubation.

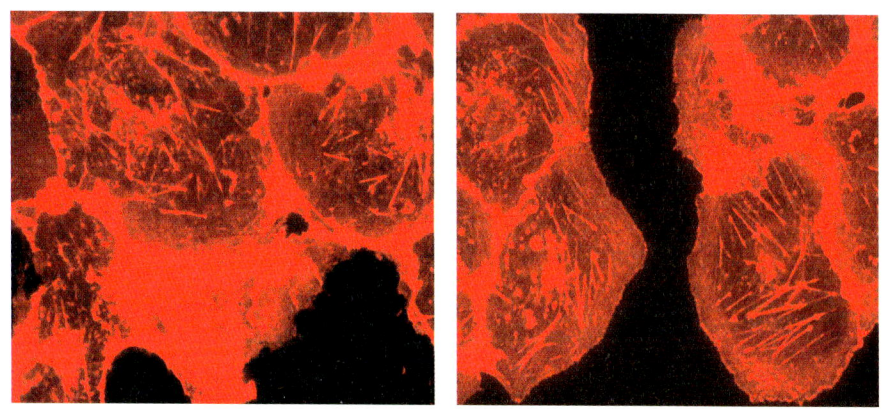

Fig. 6.16 Typical cytoskeletal structures of the COS7 monkey kidney cell line after 48 h of cell culture [358].

6.6
Concluding Remarks

To summarize, we have discussed an innovative approach for the fabrication of biocompatible hydroxyapatite bioceramics [358]. This approach capitalizes on intentional disintegration of compressed commercial HA powder into smaller sub-nano (atomic, molecular) and nano- (macromolecular, nanocluster, and nanoparticle) building units in low-pressure reactive plasma environments [358, 381, 388].

The plasma-assisted RF concurrent magnetron sputtering of HA and Ti targets can be used as an ideal practical framework of the above bioceramic synthesis concept. More importantly, by using smaller building units and lower process temperatures, it appears possible to grow highly crystalline films and substantially improve the adhesive strength of the bioceramic–implant alloy interface. Furthermore, transport of the main calcium- and phosphorus-bearing species (CaO^+ and PO_4^{3-}, respectively) to Ti-6Al-4V deposition substrates can be efficiently controlled by varying the DC bias. The different responses of oppositely charged CaO^+ and PO_4^{3-} species to the DC substrate bias makes it possible to control absolute and relative elemental concentrations of calcium and phosphorus in the films.

Thus, CaO^+ cations generated in $Ar+H_2O$ plasma discharges can be regarded as a "cause" of the elevated presence of calcium in the bioceramic, with a similar conclusion for the phosphorus-bearing anion PO_4^{3-}. In the set of experiments reviewed in this chapter, there are no specific means of surface preparation for the deposition of the building units other than conventional pretreatment in argon and maintaining substrate temperatures high enough (usually in excess of 550 °C) to promote the required degree of crystallization.

We emphasize that by using atomic, radical, and, presumably, nanocluster building units, one can obtain compact and dense films with nanoscale surface morphology (Fig. 6.6), which is believed to be ideal to support biomolecule–bioceramic interactions resulting in enhanced bone ingrowth and remodeling. The hydroxyapatite films prepared by the RF magnetron concurrent sputtering technique feature excellent bioactivity and cytocompatibility, as suggested by *in vitro* simulated body fluid and osteoblast cell culture assessments. However, specific application of the four-stage "cause and effect" approach [36] for the plasma-assisted nanofabrication of hydroxyapatite bioceramics warrants substantial theoretical and experimental efforts in the future.

7
Other Examples of Plasma-Aided Nanofabrication

In this chapter, we give a few other representative examples of successful uses of plasma-based nanotools for nanoscale assemblies that do not fall within the categories covered in Chapters 4–6. In Section 7.1, we consider an interesting example of highly controlled plasma-assisted incorporation of Er atoms into silicon carbide nanoparticle films, also synthesized by using hybrid plasma-based methods. As shown in Section 7.2, a combination of reactive plasma-assisted sputtering of solid titanium targets and plasma-enhanced chemical vapor deposition makes it possible to synthesize unique, from many points of view, nanocrystalline films made of atoms of four elements, namely Ti, Si, N, and O. These nanostructured films are extremely promising for applications as barrier interlayers in micro- and nanoelectronics. In another example, considered in Section 7.3, we discuss the possibility of tailoring the composition and other properties of nanostructured AlCN films by using a range of plasma-generated and plasma-synthesized species and the "plasma-building unit" approach. There is a direct correlation between the abundance of carbon and nitrogen-containing species in the ionized gas phase and the actual content of these elements in the AlCN nanofilms. Section 7.4 is devoted to the reactive plasma-assisted RF magnetron sputtering of highly oriented nanocrystalline AlN films at low substrate temperatures. Interestingly, by adjusting the plasma process parameters, it turns out possible to significantly improve the crystallinity and tailor many other (e.g., optical) properties of the films concerned. In Section 7.5, we give yet another example of outstanding performance of the Integrated Plasma Aided Nanofabrication Facility in the growth of nanocrystalline vanadium pentoxide films with many interesting properties that make such films indispensable for the development of a new generation of smart multipurpose coatings. In the last section of this chapter (Section 7.6), several impressive examples of the uses of low-temperature plasma-based processes for processing of materials that contain ultrasmall holes with the sizes ranging from submicrons down to nanometers are presented. These examples are certainly not exhaustive and do not pretend for a wide coverage of all amazing possibilities plasma nanotools offer for materials synthesis and processing at nano- and even subnanometer scales.

Plasma-Aided Nanofabrication. Kostya (Ken) Ostrikov and Shuyan Xu
Copyright © 2007 WILEY-VCH Verlag GmbH & Co. KGaA, Weinheim
ISBN: 978-3-527-40633-3

7.1
Plasma-Assisted Er Doping of SiC Nanoparticle Films: An Efficient Way to Control Photoluminescence Properties

In this section we consider an example of a successful application of plasma-based techniques to perform an intricate process of doping of SiC nanostructured material with atoms of a rare-earth metal, Erbium. Again, as in numerous examples of synthesis of semiconductor quantum confinement structures of various dimensionality of Chapter 5, a plasma-assisted reactive magnetron sputtering deposition proved an indispensable nanofabrication technique.

Specifically, polynanocrystalline silicon carbide (pn-SiC) films containing SiC nanoparticles and Er have been deposited on Si(111) substrates by the above plasma-assisted technique [407]. In this case, high-purity SiC and Er targets are concurrently sputtered in the Integrated Plasma-Aided Nanofabrication Facility described in detail in Chapter 3. Under certain conditions, plasma-based doping of SiC with Er effects in room-temperature photoluminescence is inefficient or impossible otherwise. Moreover, the PL peak position, intensity, and the full width at half maximum strongly correlate with the Er doping levels and plasma-based process conditions.

As usual, before we proceed to the analysis of the details and results of the experiments of Huang et al. [407], we briefly discuss the importance of doping of nanostructured semiconductors by rare-earth elements. This kind of doping is presently well known to lead to the formation of luminescent centers in semiconductor materials being doped, which make them suitable for various applications in optoelectronic devices [408–410].

Among the various rare-earth elements, erbium has attracted a particular attention because of the $4I_{13/2} \rightarrow 4I_{15/2}$ transition, which involves nonbonded 4f shell electrons of the Er^{3+} ion ($4f^{11}$) and occurs at a technologically important wavelength of 1.54 µm. This wavelength perfectly matches the absorption minimum of indirect bandgap semiconductors such as silicon and silicon-carbide-based optical fibers.

Er-doped silicon has recently attracted much interest because of its potential application in Si-based optoelectronic devices [411–413]. As we have already mentioned elsewhere in this book, silicon carbide (SiC), a semiconductor with a relatively wide energy bandgap (~ 3 eV), is an outstanding candidate material for electronic and optical devices and has advantages over other semiconductors especially for high-frequency, high-temperature and high-power applications [414–416]. Therefore, doping SiC with atoms of rare-earth metals could bring in many exciting opportunities for the development of advanced optoelectronic devices based on nanostructured SiC with a tunable bandgap [407,417].

Huang et al. [407] synthesized Er-doped SiC films on Si(111) substrates by the RF reactive magnetron sputtering from metal Er and ceramic SiC targets in a gas mixture of argon and hydrogen at low substrate temperatures (~350 °C). Erbium content (at.%) in SiC film can be controlled by varying the RF power applied to each sputtering electrode in the IPANF and also other deposition conditions. Similar to most of the cases considered in this book, the process parameters strongly affect nanoparticle sizes and other properties of the nanostructured SiC films. The details of nanofilm characterization and photoluminescence measurements are quite standard and can be found in the original work [407].

It is notable that the surface morphology of Er-doped SiC films is most strongly affected by the RF power applied to the magnetron electrode with the SiC target, substrate temperatures, and Er content in the films. The latter can be controlled by changing the RF power supply to the erbium-releasing sputtering electrode. In the experiments concerned, four different samples 1–4 have been analyzed. The deposition conditions of samples 1–4 are the same except for the RF power applied to the Er target; namely, they are 40, 20, 10, and 0 W, respectively.

The surface of Sample 1 features dispersed compact and quite separated nanoparticles, which are very much alike and in some cases are made of smaller nanoparticles. The mean nanoparticle diameter of Sample 1 is about 13.2 nm, with the particle density $\sim 7.2 \times 10^{12}/cm^2$. The XPS spectra of the four samples concerned (not shown here) suggest that the content of erbium in the films is apparently too small and it cannot be detected by the X-ray photoelectron spectroscopy. However, the elemental ratios of Si and C turn out to be approximately 46.75 at.%:48.30 at.%, which is indeed very close to the stoichiometric ratio of the SiC.

It is interesting to note that the elemental presence of silicon and carbon in this series of experiments does not change significantly with the variation of the total pressure and substrate temperature. The surface morphologies of the Samples 2 and 3 are very much alike and similar to that of Sample 1. However, the surface morphology of Sample 4 turns out to be quite different; in this case the mean nanoparticle size is ~30 nm. Thus, erbium atoms may act as initial seed nuclei of SiC nonoislands.

Figure 7.1 shows the XRD spectra of Samples 1–4. From these spectra one can make an important conclusion. Namely, with the increase of Er content, the structure of the SiC films changes from amorphous to polycrystalline. Moreover, the main peak position also shifts toward larger diffraction angles, which means that as more erbium atoms incorporate in the film, the lattice constants of Er-doped SiC become smaller.

Figure 7.2 shows the relationship between the intensity of room-temperature photoluminescence and the wavelength for the four samples of our interest here. A striking observation from Fig. 7.2 is that Sample 4 prepared without any erbium incorporation, does not luminesce!

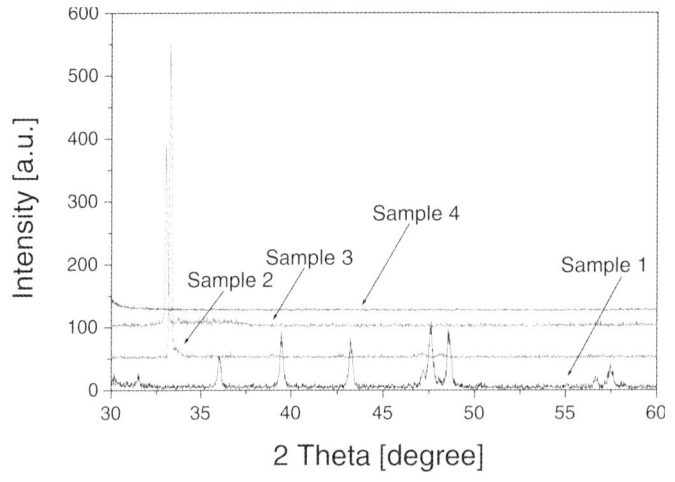

Fig. 7.1 XRD spectra of Er-doped SiC nanoparticle films [407].

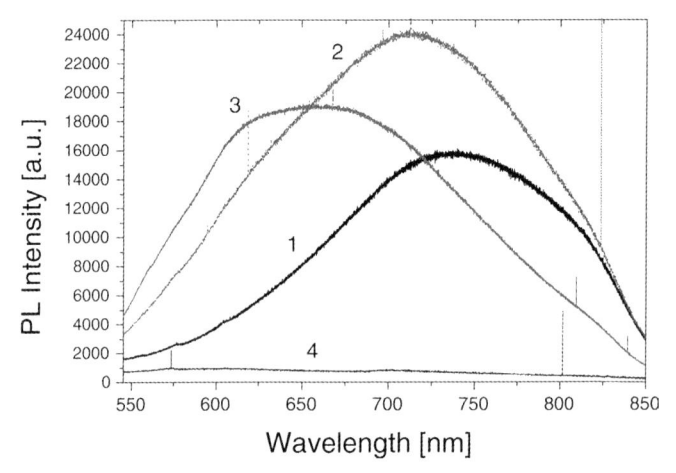

Fig. 7.2 Room-temperature photoluminescence spectra of Er-doped SiC nanoparticle films [407]. Labels 1–4 denote the sample numbers.

Table 7.1 lists the peak positions and the full widths at half maxima of erbium-doped SiC specimens. It is interesting to note that the PL peak position shifts toward shorter wavelengths and the FWHM decrease as the amount of the dopant is reduced. However, there is no clear dependence of the PL intensity and the Er content in the SiC nanofilms. This conclusion applies to Samples 1–3 with an average nanoparticle size of 13.2 nm.

The results of investigation of the effect of the nanoparticle size evidence that the photoluminescence intensity decreases and its peak position shifts to longer wavelengths when the sizes of the SiC nanoparticles become larger. Interestingly, quite similar relations between the PL intensity and Er content hold for a range of SiC nanoparticle sizes.

Tab. 7.1 Photoluminescence peak position and FWHM of Er-doped SiC films [407].

Sample number	PL peak position (nm)	FWHM (nm)
1	732	125
2	713	150
3	658	165

Huang et al. [407] also state that the photoluminescence from Er-doped SiC nanoparticle films might be attributed to size-dependent quantum confinement effects rather than the relaxation of chemical bonds. Moreover, intentionally introduced Er-based defects might be required for the thermodynamic stability of the SiC nanoparticles [418].

However, in our view, this conclusion still requires further experimental/computational evidence to be commonly accepted. Furthermore, there is still plenty of room for atomistic *ab initio* simulations and also modeling of growth kinetics of SiC nanofilms doped with atoms of rare-earth metals.

Nonetheless, pilot experiments of Huang et al. [407] introduce a new low-temperature plasma-based method of doping nanostructured Si-based films with erbium. It is imperative that the Er content in SiC nanoparticles can be precisely controlled by varying the RF power supplied to the magnetron sputtering electrode of the IPANF reactor.

In these experiments, the nanoparticle sizes sensitively depend on the plasma-based parameters, especially the RF power supply to both sputtering electrodes and the substrate temperature. The peak position, intensity, and FWHM of visible room-temperature photoluminescence are directly related to the content of the erbium dopant in the SiC nanoparticle films. Finally, this method has an outstanding potential to become widely adopted for nanofabrication of a variety of nanostructured semiconductor films with controlled levels of externally introduced dopant elements, on temperature-sensitive substrates, such as polymers and plastics.

7.2
Polymorphous (poly)Nanocrystalline Ti–O–Si–N Films Synthesized by Reactive Plasma-Assisted Sputtering

In this section, we consider another very interesting example of nanostructured materials successfully fabricated by using reactive plasma-based tools. These films are made of nanocrystals of different binary compounds embedded in an amorphous matrix. More specifically, let us consider T–N, Si–N, and Ti–Si nanocrystals in a composite Ti–Si–N–O amorphous matrix. These kinds of films are commonly termed polymorphous (*pm*) to emphasize the presence of two phases in the same material.

An example of such a material is polymorphous hydrogenated silicon (*pm*-Si:H), which is made of ultrasmall Si:H nanoclusters embedded in an amorphous hydrogenated silicon matrix [36, 38] and has already been discussed in Chapter 1. It is worthwhile to mention that *pm*-Si:H films have been successfully synthesized using reactive low-temperature plasmas of mixtures of silane and hydrogen, with optional dilution in an inert gas such as argon. These films feature excellent size uniformity and sometimes even distribution of nanocrystals within the amorphous layers [38].

Another example of polymorphous but mixed-composition films, wherein nanocrystallites of TiN are embedded in an amorphous Si_xN_y matrix, has been reported by Marcadal et al. [420]. As has been mentioned above, the films of our interest here are also based on an amorphous matrix but they contain more than one nanocrystalline "species." This is why we term the Ti–Si–O–N films *polymorphous* and at the same time, *poly-nanocrystalline* to reflect the presence of tiny nanocrystals made of binary Ti–N, Si–N, and Ti–Si compounds, even though such a terminology is not used in the original work [419].

Let us now go through the main details of the reactive plasma-aided synthesis of Ti–Si–N–O films and highlight their practical importance. The present-day microelectronic circuitry heavily relies on Ta or TaN-based barrier alloys that serve as efficient barriers for copper diffusion in dielectric–metal interconnects. However, as the feature sizes in microelectronic devices relentlessly shrink into 100 nm and sub-100 nm domain, there emerges the need to replace Ta/TaN copper diffusion barriers in deep sub-micron CMOS technology; the present-day technology of fabrication of such interlayers is mostly based on physical vapor deposition (PVD).

Titanium-based ternary materials, such as $Ti_xSi_yN_z$ alloys have recently emerged as potential candidates with the outstanding promise. Interestingly, $Ti_{34}Si_{23}N_{43}$ films appear to be a cement-like blend of nanophase TiN with a typical grain size of 2 nm intermixed with amorphous Si_3N_4 [419, 421].

Ee et al. [419] used reactive silane plasma-assisted RF magnetron sputtering to synthesize nanodevice-quality quarternary Ti–Si–N–O barrier alloys and studied the effect of plasma-based process parameters on the microstructure, phase composition, resistivity, surface morphology, and surface roughness of the films concerned. The substrates used by Ee et al. were p-type silicon wafers with 630 nm thick silica film predeposited by PECVD. The deposition of Ti–Si–N–O barrier films is conducted in the IPANF plasma reactor by RF magnetron sputtering of a titanium target in the presence of reactive silane precursor.

In the experiments concerned, no external heating or biasing was applied to the substrate surface. The gas feedstock comprised argon, nitrogen, and silane gases with the inflow rates typically varied within the range of 20–40 sccm. The main analytical tools used for the film characterization are similar to the previous sections, with the exception of the Rutherford backscattering spectrometry (RBS, 2.0 MeV ^{4}He$^+$ beam generated in a 4.0 MeV Dynamitron accelerator), which was used to measure the elemental composition of the films. In particular, this technique suggests that the elemental compositions of the three specimens are $Ti_{24}Si_{12}N_{35}O_{29}$, $Ti_{20}Si_{13}N_{35}O_{32}$, and $Ti_{16}Si_{15}N_{32}O_{37}$, hereafter referred to as specimens Si-12, Si-13, and Si-15 with the film thicknesses 215, 262, and 300 nm, respectively [419].

In addition, X-ray photoelectron spectroscopy reveals the presence of Ti–N, Si–N, Ti–Si, Ti–O, Si–O, and Si–N–O bonds, which are found after deconvolution of the Ti_{2p}, Si_{2p}, N_{1s}, and O_{1s} narrow-scan XPS peaks. Moreover, the presence of Ti–N, Si–N, and Ti–Si crystal phases is further confirmed by the XRD. Other phases, such as silicon and nitrogen forming titanium oxide, silicon oxide, and silicon oxynitride are likely to make up the amorphous matrix.

Figures 7.3 and 7.4 show high-resolution TEM graphs of the $Ti_{24}Si_{12}N_{35}O_{29}$ specimens. From Fig. 7.3, one can observe a large number of nanocrystals typically ranging from 2 to 15 nm in size. More importantly, by combining the XPS, XRD results, and the lattice spacing measurements from the HRTEM micrograph in Fig. 7.4, the nanocrystalline features can be identified as Si_3N_4(111) (top left in Fig. 7.4), Ti_5Si_4(105) (top right in Fig. 7.4), Ti_5Si_2(311) (bottom left in Fig. 7.4), and ϵ-TiN(102) (bottom right in Fig. 7.4). The remaining Ti–O, Si–O, and Si–N–O compounds are in the amorphous state [419].

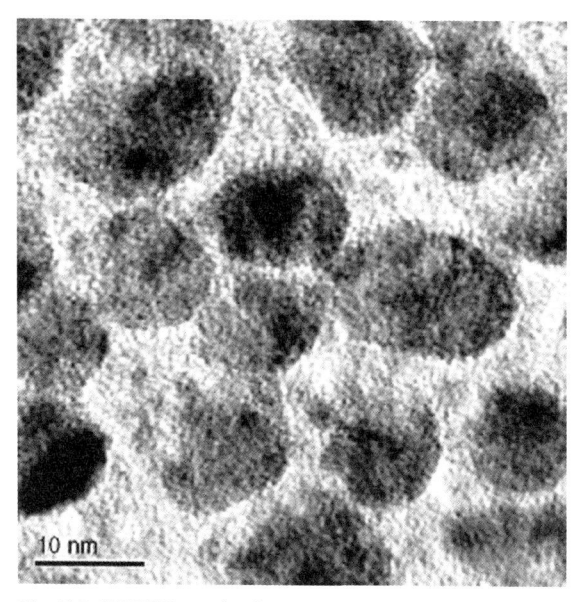

Fig. 7.3 HRTEM graph of nanocrystals embedded in amorphous matrix in the $Ti_{24}Si_{12}N_{35}O_{29}$ film [419].

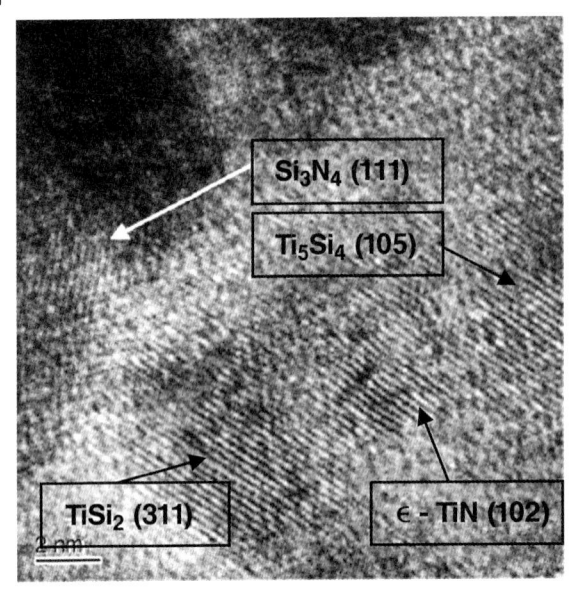

Fig. 7.4 HRTEM graph showing the presence of Ti–N, Si–N, and Ti–Si nanocrystals in the $Ti_{24}Si_{12}N_{35}O_{29}$ film [419].

Interestingly, the barrier Ti–Si–N–O films synthesized by the reactive plasma-assisted RF magnetron sputtering deposition turn out to be very smooth, with the surface roughness not exceeding 3.3 nm. Most importantly, with the very unusual type of mixed nanostructure (which we have termed polymorphous, (poly)nanocrystalline above), the films show excellent electrical stability against copper ion diffusion and as such are outstanding candidates for the future generation ULSI diffusion barrier applications in microelectronics.

7.3
Fabrication of Nanostructured AlCN Films: A Building Unit Approach for Tailoring Film Composition

Let us turn our attention to discuss how the elemental composition of plasma-synthesized nanofilms can be directly controlled by the plasma process parameters. Changes in discharge operation parameters lead to different electron energy distributions, the latter determine the rates of major electron impact elementary reactions in the ionized gas phase; as a result, production of certain building and working units is affected.

In this section, by considering the plasma-assisted DC magnetron sputtering deposition of nanostructured AlCN films, we show the way to modify the elemental composition in the films by changing relative importance of selected gas-phase chemical reactions leading to the generation of the desired building units. By following the original work [422], we introduce the most essential details of the control and diagnostics of low-frequency (~500 kHz) inductively coupled plasmas for the chemical vapor deposition (CVD) of nanostructured carbon nitride-based films.

Interestingly, there is a direct relation between the discharge control parameters, plasma electron energy distribution/probability functions (EEDF/EEPF), and elemental composition in the deposited AlCN thin films. The Langmuir probe technique is employed to monitor the plasma density and potential, effective electron temperature, and EEDFs/EEPFs in $Ar+N_2+CH_4$ discharges.

In practice, by varying RF power and gas composition/pressure one can engineer the EEDFs/EEPFs to enhance the desired plasma-chemical gas-phase reactions thus controlling the film chemical structure. It is interesting that it is possible to implement a fine tailoring of the electron energy distributions and hence the production of the desired species. Such a possibility has already been briefly mentioned in the introductory Chapter 1.

Carbon-based materials considered here have recently been of a substantial interest because of their excellent mechanical, electrical, optical, and optoelectronic properties. Various technologies, such as RF and DC sputtering, electron-beam evaporation, shock compression, and implantation have been used to synthesize carbon-based compounds and composite materials [423–425].

Among them, sources of low-temperature RF plasmas have already shown a great efficiency for the growth of nanofilms of diamond-like carbon, C_3N_4, AlN, BN, AlCN, as well as for numerous surface modification technologies. Such films prepared by the RF plasma-enhanced chemical vapor deposition (CVD) feature excellent mechanical, optical, and electrical properties [426].

However, to improve the efficiency, predictability, stability, and uniformity of the film deposition and surface modification processes, fine engineering/control and diagnostics of the plasma is required. Here, we introduce and comment on the results on control and diagnostics of low frequency (~500 kHz) inductively coupled plasmas for CVD of nanostructured AlCN films [422].

An RF discharge in $Ar+N_2+CH_4$ gas mixture, also successfully applied for synthesizing nanocomposite Al–C–O–N films [422] has been sustained in the CVD ICP reactor combined with a DC magnetron sputtering electrode (see Chapter 3 for detailed description). We recall that this facility has also been successfully used for a range of surface modification and materials synthesis application discussed elsewhere in this monograph.

One of the most attractive features of this device is that it can operate at relatively high RF powers and provide an independent control of sputtering, high rate of dissociation of molecular species, and high deposition rates. Tsakadze et al. demonstrated that varying the discharge parameters, one can properly adjust the energy and probability distribution functions of the plasma electrons and enhance/diminish rates of certain molecular/radical gas-phase reactions, and hence, the deposited film composition [422].

For the convenience of the reader, we recall the main details of the source of low-frequency (~460 kHz) inductively coupled plasmas used in the process of synthesis of ternary aluminium-enriched C–N-based compound [422]. The plasma is generated in a cylindrical, stainless steel vacuum chamber with the inner diameter $2R = 32$ cm and length $L = 20$ cm. The top plate of the chamber was a fused silica disk, 35 cm in diameter and 1.2 cm thick. High purity aluminum target is placed on a DC magnetron sputtering electrode, positioned 10 cm above the chamber bottom endplate.

Negatively biased substrate holder is fixed at approximately 6 cm above the sputtering target. The operating pressure in the Ar, N_2, and CH_4 gas mixture is typically maintained in the range $p_0 = 20$–50 mTorr. An RF-compensated single Langmuir probe is used to measure the electron energy/probability distribution functions (EEDF/EEPF), electron density, effective temperature and plasma potential. The probe can be moved radially and axially through one of the four port windows or holes in the bottom endplate of the chamber.

The optical emission from the ICP discharge has been collected using a light receiver mounted on the side facing view-port and transmitted via the optical fiber to SpectroPro-750 spectrometer (Acton Research Corporation) with a resolution of 0.023 nm. The optical emission spectra (OES) of excited neutral and/or ionized argon atoms have been investigated in the wavelength range 350–850 nm. Other details of the setup can be found elsewhere [93] (see also Chapter 2 of this book).

The easiest way to control the plasma properties in a $Ar+N_2+CH_4$ discharge is to vary RF power input P_{in}. It is instructive to investigate how P_{in} affects the electron number density and temperature, energy probability functions, species in the discharge, and film composition. In the experiment of Tsakadze et al., a Langmuir probe was located at 2 cm above the substrate holder with the samples and 4 cm away from the center of the vacuum chamber [422]. Optical emission was collected by an optical probe mounted in a side porthole with the axial position closest to that of the substrates.

Electron energy distribution functions in the discharge are shown in Fig. 7.5 for two different values of RF power. From Fig. 7.5(a), it becomes clear that the electron number density increases with the RF power. Meanwhile, the OES spectra collected *in situ* from the discharge collected reveal that the emission intensities of the majority of the plasma species also increase. As the intensities of different plasma species increase, their densities also increase leading to a stronger interaction with the samples. Moreover, higher plasma densities are favorable to enhance the sputtering yield from the Al target.

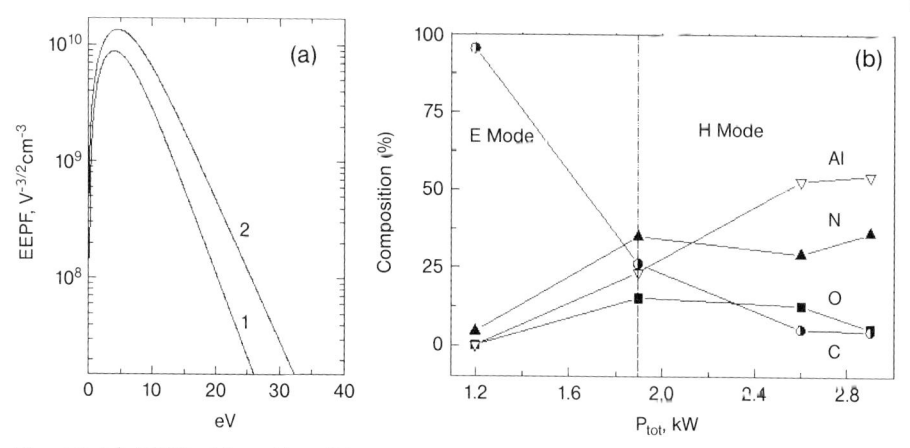

Fig. 7.5 (a): EEPF of Ar + N$_2$ + CH$_4$ 25 mTorr discharge for RF power of 1.9 kW (curves 1) and 2.6 kW (curves 2), respectively; (b): dependence of the elemental composition in the deposited film on RF power [422].

Thus, more aluminum atoms are released from the target, so that the rate of incorporation of Al into a C–N-based matrix can be augmented by properly adjusting the DC voltage on the magnetron electrode, RF power, and substrate bias. The results of XPS analysis of the surface layer shown in Fig. 7.5(b) suggest that the elemental composition of Al in the film increases from 20% to 50% with variation of P_{in} from 1.9 to 2.6 kW. This is consistent with the Langmuir probe measurements revealing higher densities of ionic species [422].

One can also control the process outcomes by varying the operating pressure and composition of the gas mixture. Using the EEDF/EEPF and reaction threshold data, the electron population able to initiate the major dissociation/ionization reactions in the plasma, is computed under different compositions of the gas mixture. In the first experiment, the discharge parameters were studied in a mixture of nitrogen and methane with 8 and 22 sccm mass flow rates, respectively.

In the second experiment, 2 sccm of argon were added to the original gas feedstock. Interestingly, a simple addition of argon gas can increase the plasma density from 5.5×10^{10} cm^{-3} to 1.5×10^{11} cm^{-3} keeping the RF power level fixed at 2.1 kW. Having selected, for simplicity, the following major gas-phase reactions:

$$N_2 + e \rightarrow N_2^+ + 2e \tag{7.1}$$

$$N_2 + e \rightarrow 2N + e \tag{7.2}$$

$$CH_4 + e \rightarrow CH_3 + H + e \tag{7.3}$$

and using the electron energy distribution functions in the discharge depicted in Fig. 7.6(a), one can calculate the plasma density and the number of electrons

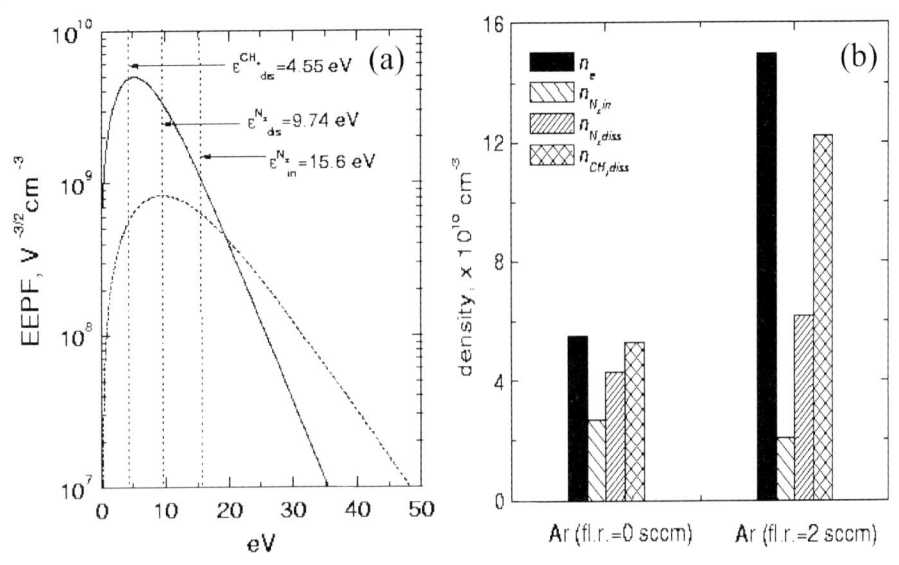

Fig. 7.6 EEPFs of N_2+CH_4 (solid line) and $Ar+N_2+CH_4$ (dashed line) discharges for the flow rates of argon, nitrogen and methane 2, 8, and 22 sccm, respectively (a). Variation of densities of the plasma and hot electrons able to dissociate/ionize N_2 and CH_4 in two discharges without and with argon inlet (b). RF power is ∼2.1 kW [422].

with energies exceeding the ionization/dissociation thresholds of reactions (7.1)–(7.3).

The estimates show that introduction of argon to the N_2+CH_4 mixture increases not only the plasma density, but also the probability of dissociation of N_2 and CH_4 molecules. In this case the XPS analysis of the elemental composition reveals an increase in nitrogen content in the film from 18 % in argon-free to 36 % in argon-enriched (additional inlet of 2–6 sccm) discharges. Addition of Ar also augments the proportion of carbon from 1.8 % to 3.2 %.

Furthermore, additional inlet of argon slightly decreases the population of electrons able to initiate ionization (7.1) of N_2 from 2.7×10^{10} to 2.1×10^{10} cm^{-3}. Meanwhile, the total number of electrons able to dissociate nitrogen molecule (reaction (7.2)) increases from 4.3×10^{10} to 6.2×10^{10} cm^{-3}. Therefore, the reaction balance in the gas phase is shifted from the formation of molecular ions N_2^+ to electron-impact dissociation of nitrogen molecule N_2.

Therefore, production of atomic nitrogen building units relative to molecular ion N_2^+ species can be enhanced by letting more argon in the discharge as the bar charts in Fig. 7.6(b) suggest. Similarly, adding argon to the mixture increases the number of electrons responsible for dissociation (reaction (7.3)) of methane molecules from 5.3×10^{10} to 1.22×10^{11} cm^{-3} [422].

Elegant experiments of Jiang [427] shed some light on the role of ionized gas-phase-borne building units in the synthesis of nanostructured AlCN films.

He recorded optical emission intensities (OEIs) of the most abundant species in the plasma discharge under different process conditions (such as the CH_4 flow rate in Fig. 7.7) and related the variation of the OEI to the changes in the intensities of relevant chemical bonds formed in the AlCN films (measured by the FTIR technique). The *in situ* OES observations provides the useful information about dominant chemical reactions in the deposition process.

In this series of experiments, atomic nitrogen, CN radicals and C_mH_n species are the key precursors of the carbon and nitrogen components in the film. On the other hand, atomic Al sputtered out from the target is an indispensable building unit for Al incorporation in the AlCN nanofilm. Interestingly, Al emission lines were not detected in the optical emission spectra [427].

As suggested by Jiang [427], the relationship between the deposition parameters and the material properties can be established by comparing the intensities of the OES signal with those of FTIR peaks, as shown in Fig. 7.7. It is clearly seen that with an increase of the CH_4 gas flow rate, both the intensities of CH species and the C–Al bond increase, while those of nitrogen species and the N–Al bond decrease.

On the other hand, the intensities of the OEI of CN species and the C–N bond initially increase and then decrease as the CH_4 flow rate increases. A similar variation trend between the OES intensities and FTIR bonding intensities have also been reported under conditions when the RF input power and/or the flow rates of N_2 changed [427].

Therefore, one can conclude that in the CN species *formed in the ionized gas phase* is indeed a building unit responsible for the C–N bond formation. On the other hand, while CH and atomic nitrogen species are combined with Al atomic species to form Al–C and Al–N bonds, respectively. Relatively high rates of formation of such bonds is probably the reason of the undetected optical emission from atomic aluminium.

Based on the above experimental results, the mechanism of the synthesis of Al-modified carbon nitride thin film has been proposed [427]. First, CN species is formed in the plasma by the chemical reaction between CH_x and N_2 [428]

$$CH_x^* + N_2^* \rightarrow CN + N + H_2 + H + e$$

These species are responsible for the C–N bond in the material, which means that C–N bond in the thin film directly comes from the CN building units in the plasma, rather than formation by chemical reaction between CH_4 and N_2 on the substrate surface.

Second, although optical emission from Al-containing species was not detected in the spectra, a similar trend between the intensities of the optical emission and the FTIR signal in Fig. 7.7 indicate that CH and N species incorporate

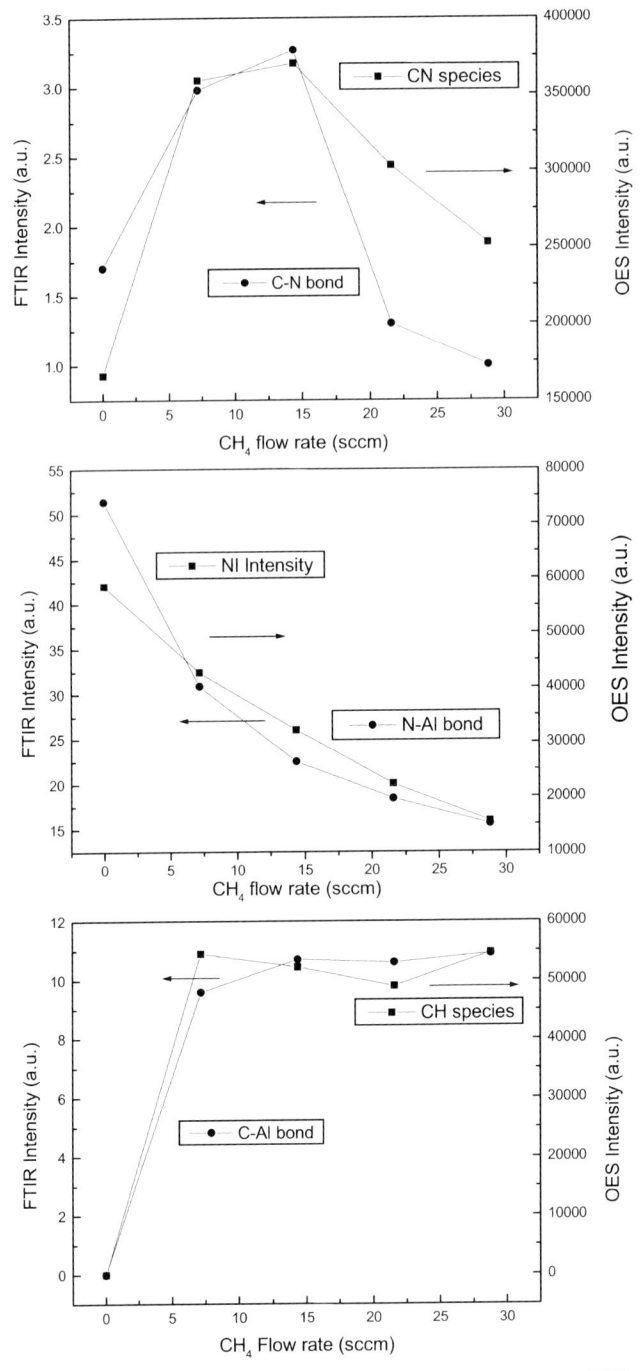

Fig. 7.7 The comparison between the FTIR intensity and the OES emission intensity in plasma-assisted DC magnetron sputtering deposition of nanostructured AlCN films [427].

into the Al–C and Al–N bonds in the AlCN film. Since the CH and N species exist as the reactants in the plasma, one can conclude that the chemical reactions forming Al–C and Al–N bonds occur on the substrate surface rather than in the ionized gas phase.

Third, in the optical spectra, CN species shows the strongest optical emission, whereas C–N bond in the FTIR spectra becomes much weaker compared to Al–N and Al–C peaks. This reveals that the affinity of species is another key factor in the deposition process. CH and N species have three and two dangling bonds that make them easily adsorbed on the substrate surface and form bonds with other atoms, whereas CN species has only a single dangling bond, which results in a low bonding probability.

On the other hand, CN species is volatile, which means that at higher temperatures or in the case of sputtering by energetic particles, CN component can be easily etched away. This is another reason for lower C–N bond strengths in the film.

Finally, the above analysis suggests that the bonding properties of nanostructured films can be optimized by controlling the building units in the plasma and in turn by adjusting the plasma process parameters. These results are yet another argument in favor of the "cause and effect" building-unit based nanofabrication approach [36].

7.4
Plasma-Assisted Growth of Highly Oriented Nanocrystalline AlN

In this section we continue showing examples of successful uses of plasma-based nanofabrication techniques. Let us turn our attention to optically transparent, highly oriented nanocrystalline AlN(002) films synthesized by using a hybrid plasma-enhanced chemical vapor deposition and plasma-assisted RF magnetron sputtering process in reactive $Ar+N_2$ and $Ar+N_2+H_2$ gas mixtures at a low Si(111)/glass substrate temperature of 350 °C [429].

In the following we will discuss how the most important process conditions, such as the sputtering pressure, RF power, substrate temperature, and partial pressure of reactive nitrogen gas feedstock can be optimized to achieve the desired structural, compositional, and optical characteristics. These films are made of high purity and near-stoichiometric AlN, which also feature a fairly uniform distribution of nanosized grains over large surface areas and also highly oriented in the (002) direction columnar structures of a typical length ∼100–500 nm with the aspect ratio ∼7–15. Moreover, the AlN films feature an excellent optical transmittance of ∼80% in the visible region of the spectrum, promising for advanced optical applications [429].

In Chapter 5, we have already discussed outstanding properties of nanostructured AlN that make this material particularly attractive for the future nanodevice applications. We have also discussed earlier growth stages of AlN nanofilms, when arrays of quantum dots can be formed. Here we consider the case when by using plasma-based techniques it becomes possible to direct the growth process toward quasi-three-dimensional columnar structures that develop at later growth stages (as compared to the quantum dots considered in Chapter 5).

It needs to be stressed that most of the existing and emerging applications demand the AlN nanofilms to be single crystalline with a smooth surface. Furthermore, a high-quality heteroepitaxy is usually required to fully realize the outstanding potential of AlN in micro- and nanoelectronics. We recall that to fabricate device-grade AlN films, techniques such as vapor phase epitaxy [220, 221], molecular beam epitaxy [221–224], chemical vapor deposition [225], ion beam nitridation [226], reactive sputtering [228], and pulsed laser deposition [231] have been extensively used, with a different degree of success depending on specific process conditions [429].

Even though advanced epitaxial techniques such as the molecular beam epitaxy (MBE), chemical vapor deposition (CVD), and metallorganic chemical vapor deposition (MOCVD) have been extensively used for the fabrication of lattice-matched nitrides of Group III semiconductors, high process temperatures often result in a substantial thermal damage and film degradation. Moreover, the ability to synthesize highly oriented nanocrystalline films, which are of special interest as growth templates for zero- and one-dimensional quantum confinement structures, still remains quite limited [429].

In order to achieve highly oriented c-axis films, lower temperature processes are required. Low temperature plasma-assisted reactive sputtering deposition is an excellent candidate owing to its apparent advantages in the ability to control the fluxes of numerous reactive species, the ease in handling and operation, and a relatively low cost. Recent advances in plasma-aided synthesis of various nanocrystalline films and aligned one- and two-dimensional nanostructures with the crystallographic growth directions controlled by the plasma process conditions discussed elsewhere in this book further justify this claim.

Likewise, as we have seen from Chapter 6, plasma-aided RF sputtering deposition provides a highly reliable and reproducible method of thin film growth and as such should be particularly suitable for the synthesis of AlN nanofilms with a smooth surface and a negligible oxygen content, the latter still remaining a major concern for the many existing fabrication techniques [431].

Mirpuri et al. recently reported on the plasma-assisted RF magnetron sputtering deposition of optically transparent, highly oriented (002) nanocrystalline AlN films on Si(111) and glass substrates [429]. Below, we will follow how the process conditions such as the RF input power, substrate tempera-

7.4 Plasma-Assisted Growth of Highly Oriented Nanocrystalline AlN

Tab. 7.2 Plasma-assisted synthesis of nanocrystalline AlN: summary of process conditions [429].

Process parameter	Values
Base pressure	below 10^{-3} Pa
N_2 flow rate	10–85 sccm
Ar flow rate	0–30 sccm
Substrate temperature	150–350 °C
RF power	200–400 W
Working gas pressure	0.4–1.4 Pa
Deposition time	0.5–3 h

ture, gas flow rates and partial pressures can be optimized to obtain the desired structural, compositional, and optical properties of the AlN films critical for their envisaged applications.

The AlN thin films of our interest here have been synthesized in the Integrated Plasma-Aided Nanofabrication Facility described in Chapter 3. Table 7.2 summarizes the range of the process parameters in the experiments of Mirpuri et al. [429].

Figure 7.8(a) shows the XRD spectra of AlN films deposited for one hour at a working gas (pure nitrogen) pressure of 0.4–1.4 Pa, substrate temperature of 350 °C, and RF power of 400 W. All the specimens featured a single hexagonal AlN(002) peak at 36.1°. The full width at half maximum (FWHM) of the AlN(002) peak has shown a strong dependence on the working gas pressure.

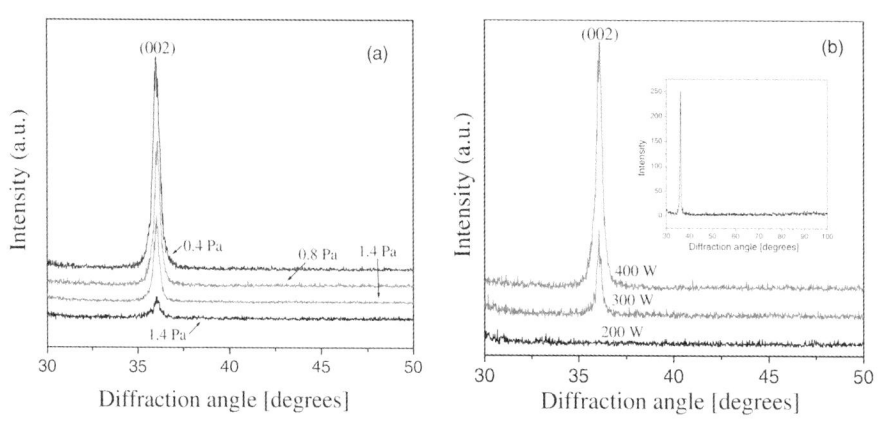

Fig. 7.8 X-ray diffraction spectra of AlN films for different working gas pressures at a RF power of 400 W (a) and different RF powers at a working pressure of 0.8 Pa (b). In both the cases nitrogen concentration and substrate temperature are 100 % and 350 °C, respectively. Inset in panel (b) displays the XRD wide scan at a RF power of 400 W and the same other parameters as in the case (b) [429].

Specifically, the FWHM has a clearly resolved minimum in the pressure range 0.8–1.1 Pa, which is indicative to somewhat better crystallization of the AlN films under such conditions. The X-ray diffractometry in the rocking curve mode also reveals that a sputtering pressure of 0.8 Pa leads to a smaller FWHM of 10.4° as compared to 18.6° at 1.1 Pa. Thus, the alignment of the growth direction is improved in the films synthesized at $p_0 = 0.8$ Pa. One can thus conclude that the optimal sputtering pressure for the growth of highly c-axis oriented AlN films is 0.8 Pa.

Figure 7.8(b) shows the XRD patterns of the samples after 1 h deposition in pure nitrogen at RF power 200–400 W, substrate temperature 350 °C, and sputtering pressure 0.8 Pa. Interestingly, the film deposited at 200 W did not show any crystalline orientation peculiar to AlN. This may be attributed to the fact that when the input power is low, the adatoms cannot gain enough energy to form crystalline structures on the substrate.

However, a pronounced diffraction peak corresponding to the (002) plane appears when the input power is increased to 300 W. The (002) preferred orientation of AlN further improves at higher RF powers, as can be seen in Fig. 7.8(b). Further XRD analysis shows that the c-axis remains the preferred growth direction within the entire temperature range considered. This growth direction becomes pronounced at substrate temperatures as low as 150 °C, when the (101) crystallographic direction emerges. Moreover, the (002) crystallographic orientation is always the preferential growth direction when the partial pressure of nitrogen p_{N_2} exceeds a half of the total pressure of the Ar+N$_2$ gas mixture [429].

Thus, the crystalline orientation is strongly affected by the relative partial pressures of nitrogen and argon gases. Nonetheless, at p_{N_2}/p_0 ratios larger than 0.5, the films appear perfectly oriented along the c-axis. Therefore, the results of the XRD analyses [429] show that the growth of highly oriented nanocrystalline AlN(002) thin film can be optimized at a sputtering pressure of 0.8 Pa, a RF power of 400 W, a substrate temperature of 350 °C and relative ratio of partial pressures of N$_2$ and Ar gases 3:1 (75 % concentration of N$_2$).

The top-view microstructure of platinum sputter-coated AlN films deposited for 0.5 and 3 h under the above optimized process conditions is depicted in Figs. 7.9(a) and (b), respectively. The surface morphology features smooth and homogeneously uniform pebble-like nanocrystalline grains (with the typical size within the 40–80 nm range) formed on the substrate with some larger (>100 nm) triangular-shaped particles on top of the layer; the latter appearing in the 3 h deposition process.

The SEM images also reveal clear grain boundaries and also voids between the crystals. This compact and almost void-free crystalline structure (viewed from the top) is indicative to preferential columnar growth, which was confirmed by the cross-sectional scanning electron microscopy.

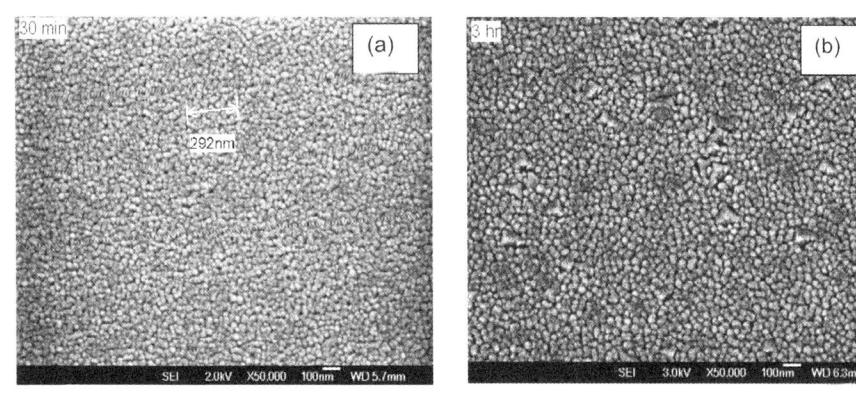

Fig. 7.9 Surface morphology of AlN film deposited for 0.5 (a) and 3 h (b) imaged by FESEM. Optimized process conditions (see text) [429].

The cross-sectional SEM images of the AlN film grown for 0.5 and 3 h are shown in Figs. 7.10(a) and (b), respectively. From Fig. 7.10(a) one can also notice a thin (a few nm thick) amorphous layer along the interface between the AlN film and the Si(111) substrate. The film grown in the 3 h long process is approximately 0.5 μm thick and shows excellent vertical alignment of the columnar structure, which is quite similar to that of typical quasi-one-dimensional quantum confinement structures considered in Chapter 5 of this monograph. A typical aspect ratio of the highly oriented crystalline columnar structures is 7–15 [429].

The composition and chemical states in the as-grown samples have been analyzed by the XPS. Typically, the percentage atomic concentrations of Al and N were quite close to each other. More importantly, the abundance of O and C was typically one order of magnitude lower than that of the main constituent elements Al and N. This contamination is very difficult to elimi-

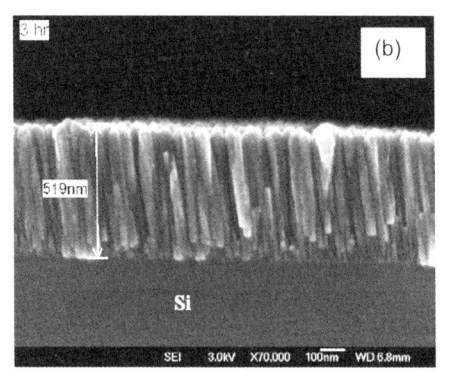

Fig. 7.10 Cross-sectional SEM micrograph of AlN film deposited for 0.5 (a) and 3 h (b) and same process conditions as in Fig. 7.9 [429].

nate completely and much higher concentrations of oxygen (owing to its large affinity to aluminium) have been reported by other authors [432, 433].

Figures 7.11(a) and (b) show the narrow scan XPS spectra of Al2p and N1s, respectively. The binding energy (BE) of the Al2p photoelectron peak in Fig. 7.11(a) is 73.9 eV, and is attributed to the Al–N bonds in AlN [434]. The peak corresponding to Al–Al bonding (with the BE ~72 eV) was not detected in the Al2p spectrum in Fig. 7.11(a). This suggests that all of the Al atoms are bonded with other elements and no metallic Al is present in the film. The N1s spectrum in Fig. 7.11(b) after deconvolution into the Gaussian/Lorentzian peaks via the nonlinear least squares routine yields 2 peaks with binding energies of 397.3 and 399.3 eV. The strongest peak at 397.3 eV is characteristic to the N1s state in the N–Al bond [433]. The almost negligible peak at 399.3 eV can be assigned to physiosorbed molecular nitrogen, a common artifact of the sputter-cleaning.

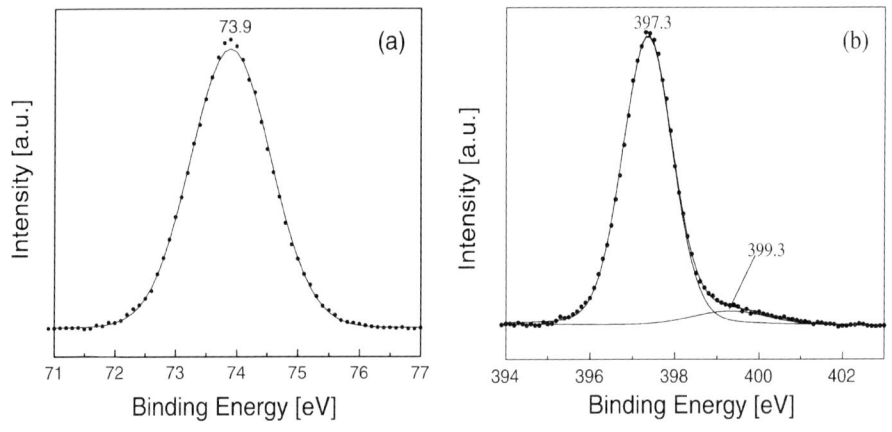

Fig. 7.11 XPS narrow scan spectra of Al2p (a) and N1s (b) recorded from AlN film deposited for 3 h [429].

The elemental composition of the AlN films was further investigated by using the EDX spectroscopy [429]. The amplitudes of the elemental peaks were used to determine the relative composition of each of the elements in the films. Interestingly, the nitrogen content consistently increases with the RF power and becomes optimum at $P_{rf} = 400$ W. More importantly, the EDX analysis suggests that an increase of the input power from 200 to 300 W leads to a drastic improvement in the [N]/[Al] ratio toward 1. Thus, near-stoichiometric and highly oriented AlN films can indeed be synthesized at a relatively high RF power level of 400 W.

The bonding states of AlN were further studied by means of Raman and FTIR spectroscopy. The Raman spectroscopy suggests that the AlN thin film is formed in the pressure range between 0.8 and 1.1 Pa, as suggested by clearly resolved characteristic AlN phonon modes at 657 cm^{-1} (E2 high) and 910 cm^{-1} (E1 LO) [435]. A complementary FTIR analysis also confirms the formation of Al–N bonds in the films synthesized under the optimized process parameters.

Indeed, a strong peak at 682 cm^{-1} corresponds to the E1 transverse-optical (TO) vibration mode of the infrared-active Al–N bond [429, 435]. Notably, the E1(TO) linewidth narrows as the power increases, which is indicative of an improved quality of AlN specimens synthesized at higher RF powers. This is consistent with the results obtained from the XRD spectra at different power (Fig. 7.8(b)) which suggest that no crystalline AlN film form at $P_{rf} = 200$ W and that a further increase in the intensity of the (002) peak is proportional to the input power.

The FTIR results also suggest that when the partial pressure of nitrogen is less than a half of the total gas pressure ($p_{N_2}/p_0 < 0.5$), no phonon modes characteristic to AlN can be detected. Furthermore, the FWHM of the peak related to the E1 (TO) phonon mode is noticeably larger in pure nitrogen compared to the case when $p_{N_2}/p_0 = 0.75$. Thus, the film quality is indeed better when the relative percentage of N_2 is 75%, which further justifies the appropriateness of the choice of the optimum process conditions [429].

The nanocrystalline AlN films deposited on glass substrates also show excellent optical properties. A comparison of the spectral characteristics of the glass and AlN-coated glass specimens suggests that the AlN film has a lower transmittance of ∼8-14% (compared to the glass substrate) in the visible region (400 to 780 nm) of the spectrum [429]. The main cause of the incident light loss is due to surface reflection whereas a very small amount of light is absorbed. The onset of the band edge and interband absorption results in a reduced transmittance of the AlN sample. Since optical coatings are required to be visible in the whole of the visible spectrum, this feature is undesirable from the optical applications point of view.

In order to improve the optical transmittance of the AlN films, a small percentage (∼5%) of H_2 gas can be added to the Ar+N_2 gas mixture. The addition of hydrogen results in the growth of AlN:H (hydrogenated AlN) films which, unlike the AlN films, exhibited a better transparency in the 300–600 nm spectral range. In this case, there is a significant reduction in the near-band-edge absorption [436]. This effect can be explained in terms of the reduced density of states at the band edges of the crystalline AlN material as a consequence of the passivation of the dangling bonds due to the addition of hydrogen in the films.

To conclude this section, the Ar+N_2 plasma-assisted RF magnetron sputtering of aluminium at a low temperature of 350 °C enabled a successful synthesis of optically transparent, highly oriented (002) nanocrystalline AlN films, which was a major challenge for other fabrication techniques. Remarkably, single-crystalline, highly oriented, chemically pure, near-stoichiometric AlN films can be synthesized at a sputtering pressure of 0.8 Pa, RF power of 400 W and relative partial pressure of nitrogen in the Ar+N_2 mixture of 75%. Moreover, the AlN(200) films exhibit an excellent optical transmittance of ∼80% to visible light, making it also promising for optical applications.

7.5
Plasma-Assisted Synthesis of Nanocrystalline Vanadium Oxide Films

Another example of successful application of low-temperature plasma nanotools is the synthesis of nanocrystalline vanadium pentoxide films for advanced optical applications. Lim et al. used the plasma-assisted reactive RF magnetron sputtering deposition technique to fabricate vanadium oxide films on glass, silica, and silicon substrates [437]. Similar to the fabrication of highly oriented nanocrystalline AlN films of the previous section, the process conditions have also been optimized to synthesize phase-pure vanadium pentoxide (V_2O_5) featuring: (i) a nanocrystalline structure with the predominant (001) crystallographic orientation, (ii) surface morphology with rod-like nanosized grains, and (iii) very uniform coating thickness over large surface areas. Moreover, the V_2O_5 films also show excellent and temperature-independent optical transmittance in a broad temperature range (20–95 °C).

We now consider this successful application of the plasma-based nanofabrication technique in more detail following the original work [437]. Vanadium oxides VO_2 and V_2O_5 have been studied extensively owing to their unusual properties that make them particularly attractive for various applications ranging from data storage [438] to smart window coatings [439].

Vanadium pentoxides have been widely investigated because of their superior electrochemical performance in various functional films and nanostructures, such as smart windows for solar cells [440], high aspect ratio nanowires [441], cathode coatings in high-capacity lithium batteries [442], reactive catalyst nanoparticles [443], and other films for electrochromic devices [444, 445] and optical switches [446]. In the last decade, various thin film growth techniques including sputtering [445, 447–450], atmospheric pressure chemical vapor deposition (APCVD) [451], and pulsed laser deposition [452, 453] have been adopted to deposit vanadium oxide thin films on various substrates.

However, these techniques in most cases fall short to produce highly crystalline vanadium oxides and substantially lack the ability to satisfactorily control the most important coating parameters such as the phase composition, crystallinity, crystallographic orientation, optical transmittance, thermal and mechanical stability, and several others. Moreover, these problems become even less manageable in the case of nanocrystalline or nanostructured V_2O_5 [441].

In a recent work, the advanced plasma-assisted RF magnetron sputtering deposition technique has been successfully used to synthesize high-purity nanocrystalline vanadium pentoxide with preferential (001) crystallographic orientation and excellent optical transmittance in a broad range of temperatures. These results are confirmed by multiple analytical tools including Raman spectroscopy, ultraviolet-visible spectroscopy (UV/VIS), X-ray diffractometry (XRD), and X-ray photoelectron spectroscopy (XPS).

Vanadium oxide films can be deposited on glass, amorphous silica, and silicon substrates by using the IPANF plasma nanofabrication facility described in detail in Chapter 3 (see also Ref. [157]). The distance between the vanadium target and the deposition substrate is 60 mm. To sustain plasma discharges in oxygen–argon gas mixtures, the RF power input level is set at 400 W [437]. The effect of the growth process parameters, such as the substrate temperature, oxygen flow rate, and the working gas pressure in the plasma reactor, on the composition, structural and optical characteristics of the vanadium oxide films has been investigated.

In the series of experiments of our interest here, the deposition substrates (glass, silica, and silicon) are externally heated and thermo-stabilized within the temperature range from 370 to 550 °C. The oxygen inflow rate is varied from 0.3 to 5 sccm. At a fixed oxygen influx, the total pressure of the working gas feedstock is controlled by additionally letting high-purity (99.99%) argon. The range of total pressures in the experiments of Lim et al. is maintained within the range 10–50 mTorr [437].

To achieve the desired characteristics of the vanadium oxide films, the plasma-based process conditions can be optimized by adjusting the substrate temperature, oxygen flow rate, and working pressure in the deposition reactor. Likewise, the film properties turn out quite sensitive to substrate properties, especially the surface morphology and texture as well as the growth temperature.

The results of the Raman spectroscopy analysis of the vanadium oxide thin films synthesized at different process conditions suggest that if the substrate temperature T_s exceeds 450 °C, the Raman spectra feature eight well-resolved bands at 147, 197, 285, 305, 405, 483, 699, and 999 cm^{-1}. The narrow band at 147 cm^{-1} is due to vibrations in a V–O–V atomic chain; the band at 999 cm^{-1} is peculiar to V=O stretching modes. All eight bands perfectly match the features of the reference Raman spectrum of commercially available pure V_2O_5 powders; thus, the predominant phase in the experiments of Lim et al. [437] is indeed pure vanadium pentoxide.

Moreover, a high resolution of the Raman spectra is indicative of a high degree of crystallinity and phase purity in the films concerned. Noteworthy, all eight vanadium pentoxide "fingerprint" bands in the Raman spectra resolve when the substrate temperature is set above a certain threshold. A surface temperature of 450 °C turns out to be sufficient to obtain high-purity V_2O_5 coatings [437].

Apparently, higher surface temperatures are favorable for the efficient crystallization, which happens when deposited ionic/atomic/radical building units (with some of them becoming adatoms on the surface and some inserting directly from the ionized gas phase) have a sufficient energy (and also lifetime) to visit the available bonding cites and properly insert into a reconstructed surface and relax in a stable position [36, 437]. However, in several cases, all vanadium pentoxide "fingerprints" appear in Raman spectra of specimens synthesized at lower deposition temperatures, such as 410 and even 370 °C. This becomes possible by properly manipulating the oxygen inflow rates and additionally increasing the working pressure.

For example, by adjusting the supply of oxygen to the growth surfaces one can synthesize two structural types of vanadium oxides with distinctly different Raman spectra and also control the film crystallinity at a lower substrate temperature of 410 °C. When the oxygen flow rate is low, the coatings contain mixed phases, with quite different contents of pure vanadium and two VO_2 and V_2O_5 vanadium oxide phases.

On the other hand, when the oxygen flow rate increases to 0.3 sccm, the vanadium pentoxide phase becomes dominant. Notably, high-purity V_2O_5 films can be successfully grown at oxygen flow rates exceeding 0.3 sccm. Interestingly, the intensity of the 996 cm^{-1} Raman peak (associated with the V=O bonds in V_2O_5) increases with the O_2 influx. This is quite natural to expect since oxygen deficiency in the films usually leads to the formation of vanadium oxides with lower valence vanadium states, such as VO_2.

However, crystalline features of the films in this case become pronounced only at oxygen flow rates 0.5 sccm and above [437]. This can be attributed to the enhanced crystallization due to highly reactive atomic and anionic oxygen, which can act as efficient crystallization agents [441].

Meanwhile, when more oxygen is let in the chamber, the Raman spectra change dramatically [437]. Specifically, all eight major "fingerprints" of vanadium pentoxide appear in the spectra when the working pressure becomes higher than 35 mTorr. Moreover, when the pressure reaches 50 mTorr all the major eight spectral bands become well resolved, with the dominant band at 147 cm^{-1}.

Thus, vanadium oxides synthesized at higher argon gas pressures, feature stronger vibrations of two V atoms bonded with oxygen, which may be indicative of a relative increase of the VO_2 phase content in the film. Therefore, at $T_s = 410$ °C, the optimum working pressure for the synthesis of high-purity vanadium pentoxide is in the range between 35 and 50 mTorr.

The effect of the working gas pressure becomes clear if one notes that larger numbers of argon atoms/ions in the reactor chamber effectively increase collision rates of vanadium atoms sputtered from the target and also those of atomic/ionic oxygen. This leads to somewhat lower deposition rates [437]. On the other hand, higher argon pressures result in stronger fluxes of argon ions onto the substrate, which in turn lead to higher surface temperatures and, hence, more efficient crystallization of vanadium oxide films. Thus, the working pressure needs to be properly adjusted for the highly controlled synthesis of crystalline V_2O_5.

The results of the analysis of the optical transmittance of the vanadium oxide films suggest that different samples feature quite different optical transmittance and hence vary in color quite significantly. There is a strong correlation between the experimental conditions used to prepare the various samples, the resulting colors, and the results of the Raman analysis.

The optical transmittance of the samples is measured in the near-UV and near-infrared ranges, by using UV/Vis spectrophotometer within the range of temperatures from the room temperature up to 95 °C. For this analysis, the specimens with the dominant vanadium pentoxide phase and improved crystallinity are used [437]. We recall that such requirements can be met by adopting the following optimized process conditions: substrate temperature above 450 °C, oxygen flow rate above 0.5 sccm, and working gas pressure in the range 35–50 mTorr.

Under such conditions the films appear grayish yellow in color, and do not show any significant (less than 1%) change in the optical transmittance even when exposed to high temperatures. On the other hand, the films synthesized at different process conditions (lower substrate temperatures or working gas pressures) can show quite a significant (more than 5%) decrease in their optical transmittance when the temperature increases up to 95 °C. It is notable that in such cases the V_2O_5 "fingerprints" in the Raman spectra fade quite significantly. Therefore, an almost negligible change in the optical transmittance within the broad temperature range is actually observed for the films with the high vanadium pentoxide phase content.

The elemental composition and phase content have been analyzed by using the X-ray photoelectron spectroscopy [437]. Figure 7.12 (left panel) shows a narrow scan $2p_{3/2}$ XPS spectrum of vanadium in the film synthesized at $T_s = 410\,°C$, $p_0 = 18$ mTorr, and O_2 inflow rate of 0.3 sccm. The two peaks with binding energies of 517.6 eV (a much stronger peak) and 516.3 eV obtained after de-convolution of the V $2p_{3/2}$ spectrum are clearly seen in the spectrum. The former peak characterizes the binding of vanadium with oxygen in vanadium pentoxide and appears to be much stronger and has a larger peak area compared to the 516.3 eV peak representative to VO_2.

Therefore, the XPS analysis confirms that the vanadium pentoxide appears to be a dominant phase even at "nonoptimized" process conditions (such as the substrate temperature 410 °C in this example). Nevertheless, the presence

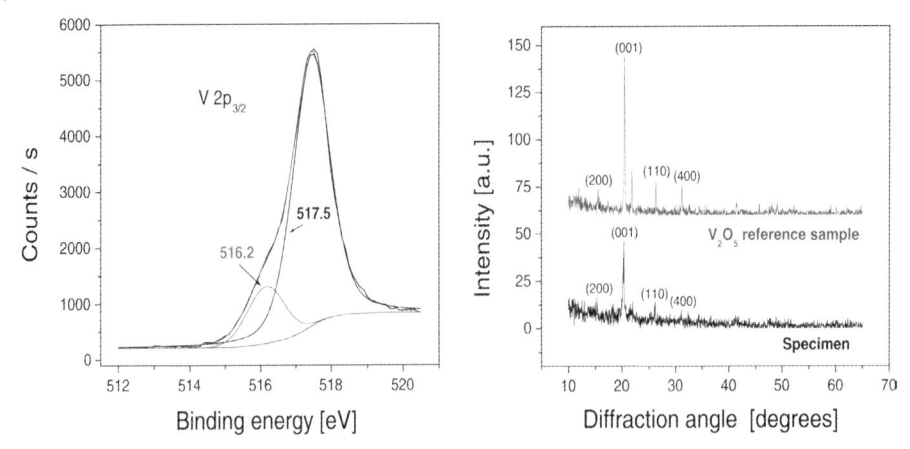

Fig. 7.12 Narrow scan V 2p$_{3/2}$ XPS spectrum of the film deposited on a Si substrate at oxygen flow rate 0.3 sccm, substrate temperature 410 °C, and working gas pressure 18 mTorr (left panel); XRD spectra from the specimen deposited on the amorphous silica substrate (same conditions) and V$_2$O$_5$ reference sample (right panel) [437].

of the VO$_2$ still remains nonnegligible, which explains a minor decrease in the optical transmittance shown by the UV/VIS analysis [437].

Crystallinity of the vanadium pentoxide films has been investigated by the X-ray diffractometry [437]. Interestingly, the X-ray diffraction patterns convincingly show that the V$_2$O$_5$ thin films can be grown on amorphous silica substrates with a specific preferential crystal orientation. The XRD spectra of the specimens perfectly match those from the orthorhombic V$_2$O$_5$ reference sample (Fig. 7.12 (right panel)).

As can be seen from Fig. 7.12, both spectra feature four major peaks at 2θ = 15.4, 20.3, 26.2, and 31.0° related to the main crystallographic orientations (200), (001), (110), and (400) of vanadium pentoxide [437]. More importantly, it turns out possible to synthesize V$_2$O$_5$ films with a predominant crystallographic orientation, being (001) in the case considered. In fact, this is one of the possible pathways to fabricate high-aspect-ratio vanadium pentoxide nanowires [441].

The surface and cross-sectional morphology of the as-grown films has been investigated by the field emission scanning electron microscopy, with the representative results shown in Fig. 7.13. Cross-sectional scanning electron micrographs similar to the one shown in Fig. 7.13 (left panel) suggest that the film thickness of a selected sample at 8 different points varies in a very narrow range from 383 to 422 nm. The SEM micrograph in Fig. 7.13 (right panel) shows clearly resolved crystalline features (grain boundaries, sharp edges, multiple facets) of a nanocrystalline V$_2$O$_5$ film on a Si substrate. One can notice a large number of uniform-size rod-like nanometer-sized grains ran-

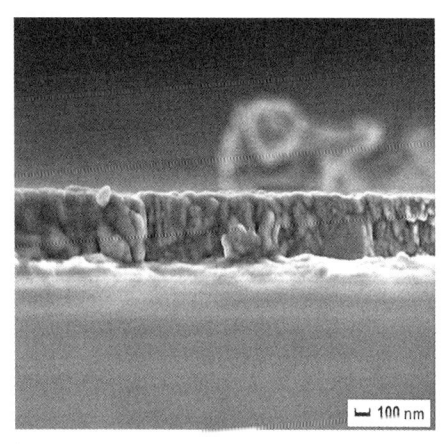

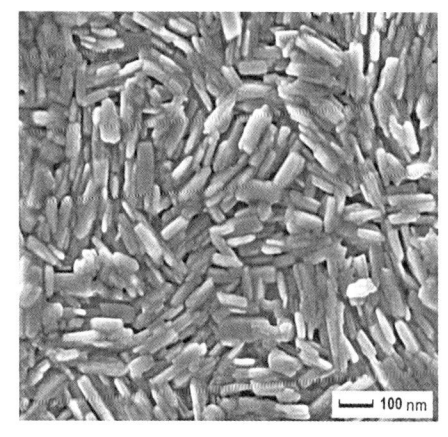

Fig. 7.13 Representative cross sectional electron micrograph (left panel) and surface morphology (right panel) of nanocrystalline vanadium pentoxide films imaged by field emission scanning electron microscopy [437].

domly oriented about the surface. Thus, the V_2O_5 films concerned are poly-nanocrystalline, with the preferential crystallographic orientation (001).

In applications, the performance and reliability of polycrystalline films are strongly affected by the average grain size and the distributions of the grain sizes and crystallographic orientations. Therefore, these important properties need to be controlled by the process conditions, which in turn affect the grain growth. The resulting surface morphology depends on the film thickness, working gas pressure, substrate temperature, surface charges, thermal expansion coefficients of the film and the substrate, mechanical properties of the film, as well as some other factors.

For the convenience of the reader, we now summarize the features of the nanocrystalline vanadium pentoxide films synthesized by using the advanced plasma-assisted RF magnetron sputtering implemented in the IPANF facility described in Chapter 3 of this monograph. In this way, nanocrystalline vanadium oxide coatings of Si, SiO_2, and glass substrates have been successfully fabricated and characterized [437]. By varying the process parameters, it turns out possible to optimize the crystallinity, crystal size and shape, surface morphology, phase composition, and optical transmittance of the vanadium oxide films.

In particular, when the substrate temperature, oxygen flow rate, and working pressure in the plasma reactor exceed 450 °C, 0.5 sccm, and 35 mTorr, respectively, nanocrystalline vanadium pentoxide becomes a dominant phase in the coating and significantly outweighs the other abundant, VO_2 phase. By adjusting the flow rates of chemically active oxygen (atomic/ionic oxygen can act as a crystallization agent [36]) and inert argon (i.e., manipulating the working pressure), one can also achieve quite similar results at even lower substrate temperatures, such as 410 °C.

It is remarkable that V_2O_5 optical coatings also exhibit excellent thermal stability, with their optical transmittance showing only a slightest ($\sim$1 %) decrease at high temperatures, and most likely due to the nonnegligible presence of the VO_2 phase. The deposition rates are typically in the $\sim$100 nm/h range.

We stress that this plasma-based technique allows one to deposit vanadium pentoxide coatings with excellent thickness uniformity (in the case considered the nonuniformities do not exceed 4 %) and clearly predominant (001) crystallographic orientation. Future research should aim at improving the grain size/shape uniformity and crystalline orientation and also synthesize highly crystalline V_2O_5 nanorods with a high aspect ratio.

Finally, numerous exciting possibilities to control the nanofilm properties by properly manipulating the plasma process parameters (with only a small number discussed in this monograph!) make us optimistic that the yet elusive goal of deterministic plasma-aided synthesis of smart nanocrystalline vanadium pentoxide films will become a reality in the near future.

7.6
Plasma-Treated Nano/Microporous Materials

We now turn our attention to another interesting example showing how nanometer-size porous features can be treated and even synthesized by using low-temperature plasmas. First, a successful application of inductively coupled plasmas for the deposition of carbon-based molecular sieve membranes for gas separation will be considered in detail followed by an overview of other applications of plasma-based tools and techniques for the fabrication and/or treatment of materials with ultrasmall pores/voids.

To begin our discussion of applications of plasma-enhanced chemical vapor deposition for the synthesis of nano-/microporous membranes, we first stress that the pore/void sizes in such membranes need to be selected for the envisaged applications. For example, if gas separation is a targeted application, the membranes need to exhibit molecular sieving capabilities [454].

In other words, their pore/void sizes should be comparable with the dimensions of gas molecules to be separated. For example, one needs to separate hydrogen H_2 and nitrogen N_2 molecules with kinetic diameters $d_{H_2} = 0.29$ and $d_{N_2} = 0.364$ nm, respectively [455]. In this case one can expect that a pore with a feature size $d_{H_2} < d < d_{N_2}$ will filter larger nitrogen molecules and let smaller hydrogen molecules pass through the membrane. Thus, by engineering pore/void sizes within the range $\sim$0.29–0.36 nm, one can completely separate hydrogen and nitrogen gases.

However, in reality it is very difficult to create so uniformly and narrowly distributed sub-nano-sized pores. Even suitable naturally existing materials such as crystalline zeolite and amorphous diamond-like carbon (DLC) feature nonuniform pore sizes, which makes it difficult to separate different gas molecules completely. This problem becomes even more difficult in situations when the dimensions of molecules being separated are close. This is the case when oxygen molecules O_2 with a kinetic diameter of 0.346 nm [455] need to be separated from nitrogen ones.

In this case the acceptable pore sizes should belong to a very narrow range between 0.346 and 0.36 nm. Therefore, at given nanopore sizes the selectivity of H_2/N_2 separation is expected to be higher than that of O_2/N_2. Another issue is the ability of the gas to efficiently permeate through the membrane, which effectively leads to the requirement of the presence of a larger numbers of clear pores/channels within the material.

As to the appropriate choice of a suitable material, this is an interesting example when amorphous materials may have advantages over nanocrystalline ones. Indeed, cracks along grain boundaries have been reported as critical factors that severely undermine the gas separation capabilities of nanocrystalline zeolite [454]. Diamond-like carbon is an excellent candidate for a high-performance material suitable for carbon molecular sieve membranes.

It is is worth emphasizing that plasma-based tools proved very efficient in synthesizing dense DLC films owing to efficient film structurization and densification controlled by energetic ion bombardment [422, 456, 460]. The pore size distributions and porosity can be controlled by varying the deposition and postprocessing conditions [454].

More importantly, plasma-synthesized DLC films often feature tiny pores with the sizes characteristic to molecular dimensions [456]. The quality and size distribution of the pores are also very sensitive to the precursor gas. For example, silicon-containing monomers generated as a result of plasma-assisted dissociation of hexamethyldisiloxane (HMDSO) facilitate the formation of highly mobile siloxane bonds on the surface and eventually to a better permeance in gas separation applications [457–459].

Stability and performance of gas separation membranes can be substantially improved by depositing diamond-like carbon films on microporous supports such as Al_2O_3 commonly used as a template in the synthesis of carbon nanotubes and other related nanostructures. However, fabrication of high-quality nanoporous membranes or microporous supports by using conventional chemical routes represents a significant challenge due to numerous problems with surface cracking, film delamination, insufficient mechanical strength and some other problems. [454]. Currently, conventional polymer gas separation membranes fabricated by various chemical methods have almost reached unsurpassed upper limits, both in selectivity and permeability [454].

Wang and Hong used a remote inductively coupled plasma-based technique to synthesize high-performance carbon-based molecular sieves [454]. We now comment on the choice of the plasma source arrangement used in this series of experiments. In fact, this is a good example of deterministic choice of the building units participating in the synthesis process.

More specifically, a dielectric tube (where a plasma glow discharge is sustained) with a helical inductive coil wound around it, is intentionally moved away from the deposition substrate to eliminate (e.g., self-bias) electric fields in the near-substrate sheath. In this case the effects of ion bombardment that lead to a significant film densification [456] are reduced and plasma-generated neutral radicals become the main building units in the deposition process. In this remote plasma-based process only the radicals are likely to diffuse out of the plasma to deposit on the substrate unless a negative surface bias is applied to extract positively charged ions from the plasma [454].

This arrangement enables one to avoid the formation of overdense diamond-like films, which, as has been mentioned above, is undesirable for molecular separation applications. Moreover, the use of a funnel-type gas shower made it possible to direct the gas feed and control the radical residence times to eliminate their polymerization in the ionized gas phase, which often leads to the growth of nano-sized particles described in Chapter 1 of this book.

In terms of the reactive gas used, the best performance was shown by the nanoporous films synthesized from the HMDSO precursor. These films turn out less dense compared to those fabricated by using pure CH_4.

Interestingly, a combination of pretreatment of porous alumina with highly energetic (accelerated up to 300 eV), remote plasma deposition of carbon-based nanoporous films, followed by high-temperature pyrolysis, made it possible to significantly improve the selectivity and permeance of molecular sieves compared to the results achieved by conventional chemical methods. Indeed, the H_2/N_2 selectivity can be increased to 30–45 with an extremely high permeance of 2×10^{-6} mol/(m^2·s·Pa) at 423K [454]. On the other hand, the O_2/N_2 selectivity reached as high as 3.8, which further evidences an excellent uniformity of nanopore sizes within the membrane.

It is important to note that the selectivity of the process is mostly determined during the plasma deposition stage and did not change during the postprocessing. However, pyrolysis at high (∼700–800 K) temperatures significantly improves the permeance of the membrane. Even though there is no specific explanation for this effect given in the original work, we can speculate that a substantially improved permeation can presumably be attributed to smoothening of nonuniform carbon-based deposits that block pore openings, the latter being a quite common problem in the plasma treatment of microporous materials [461]. Therefore, Wang and Hong [454] proposed a new and efficient method for the fabrication of carbon-based molecular sieve membranes by combining the surface treatment with high-energy ions, remote inductively coupled HMDSO plasma deposition, and the subsequent pyrolysis at high temperatures.

We now briefly go through some other examples showing that the use of plasma-based techniques can be instrumental in synthesizing or postprocessing functional materials with micro-/nanoporous features. Microstructure and permeation performance of organosilicon RF plasma-polymerized membranes show a strong dependence on the "hardness" of the plasma conditions characterized by the ratio of the RF input voltage to the flow rate of precursor monomer (N_2, H_2, O_2, CO_2, and CH_4) [462]. If the plasma conditions are "hard" (high degree of precursor dissociation), the permeation appears to be in the diffusion-controlled regime (larger pores), whereas under "soft" plasma conditions (low degree of precursor dissociation), the pore size decrease and the gas separation is controlled by the Knudsen-like transport phenomena [462].

Plasma polymerization can also be used to produce tailored coating for microporous materials and nanotubules [463]. Likewise, surface properties of biomaterials can easily be modified by glow-discharge processing for very diverse biomedical applications, such as coating of microporous polypropylene oxygenerator with thin polymeric films [464]. Reactive (fluorocarbon- or oxygen-based) plasma-assisted processes have also been successfully used to modify porous low-k dielectric materials [465]. The resulting microstructure depends sensitively on the competition between the deposition and removal of plasma-generated building units, which turn out completely different in the cases of micro- and mesoporous materials.

Plasma polymerization can also be used for fast and efficient pore sealing of microporous low-k dielectrics [466]. It is imperative to identify the minimum plasma process conditions (such as the input power and bias voltage) to avoid damage of such materials. Meanwhile, plasma-enhanced chemical vapor deposition turns out to be extremely efficient in the coating of inner surfaces of porous alumina tubes by silica-like membranes [467]. Deposition rates of such coatings need to be kept reasonably low to obtain the microstructure suitable for hydrogen gas separation.

Appropriate tailoring the plasma process parameters (e.g., minimizing the ion bombardment) is a necessary prerequisite for the efficient polymerization of silicon organic gas-selective membranes on a microporous support [468]. Plasma treatment can also be used for modification of composition of functional coatings of microporous membranes. For example, exposure of silicon-coated Vycor® membranes results in the formation of silicon dioxide upper layer featuring a large number of pores ranging in size from microns down to a few tens of nanometers [469]. This modified membrane can be used for gas/vapor separation, reverse osmosis and the low molecular weight end of nanofiltration.

It is also worth mentioning that separation of red blood cells from whole blood was made a reality as a result of efficient transport of blood-plasma through the (gaseous) plasma-modified microporous membrane [470]. Finally, the examples discussed above show an outstanding potential of plasma-based nanotools in high-precision tailored synthesis and postprocessing of functional porous materials for a variety of advanced applications.

8
Further Examples, Conclusions, and Outlook

As is often said, all good things eventually come to an end, sooner or later. So is our success story about the use of properly chosen and developed plasma facilities for nanoscale synthesis of an overwhelming variety of nanoassemblies, ranging from zero-dimensional quantum dots to "bulk" nanofilms and three-dimensional nanostructures and nanopatterns. The most amazing thing is that the range of nanoassemblies and other relevant application in reality is, in fact, much wider than we managed to introduce in this monograph.

To emphasize this, in Section 8.1, we continue with providing, this time even more diverse, examples of successful applications of low-temperature plasmas for nanoscale processing. However, limited size of this monograph prevents us from trying to embrace the entire range of relevant processes, techniques, and tools that are based on or benefit in any way from the presence of low-temperature plasmas. Rather than doing this, we aim at making the reader belief that the actual possibilities of plasma nanotools are virtually infinite provided that one knows how to use them effectively and with the maximum efficiency.

One might have found our narration of the pathway from the development of advanced and sophisticated plasma sources to high-precision and highly controllable synthesis of various nanoassemblies a bit amazing in a sense that, although mentioned in several places in this monograph, many current and emerging problems and challenges of the plasma-aided nanofabrication have been sidestepped. One of the reasons we did that is not to reduce the enthusiasm many researchers might get (and potentially embark in the plasma-based nanotech research and development) after reading this book. However, we understand that real challengers always prefer to start from learning the existing problems to realize that the field is still full of opportunities, fundamental mysteries, and technical puzzles. This is one of the main reasons why we included a relatively brief discussion (Section 8.2) of some of the important problems and challenges that exist in this exciting research area and await their solutions.

As can be seen from Section 8.3, plasma-aided nanofabrication does have an enormous potential for the future development. In fact, the amazing possi-

Plasma-Aided Nanofabrication. Kostya (Ken) Ostrikov and Shuyan Xu
Copyright © 2007 WILEY-VCH Verlag GmbH & Co. KGaA, Weinheim
ISBN: 978-3-527-40633-3

bilities to synthesize exotic bits of nanoscale matter by using low-temperature plasmas disclosed in this monograph, is just the beginning of the even more fascinating and exciting journey toward the fully deterministic plasma-aided nanofabrication and eventually to atomic-precision plasma processing.

8.1
Further Examples of Plasma-Aided Nanofabrication

In most examples given in this monograph, silicon- and carbon-based nanoassemblies are considered. In fact, these two materials still remain in the spotlight of modern nanoscience and nanotechnology. Plasma-assisted nanoscale processing is not an exemption and silicon- and carbon-based materials have recently been among the most widely studied topics in the area.

We can state without any doubt that the results available to date are sufficient to support the importance of the proper choice of the right plasma, plasma source, and process parameters to achieve highly controlled synthesis of the desired nanosized objects. Below, we will outline some other representative examples of the application of nonequilibrium low-temperature plasmas to fabricate various nanoscale assemblies and comment on the most important aspects relevant to the choice of the plasma and process parameters.

In the first example, we consider the plasma-assisted growth of charged nanoclusters, which can be used as effective building units of highly crystalline TiO_2 nanofilms fabricated on unheated substrates by means of low-temperature plasma-assisted DC magnetron reactive sputtering deposition [69, 70]. It is interesting that this nanofabrication technique uses the charged cluster theory [37], which has been highly successful for the growth of high-quality crystalline films in a number of neutral- and ionized gas-based processes.

Amazingly, in an ionized gas environment, this nanocluster-based approach can become even more precise and effective owing to unique charging effects not available otherwise in neutral gas-based processes. What usually happens with the plasma-grown nanoclusters can best be understood by considering the three main stages of the process.

First, small (usually in the 1–3 nm size range) amorphous nanoclusters are generated in the gas phase. Second, through intense interactions with microscopic electron and ion currents and fluxes of numerous reactive species the nanoclusters acquire electric charge in the plasma.

The sign and the amount of the electric charge critically depend on the nanocluster size. However, how exactly do they depend on the size, still remains essentially an open question. This important issue of nanomaterials synthesis by using plasma-generated nanoclusters is outside the scope of this book.

Third, depending on the rates of nanocluster charge dissipation upon their landing on the solid surface, the structures assembled can be very different. Such rates strongly depend on the conductivity, polarizability, and some other properties of the substrate material.

Some of these factors have already been discussed in great details elsewhere [36, 37]. In the example of plasma-aided synthesis of nanocrystalline titanium dioxide films, the first requirement is essential for epitaxial recrystallization of ultrasmall TiO_2 nanoclusters on the surfaces of bulk nanostructured titanium dioxide. We recall that when the nanocluster size is very small (e.g., in the 1–3 nm range), their crystalline structure is not firmly fixed and somewhat strained. Moreover, as was discussed in Section 1.2, such small objects have significantly lower melting points, which is again to the benefit of faster contact epitaxy.

The second requirement (nanocluster charge) is essential in the processes where dense and compact nanostructures need to be fabricated. Indeed, like charges of the same-size nanoclusters prevent their uncontrollable agglomeration and/or coalescence in the gas phase. In this case, the nanoclusters can deposit onto and incorporate into the reconstructed surface one by one, without significantly affecting other clusters.

Charge dissipation during the nanocluster landing process is yet another phenomenon that needs better understanding. Depending on charge dissipation rates, different amounts of surface charge can build up and affect the deposition/epitaxy processes. If the dissipation rates are high, the surface most likely has a specific potential determined by the plasma process conditions. On the other hand, if such rates are low, uncontrolled accumulation of electric charge can result in significant variations of the surface potential, the latter affecting the deposition of the nanocluster building blocks and their mobility on charged plasma-exposed surfaces.

Barnes et al. recently used low-temperature $Ar+O_2$ plasmas under low-pressure conditions to assemble, in the ionized gas phase, TiO_2 nanoclusters in the 1–3 nm range [69, 70]. A carbon-coated copper TEM grids positioned at different distances (from 50 to 250 mm) from a titanium sputtering target was used to collect the nanoclusters, which were subsequently imaged by transmission electron microscopy.

Remarkably, the TiO_2 clusters collected closer (at 50 mm) to the sputtering target contained a much smaller numbers of atoms (i.e., less than 2 nm in size and containing ~400 atoms) than those captured upon the grid located 250 mm away from the target (which were larger than 3 nm in size and contained ~1400 atoms) [69, 70].

These experiments prove that clustering occurs in the plasma phase and lasts while the nanoclusters are transported through the vacuum chamber toward the deposition substrate. Therefore, by moving the solid substrate closer to the sputtering target, one can collect smaller nanocluster building units. On the other hand, when the substrate is located further from the nanocluster source, the TiO$_2$ films would incorporate larger structural units. Interestingly, smaller clusters produce faceted crystalline anatase film, whereas larger clusters result in mostly amorphous films [69, 70].

This result is consistent with the size-dependent phase composition of TiO$_2$ clusters, with the amorphous-to-crystalline transition in the size range between 1.5 and 3 nm [36, 69]. It is also relevant to mention that the value of the DC power does not significantly affect the cluster generation and film properties.

However, RF power strongly affects the nanocluster BU-based synthesis of TiO$_2$ films in TiCl$_4$+O$_2$ parallel-plate RF plasmas [68]. In this case, the nanocluster size, charge, and the film structure change with the RF power input variation [36]. At low input powers ($\sim$90 W), the TiO$_2$ films contain random fractal and agglomerated structures and show relatively poor adhesion to the substrate. However, at higher powers ($\sim$180 W), the films appear to be better ordered and denser; in this case the content of the crystalline phase is higher.

We emphasize that such different effects of the input power in the two plasma-assisted sputtering [69, 70] and in plasma-enhanced chemical vapor deposition [68] processes suggest a higher reactivity of titanium tetrachloride-based chemically active plasmas; such reactivity appears to be a major factor in the generation of titanium dioxide nanoclusters in the plasma.

In PECVD experiments, the sizes of the nanocluster building units also turn out quite different when the RF power is varied. Specifically, the sizes are 10–15 nm at 90 W RF input power and only 7–12 nm at 180 W [68]. More importantly, Barnes et al. observed nanocluster agglomeration at 90 W; here we comment that the latter could be avoided by running the plasma discharge at higher RF power densities.

Another interesting example of how different building units in different plasma environments can result in totally different properties of nanofilms has already been considered in Chapter 6. We recall that two different sorts of low-temperature plasmas (plasma spraying in a thermal plasma jet and RF magnetron sputtering deposition) performed very differently in the plasma-assisted fabrication of hydroxyapatite bioceramic coatings on Ti–6Al–4V orthopedic alloys widely used by the biomedical industry for hip joint and dental implants [36]. This is striking example when low-pressure nonequilibrium plasma promoted throughout this monograph, performs much better than a thermal plasma!

We recall that the thermal plasma spraying [366, 367] uses thermal plasma jets that carry micro-dispersed nano- or micron-sized powder particles toward the deposition substrates. However, as was shown in Chapter 6, despite apparent simplicity, moderate deposition rates, and relatively low cost, this technique is unlikely to meet current standards of biomedical industry.

Physically, this happens because under typical operating conditions micron-sized nanopowders usually land onto metal surfaces without any significant decomposition into smaller nano- and subnanometer-sized fragments. This is the reason why hydroxyapatite bioceramics prepared by the thermal plasma spray method feature so high amorphous phase content, rough and irregular surface morphology and also porous microstructure [366, 367].

As was discussed in Chapter 6, thermal plasma-synthesized hydroxyapatite coatings show disastrous *in vitro* and *in vivo* performance. On the other hand, by using low-temperature nonequilibrium plasmas in the argon-assisted RF magnetron sputtering process [358], it turns out possible to fabricate high-quality hydroxyapatite films, which are almost completely free from the above drawbacks. Moreover, this technique enables one to substantially improve the adhesive properties of the bioceramic-metal implant interface, which still remain a matter of a major concern in some thermal plasma-based methods.

In this example a thermally nonequilibrium plasma has a substantial competitive advantage over thermal plasmas. Thus, when considering the most appropriate plasma facility for nanofabrication of nanocrystalline hydroxyapatite bioceramics, one should choose a nonequilibrium plasma-based RF magnetron sputtering facility rather than a thermal plasma-based plasma spray.

There are many other examples of excellent performance of plasma-based nanofabrication tools and techniques [36]. One of the most interesting examples is the plasma-assisted deposition of superhard nc-$Al_xTi_{1-x}N/a$-Si_3N_4 nanocomposite material. This nanocomposite features a very high hardness (comparable to that of natural diamond) and this is why it is widely used by the tooling industry in applications demanding the extended lifetimes of superhard coatings, such as high-speed masonry drill bits [471]. In a sense, this class of nanocomposite materials is quite similar to the polymorphous films considered in Chapters 1 and 7. As reported in the original article [471], $Al_xTi_{1-x}N$ nanocrystals are enclosed by an amorphous Si_3N_4 matrix. It was also noted that the best hardness of the nanocomposite can be achieved when each of the $Al_xTi_{1-x}N$ nanocrystals is covered by a monolayer of amorphous silicon nitride [472]. A future challenge for the numerical simulation and experimental work is to explain this intricate and highly balanced growth, which can only happen when the growth rates of the $Al_xTi_{1-x}N$ nanocrystals and the amorphous silicon nitride are approximately the same. In fact, the rates of deposition of the amorphous phase should be just sufficient to coat the continuously growing $Al_xTi_{1-x}N$ nanocrystals with an amorphous monolayer [36].

It is worth emphasizing that the range of nanofabrication processes, devices, materials, nanostructures, tools, and techniques that use low-temperature plasmas, is virtually infinite and it would be a futile attempt to try to cover all of them in a single monograph. Below, by using a summary given in a recent review article [36], we will briefly discuss only a few emerging applications of the (mostly reactive) low-temperature plasmas in nanotechnology and give relevant references for further reading in this extremely hot and exciting area of research.

It is perhaps a common knowledge that chemically active plasmas are widely used in semiconductor micromanufacturing for ultrafine selective etching of submicron and nanofeatures (currently as small as ∼90 nm, with an expected shrinkage down to ∼40–50 nm in the near future) of ultra large scale integrated (ULSI) circuits and devices [88, 473]. Interestingly, reactive chemical etching processes greatly benefit from the presence of reactive species ("working units") in a plasma.

For example, reactive cations (positive ions) accelerated in the near-substrate sheath, are very useful working units for highly anisotropic reactive chemical etching. This ion-assisted etching can be very anisotropic and is carried out in the direction normal to the substrate surface; in this way one can substantially reduce the undesired overetch of sidewalls of the microelectronic features, which is almost unavoidable in many conventional wet chemical etching techniques. It is interesting to mention that hydrogen etching of graphite can be used to synthesize diamond films and carbon nanotips (considered in Chapter 4), even without supplying any hydrocarbon gas feedstock from the plasma [474].

It is also interesting to mention that one of the most recent important advances in this direction is to use prefabricated ordered arrays of nanostructures for pattern transfer to solid surfaces [36]. Related examples include synthesis of Si nanopillar arrays in inductively coupled plasmas by using a lithographic mask made of nickel nanodots [475] and nanoassembly of GaN-based nanorod light-emitting diodes using reactive ion etching, also in inductively coupled plasmas, through self-assembled nickel nanomasks [476].

It is worth emphasizing that a combination of high-precision plasma-based chemical etching with prepatterning (e.g., lithographic) techniques holds an outstanding promise for even further miniaturization of the features of microelectronic circuitry [36]. Recently, nanolithography enhanced by chemically active oxygen plasma etching enabled an unprecedented pattern transfer in nanofabrication of a grating with a period of only 23 nm on diamond substrates [477]. To our opinion, this precision even exceeds current possibilities of many existing electron beam lithography techniques without plasma-enhanced features.

As was already mentioned in Section 7.6, templates with nanosized pores can be effectively used for nanopattern transfer onto solid substrates. This nonlithographic approach has been recently used by Menon et al. to create intricate nanoporous patterns on a range of substrates by using nanoporous alumina as an etching mask [36, 478].

Another emerging area of plasma-aided nanofabrication, not detailed in Section 7.6, is a precision treatment of porous materials with micron- and nanosized pore features [36, 479, 480], such as high-aspect ratio trenches in silicon wafers, templated mesoporous materials (e.g., aluminophospates) [481], carbon-based molecular sieves for gas separation [454] (see also Section 7.6 of this monograph), microporous polymeric cytoscaffolds for bio-implants and tissue engineering, and several others [36].

The amazing variety of subnanometer-sized building units that can be generated in low-temperature plasmas is extremely favorable for the assembly of complex nano-objects that feature very different structural organization and elemental composition; moreover, without such building units this is quite difficult to implement by using other, nonplasma-based, methods [36].

Furthermore, low-temperature plasmas have been recently successfully used for highly efficient assembly of self-organized GaN quantum dot assemblies on ternary $Al_xGa_{1-x}N$ [482]; periodic two-dimensionally arrayed nanocolumns made of quarternary compound InGaAsP [483]; various low-dimensional semiconductor structures such as Ga_2O_3 and Al_2O_3 nanowires [484], germanium quantum dots on silicon oxide [86], and several others [36].

Another interesting feature of plasma-based nanofabrication processes is that plasma-generated species with sufficient energy can incorporate into various nanoassemblies giving them new and unusual properties. For example, silicon nanocrystals or quantum dots can be embedded in silicon nitride matrix to fabricate light-emitting diodes and other nanodevices [36, 485, 486].

Yet another attractive feature of plasma nanotools is an outstanding flexibility of low-temperature plasmas in the ability to synthesize and combine (in the same vacuum environment) quite different nanostructures in the same device. Recently, tetramethyl silane plasmas have been used for highly controllable, chemically active plasma-aided nanofabrication of nanocantilevers made of crystalline β-SiC nanowires in silicon oxide nanocones [487]. What is even more amazing is that by using plasma-aided nanofabrication, one can assemble new silicon–carbon hetero-nanostructures comprising multiwalled carbon nanotubes and tapered silicon nanowires [488]. These unique nanostructures herald a new era of integration of carbon and silicon nanofeatures in nanoelectronic devices [36].

As was mentioned in a recent review [36], low-temperature plasmas have always been, and will remain in the future, a very efficient medium for advanced synthesis and processing of novel nanopowder and nanofilm materials. Most recent examples include manufacturing of polymer-coated silver nanopowder in microwave plasma chemical vapor deposition [489] and sintering of nanocomposite Cu-TiB_2 powders [490]. On the other hand, chemically active plasmas of gas mixtures of vinyltrymethylsilane and carbon dioxide have been successfully used for synthesis of nanocomposite low-k SiCOH films [36, 491].

We hope that the above brief overview of the most recent impressive achievements in the area have further convinced the reader that plasma-assisted tools and techniques are indeed extremely useful in nanotechnology; we recall that this area has an estimated market projection exceeding US$2.5 trillion by the year 2014 [492]. Let us now briefly review, from the fundamental point of view, some benefits and challenges of using plasma nanotools.

8.2
On Benefits and Problems of Using Plasma Nanotools

In this section, we attempt to discuss some of the most obvious benefits of using (mostly thermally nonequilibrium) low-temperature plasmas for nanoassembly purposes. As we have seen from the numerous examples from the previous sections, the number of nanostructures, nanofilms, nanopatterns, and other nanoassemblies that can be synthesized in plasma environments is virtually infinite. In each of the processes, there are specific advantages and disadvantages of using the plasma. Below, we will focus on plasma-synthesized carbon-based nanoassemblies and begin our discussion from vertically aligned nanotip-like structures considered in Chapter 4.

By using advanced plasma nanofabrication facilities such as the IPANF of Chapter 3, one can control and perfect the main properties of the carbon nanostructures. The best example in this regard would be single-crystalline high-aspect ratio (needless to say, vertically aligned!) carbon nanotips discussed in detail in Sections 4.3 and 4.4. These and other similar structures of Chapter 4 have been synthesized quasideterministically by following the four main nanofabrication steps of the "cause and effect" approach [36].

However, many important questions still remain open. For example, what is the cause of such pronounced vertical alignment of carbon nanotubes and carbon nanotips grown by the plasma-assisted techniques? Within the framework of the building unit-based "cause and effect" approach, the answer critically depends on the dominant building units [36]. We recall that in the near-substrate plasma sheath the electric field **E** is nonuniform. Moreover, it is focused on the growth spots from the initial stage of the growth island formation and drives the charged building units facilitating their stacking into nanoassemblies directly from the gas phase [36].

Meanwhile, strong electric fields in the vicinity of vertically aligned nanostructures polarize neutral (e.g., nanocluster) building units and align them to stack in the nanoassemblies, also with a higher probability of approaching from the top. Thus, directed precipitation of both charged and neutral building units favors the assembly of the structures along the direction of the sheath electric field **E**, on the unit-by-unit basis.

However, stacking of neutral species with poor polarization response does not necessarily happen in the direction of **E**. In this case the electrostatic force creates stress nonuniformly distributed over the interface between the catalyst nanoparticle and the carbon nanotube structure grown through either the "tip" or "base" mechanisms [36]. As a result, the carbon material precipitation rates become different in the areas with different stress, and vertical growth is dynamically maintained [81]. However, how specifically the stress nonuniformity over the interface between the catalyst nanoparticle and the carbon nanotube translates into preferential stacking of the plasma-generated species, still remains an essentially open question and requires an adequate explanation.

Another open question is why carbon nanotubes grown via the "tip" mechanism usually appear multiwalled (the most common kind of reactive plasma-grown carbon nanotubes), whereas the base-grown nanotubes are usually single-walled [36, 72]. To understand this, one should bear in mind that nanoassembly of single-walled carbon nanotubes requires *very low* supply of carbon species to the surface of the Ni/Fe/Co catalyst nanoparticle; on top of that, higher hydrocarbons should be suppressed [493]. It appears that such conditions are quite difficult to meet in chemically active plasma-assisted processes [36].

From the above considerations, one can conclude that the low-temperature plasma-assisted techniques can have a competitive edge over other techniques (e.g., thermal chemical vapor deposition) when at least one of the following is required [36]:

- control of densities and fluxes of the required species in the gas phase, which is difficult (if possible at all) to do on the surface of catalyst nanoparticles; this is extremely difficult to implement in the growth process of nanotube-like structures;

- electric field-guided delivery of preselected (desired) building units directly to nanoassembly sites from the ionized gas phase;

- specific substrate activation by the fluxes of neutral and charged species from the gas phase;

- preferential growth and alignment direction, such as the direction of the DC electric field in the plasma sheath.

Nonetheless, the actual role of the plasma in the growth of carbon nanotubes is still a subject of intense discussions in the research community. For example, the most widely accepted view is that since the dissociation of precursor gas on the surface of catalyst particles is sufficient for carbon material precipitation, the ability of the plasma to dissociate the gas feedstock into reactive radicals should not be a factor in the growth of carbon nanotubes [72].

To this end, one should clearly understand the consequences of strong dissociation of the feedstock gas in the plasma, which is usually quite inefficient in a neutral gas. On one hand, extra radical species produced in the gas phase reduce the need of their production on the catalyzed surface. Therefore, this process can be energetically favorable; in fact, this can be one of the reasons of lower substrate temperatures required to synthesize carbon nanostructures by plasma-assisted methods [36].

Another argument in favor of the importance of the gas-phase decomposition of working gas is the possibility of plasma-assisted growth of multiwalled carbon nanotubes on catalyst-free silicon surfaces [36, 494, 495]. From now, it becomes clear that carbon-bearing species indeed can precipitate onto nanoassemblies directly from the ionized gas phase.

Another argument in favor of plasma-based fabrication techniques is that chemical vapor deposition of carbon nanotubes usually requires high substrate temperatures, which in most cases exceed 600 °C; in this case the nanotube growth on unheated substrates is quite unlikely [72]. By using low-temperature plasmas, growth temperatures of carbon nanotubes as low as 120 °C have been reported [75].

Moreover, as we have seen from Chapter 4, various carbon nanostructures can be grown in thermally nonequilibrium plasmas without any external heating of the substrate. In this case, neutral component of the weakly ionized plasma environment is responsible for heating of the metal catalyst layer [36].

Another interesting point to discuss is whether it is beneficial or not to pursue high rates of species generation. If the species are highly reactive, there is a high chance of plasma polymerization and formation of macromolecules and long polymeric chains. It is also common that highly reactive plasmas feature high densities of multiple chemically active species. On one hand, this feature is clearly undesirable for the synthesis of delicate nano-sized objects such as nanodots or nanowires, which requires highly controlled supply of small (usually in submonolayer quantities) numbers of building material.

On the other hand, if one aims to synthesize large numbers of nanoparticles, it is worthwhile to use highly reactive plasmas and capitalize on chemical activity of the gas feedstock and plasma polymerization processes. Thus, using highly reactive gases such as ethylene, acetylene, or propylene is in most cases beneficial to increase the yield of nanoparticle production. For example, nanoparticle generation and nanostructure growth is more efficient in acetylene-based than methane-based plasmas [36, 66].

Despite impressive recent advances in plasma-assisted nanofabrication, there are still many hurdles that need to be overcome before it can become fully deterministic, cost-efficient and as such widely accepted by the research community and industry. Indeed, there are numerous future challenges along this pathway. For example, the process accuracy critically depends on the ability to generate, monitor and manipulate, in a controlled manner, numerous building units in the ionized gas phase.

From this point of view, one of the main challenges is to develop suitable and highly efficient plasma sources equipped with advanced process control and monitoring equipment, including moveable, flexible and temperature-controlled substrate stages, film parameter monitors (e.g., ellipsometers), plasma species and fluxes diagnostic tools (e.g., ion flux monitors, mass-spectrometers, etc.) and other instrumentation.

One of the greatest challenges is to experimentally monitor the entire range of building units generated in the plasma. For example, presently existing diagnostic methods enable one to adequately trace and characterize the building units at the smaller- and larger size ends of the building unit continuum but often fall short in the size range of small nanoclusters. Interestingly, such nanoclusters with the properties very different from those of bulk materials, have been considered among the most important building units of various nanoassemblies, thin films, and bulk materials [36]. Future research efforts in the direction of plasma-grown nanocluster-assembled films should involve comprehensive modeling of the nanocluster formation and charging process in the experiments concerned [36, 68–70].

It is notable that an elegant approach [68–70] can be used for direct *in situ* sampling and characterization of small (in the few nm size range) nanoclusters in the ionized gas phase, which is one of the most difficult challenges nowadays because of these extremely small nanocluster sizes. Some other (mostly indirect) methods of nanocluster detection and characterization in the plasma have been presented elsewhere [51, 53].

Therefore, building units of this size range still remain a major challenge for the development of suitable *in situ* diagnostic techniques and control tools. Another challenge is to control the energetics of insertion of radical/molecular units into nanoassemblies by varying the electric fields in the plasma sheath; this can be achieved by optimizing the discharge and process parameters.

In Chapter 6, we considered an example of an outstanding performance of thermally nonequilibrium low-temperature plasma in fabrication of bioimplant-grade hydroxyapatite bioceramic coatings. Even though the results of RF-magnetron sputtering deposition appear to be superior compared to thermal plasma spray-based technique, there are still a few open questions that need to be answered before the process can be fully understood. For example, since high temperatures are usually beneficial for effective crystallization, then why in cold nonequilibrium plasmas one obtains highly crystalline HA coatings whereas in hot thermal plasmas the films appear amorphous? This is just another example of an issue that still awaits its conclusive answer. Perhaps, it can be answered by comparatively examining the properties of nonequilibrium and thermal plasmas by using the basic information of Section 1.4, or maybe more research should be done to elucidate the answer. We thus leave this issue to the interested researcher.

There are many other challenges, open questions, and unresolved problems related to plasma applications in nanotechnology. One of such problems is charging of ultrasmall nanoclusters already mentioned in the introductory Chapter 1. In this case, conventional "macroscopic" models commonly used to describe electrostatic charging of larger (e.g., ∼100 nm-sized) nanoparticles, become invalid and there is a strong current demand for the development of more reliable and accurate charging models based on *ab initio* quantum mechanic treatment. This can have important practical implications. For example, accurate knowledge of the nanocluster charge can help one to explain the actual role of the low-temperature plasmas in the nanocluster-assembled highly crystalline films, such as TiO_2.

The number of existing problems, puzzles, and challenges is indeed very large and each new nanostructure and new plasma-based process generate even more to resolve. However, these problems are in most cases fascinating and become even more exciting when the knowledge on plasma-made nanoassemblies broadens and deepens. After a number of years experience in this field, we guarantee that anyone who will take the challenge will enjoy it. This research area is full of exciting and challenging opportunities that can potentially lead to major discoveries. It is our advice to a undergraduate and graduate students and junior researchers not to miss this opportunity! On this note we will now proceed to the final remarks and outlook for the future.

8.3
Outlook for the Future and Concluding Remarks

As was mentioned earlier [36], the number of nanoassemblies, as well as of approaches to fabricate them, is virtually endless. One should be very clear that in most cases nanofabrication techniques are process- and nanoassembly specific. However, low-temperature thermally nonequilibrium plasmas of our interest in this monograph indeed have a number of attractive features that make such plasmas a versatile nanofabrication tool of the "nano-age."

These features can be examined by using a conceptual framework, such as the building unit-based "cause and effect" (this approach encompasses the generation of the required building units in the ionized gas phase, plasma-assisted preparation of and transport of the BUs toward the deposition surfaces, and, finally, the stacking of the species into the required nanoassembly patterns [36]) or any other suitable practical approach (e.g., "trial and error" with the number of choices limited by some relevant practical considerations such as suitability of the available plasma source or nanofabrication facility).

In either way, one can elucidate many unique features of the selected reactive plasma-based nanofabrication methods and techniques.

Looking beyond the applications of low-temperature plasmas at nanoscales, we note that the plasma intrinsically is prone of a variety of atomic and ionic building units, and as such, holds a promise for the future-generation processing of matter at *atomic* scales [36]. To this end, recent reports on a high-precision, highly conformal plasma-aided atomic layer deposition sounds very optimistic and encouraging [496].

We hope that this topic will attract the interest and critical comments of a wide interdisciplinary community and result in dedicated experiments to establish direct and conclusive correlations between the abundance of very specific building units in the ionized gas phase, the parameters of transition areas between the plasma bulk and substrates, the energetics of the building units upon deposition/stacking, and the key properties of the nanoassemblies and translate these correlations into clear technical specifications and parameters of plasma processes and nanofabrication facilities. In the examples discussed above, the choice of the appropriate plasma and plasma sources and processes was in most cases adequate although may not necessarily ideal.

In this monograph, one can find numerous practical recipes for the optimal choice of the plasma sources and the plasma/process parameters to fabricate a range of practically important nanoassemblies. However, we did not even aim to provide exhaustive recipes for a very broad range of low temperature plasma-assisted nanoassembly processes. More importantly, the issues and approaches introduced in this monograph should not only be perceived as a stand-alone unitary concept (based on the plasmas and plasma sources *per se*) but as a multidimensional conceptual framework that can be used not only for plasma-based nanofabrication, but also for in a broader nanoscience and nanotechnology context.

Finally, we believe that low-temperature (not only thermally nonequilibrium plasmas of our interest here) plasma-aided nanofabrication has tantalizing prospects to become one of the benchmark technologies of the "nano-age" and beyond.

References

1 *Advances in Low Temperature RF Plasmas: Basis for Process Design*, edited by T. Makabe (Elsevier, Amsterdam, 2002).

2 *Ionized Physical Vapor Deposition*, edited by J. A. Hopwood (Academic Press, San Diego, 2001).

3 *Handbook of Plasma Immersion Ion Implantation and Deposition*, edited by A. Anders (Wiley Canada, Montreal, 2000).

4 *Dusty Plasmas: Physics, Chemistry, and Technological Impacts in Plasma Processing*, edited by A. Bouchoule (Wiley, Chichester, UK, 1999).

5 F. F. Chen and J. P. Chang, *Principles of Plasma Processing: A Lecture Course* (Kluwer, Amsterdam, 2002).

6 *Nanoparticles and Nanostructured Films: Preparation, Characterization and Applications*, edited by J. H. Fendler (Wiley-VCH, Weinheim, 1998).

7 *Nano-Architectured and Nanostructured Materials: Fabrication, Control and Properties*, edited by Y. Champion and H.-J. Fecht (Wiley-VCH, Weinheim, 2004).

8 G. Schmid, *Nanoparticles: From Theory to Application* (Wiley, New York, 2004).

9 S. Reich, *Carbon Nanotubes: Basic Concepts and Physical Properties* (Wiley, New York, 2004).

10 *Encyclopedia of Nanoscience and Nanotechnology*^{TM}, edited by H. S. Nalwa (American Scientific Publishers, New York, 2004).

11 S. V. Vladimirov, K. Ostrikov, and S. Samarian, *Physics and Applications of Complex Plasmas* (Imperial College Press, Singapore, London, 2005).

12 International Technology Roadmap for Semiconductors, http://www.itrs.net

13 S. Iijima, Nature (London) **354**, 56 (1991).

14 F. F. Chen, *Introduction to Plasma Physics and Controlled Fusion* (Plenum, New York, 1984).

15 M. A. Lieberman and A. J. Lichtenberg, *Principles of Plasma Discharges and Materials Processing* (Wiley, New York, 1994).

16 A. Fridman and L. A. Kennedy, *Plasma Physics and Engineering* (Taylor & Francis, New York, 2004).

17 D. A. Gurnett and A. Bhattacharjee, *Introduction to Plasma Physics With Space and Laboratory Applications* (Cambridge University Press, Cambridge, New York, 2005).

18 A. von Engel, *Ionized Gases* (Clarendon, Oxford, UK, 1955).

19 A. M. Howatson, *An Introduction to Gas Discharges* (Pergamon, Oxford, UK, 1976).

20 Y. P. Raizer, *Gas Discharge Physics* (Springer, Berlin, 1991).

21 I. Levchenko, K. Ostrikov, M. Keidar, and S. Xu, J. Appl. Phys. **98**, 064304 (2005).

22 I. Levchenko, K. Ostrikov, M. Keidar, and S. Xu, Appl. Phys. Lett. **89**, 033109 (2006).

23 E. Tam, I. Levchenko, and K. Ostrikov, J. Appl. Phys. **100**, 036104 (2006).

24 I. Levchenko, K. Ostrikov, and E. Tam, Appl. Phys. Lett. **89**, 223108 (2006).

25 M. C. Roco, S. Williams, and A. P. Alivisatos, *Nanotechnology Research Directions: Vision for Nanotechnology Research and Development in the next Decade* (Kluwer, Amsterdam, 1999). See also: US National Nanotechnology Initiative, http://www.nano.gov.

26. R. P. Feynman, *There is Plenty of Room at the Bottom*, Paper presented at the American Physical Society Annual Meeting, 29 December 1959. Reprinted in: *Miniaturization*, edited by H. D. Gilbert (Reinhold, New York, 1961). See also http://www.zyvex.com/nanotech/feynman.html.

27. E. Harvey, B. Innes, G. Smith, C. Jagadish, M. Barber, M. Lu, S. Longstaff, F. Caruso, and R. Rose, *Options for a National Nanotechnology Strategy*, Report of the National Nanotechnology Strategy Taskforce (Ministry of Industry, Tourism and Resources of Australia, Canberra, 2006).

28. M. J. Pitkethly, Nano Today **6**, 36 (2003).

29. C. P. Poole and F. J. Owens, *Introduction to Nanotechnology* (Wiley, New York, 2003).

30. V. Shchukin, N. N. Ledentsov, and D. Bimberg, *Epitaxy of Nanostructures* (Springer, Berlin/Heidelberg, 2003).

31. P. Mulvaney, MRS Bulletin **6**, 1009 (2001).

32. M. S. Dresselhaus, G. Dresselhaus, and P. C. Eklund, *Science of Fullerenes and Carbon Nanotubes* (Academic Press, New York, 1996).

33. G. Seifert, Nature Mater. **3**, 77 (2004).

34. A. Kasuya et al., Nature Mater. **3**, 99 (2004).

35. B. Gilbert, F. Huang, H. Zhang, G. Waychunas, and J. F. Banfield, Science **305**, 651 (2004).

36. K. Ostrikov, Rev. Mod. Phys. **77**, 489 (2005).

37. N. M. Hwang and D. Y. Kim, Int. Mater. Rev. **49**, 171 (2004).

38. N. Chaabane, A. V. Kharchenko, H. Vach, and P. Roca i Cabarrocas, New J. Phys. **3**, 37.1 (2003).

39. N. Chaabane, P. Roca i Cabarrocas, H. Vach, J. Non-Cryst. Solids **338–340**, 51 (2004).

40. A. Fontcuberta i Morral and P. Roca i Cabarrocas, Thin Solid Films **383**, 161 (2001).

41. A. Fontcuberta i Morral, P. Roca i Cabarrocas, and C. Clerc, Phys. Rev. B **69**, 125307 (2004).

42. Y. Poissant, P. Chatterjee, and P. Roca i Cabarrocas, J. Appl. Phys. **94**, 7305 (2003).

43. P. Roca i Cabarrocas, S. Hamma, S. N. Sharma, G. Viera, E. Bertran, and J. Costa, J. Non-Cryst. Sol. **227–230**, 871 (1998).

44. P. Roca i Cabarrocas, N. Chaabane, A. V. Kharchenko, and S. Tchakarov, Plasma Phys. Control. Fusion **46**, B235 (2004).

45. V. Suendo, A. Kharchenko, and P. Roca i Cabarrocas, Thin Solid Films **451–452**, 259 (2004).

46. G. Viera, M. Mikikian, E. Bertran, P. Roca i Cabarrocas, and L. Boufendi, J. Appl. Phys. **92**, 4684 (2002).

47. S. Thompson, C. R. Perrey, C. B. Carter, T. J. Belich, J. Kakalios, and U. Kortshagen, J. Appl. Phys. **97**, 034310 (2005).

48. M. Tanda, M. Kondo, and A. Matsuda, Thin Solid Films **427**, 33 (2003).

49. M. Shiratani, S. Maeda, K. Koga, and Y. Watanabe, Jpn. J. Appl. Phys., Part 1 **39**, 287 (2000).

50. M. Shiratani, K. Koga, and Y. Watanabe, Thin Solid Films **427**, 1 (2003).

51. L. Boufendi, W. Stoffels, and E. Stoffels, in *Dusty Plasmas: Physics, Chemistry and Technological Impacts in Plasma Processing*, edited by A. Bouchoule (Wiley, New York, 1999), p.181–303.

52. S. V. Vladimirov and K. Ostrikov, Phys. Repts **393**, 175 (2004), and the references therein.

53. R. Ghidini, C. H. J. M. Groothuis, M. Sorokin, G. M. W. Kroesen, and W. W. Stoffels, Plasma Sources Sci. Technol. **13**, 143 (2004).

54. C. Hollenstein, Plasma Phys. Control. Fusion **42**, R93 (2000).

55. A. Bapat, C. R. Perrey, S. A. Campbell, C. B. Carter, and U. Kortshagen, J. Appl. Phys. **94**, 1969 (2003).

56. U. V. Bhandarkar, M. T. Swihart, S. L. Girshik, and U. Kortshagen, J. Phys. D: Appl. Phys. **33**, 2731 (2000).

57. S. M. Suh, S. L. Girshik, U. R. Kortshagen, and M. R. Zachariah, J. Vac. Sci. Technol. A **21**, 251 (2003).

58. A. Bapat, C. Anderson, C. R. Perrey, C. B. Carter, S. A. Campbell, and U. Kortshagen, Plasma Phys. Control. Fusion **46**, B97 (2004).

59. Z. Shen, T. Kim, U. Kortshagen, P. McMurry, and S. Campbell, J. Appl. Phys. **94**, 2277 (2003).

60. L. Mangolini, E. Thimsen, and U. Kortshagen, Nano Lett. **5**, 655 (2005).
61. A. S. Barnard and P. Zapol, J. Chem. Phys. **121**, 4276 (2004).
62. D. M. Gruen, P. C. Redfern, D. A. Horner, P. Zapol, and L. A. Curtis, J. Phys. Chem B **103**, 5459 (1999).
63. D. M. Gruen, MRS Bulletin **6**, 771 (2001).
64. S. Stoykov, C. Eggs, and U. Kortshagen, J. Phys. D: Appl. Phys. **34**, 2160 (2001).
65. G. Gebauer and J. Winter, New J. Phys. **5**, 38.1 (2003).
66. S. Hong, I. Stefanovic, J. Berndt, and J. Winter, Plasma Sources Sci. Technol. **12**, 46 (2003).
67. E. Kovacevic, I. Stefanovic, J. Berndt, and J. Winter, J. Appl. Phys. **93**, 2924 (2003).
68. M. C. Barnes, A. R. Gerson, S. Kumar, L. Green, and N. M. Hwang, Thin Solid Films **436**, 181 (2003).
69. M. C. Barnes, A. R. Gerson, S. Kumar, and N. M. Hwang, Thin Solid Films **446**, 29 (2004).
70. M. C. Barnes, S. Kumar, L. Green, N. M. Hwang, and A. R. Gerson, Surf. Coat. Technol. **190**, 321 (2005).
71. D. Hash, and M. Meyyappan, J. Appl. Phys. **93**, 750 (2003).
72. M. Meyyappan, L. Delzeit, A. Cassell, and D. Hash, Plasma Sources Sci. Technol. **12**, 205 (2003), and the references therein.
73. C. L. Tsai, C. W. Chao, C. L. Lee, and H. C. Shih, Appl. Phys. Lett. **74**, 3462 (1999).
74. C. L. Tsai, C. F. Chen, and L. K. Wu, Appl. Phys. Lett. **81**, 721 (2002).
75. S. Hofmann, C. Dukati, J. Robertson, and B. Kleinsorge, Appl. Phys. Lett. **83**, 135 (2003).
76. K. B. K. Teo, D. B. Hash, R. G. Lacerda, N. L. Rupesinghe, M. S. Bell, S. H. Dalal, D. Bose, T. R. Govindan, B. A. Cruden, M. Chhowalla, G. A. J. Amaratunga, M. Meyyappan, and W. I. Milne, Nano Lett. **4**, 921 (2004).
77. C. Bower, W. Zhu, S. Jin, and O. Zhou, Appl. Phys. Lett. **77**, 830 (2000).
78. L. Delzeit, I. McIninch, B. A. Cruden, D. Hash, B. Chen, J. Han, and M. Meyyappan, J. Appl. Phys. **91**, 6027 (2002).
79. M. Chhowalla, K. B. K. Teo, C. Ducati, N. L. Rupersinghe, G. A. J. Amaratunga, A. C. Ferrari, D. Roy, J. Robertson, and W. I. Milne, J. Appl. Phys. **90**, 5308 (2001).
80. V. I. Merkulov, A. V. Melechko, M. A. Guillorn, D. H. Lowndes, and M. L. Simpson, Appl. Phys. Lett. **76**, 3555 (2000).
81. V. I. Merkulov, A. V. Melechko, M. A. Guillorn, D. H. Lowndes, and M. L. Simpson, Appl. Phys. Lett. **79**, 2970 (2001).
82. A. V. Melechko, V. I. Merkulov, T. E. McKnight, M. A. Guillorn, K. L. Klein, D. H. Lowndes, and M. L. Simpson, J. Appl. Phys. **97**, 041301 (2005).
83. J. F. AuBuchon, L-H.Chen, and S. Jin, J. Phys. Chem. B **109**, 6044 (2005).
84. I. Levchenko and K. Ostrikov, J. Phys. D: Appl. Phys., **40**, 2308 (2007).
85. M. Yan, H. T. Zhang, E. J. Widjaja, and R. P. H. Chang, J. Appl. Phys. **94**, 5240 (2003).
86. J. Shieh, T. S. Ko, H. L. Chen, B. T. Dai, and T. C. Chu, Chem. Vapor Depos. **10**, 265 (2004).
87. Y. J. T. Lii, *Etching*, in *ULSI Technology*, edited by C. Y. Chang and S. M. Sze (McGraw-Hill, New York, 1996), pp. 329–370.
88. G. S. Oehrlein, *Plasma Processing of Electronic Materials* (Springer, Berlin, 2003).
89. K. Ostrikov, Sing. J. Phys. **19**, 1 (2003), and the references therein.
90. H. Sugai, T. H. Ahn, I. Ghanashev, M. Goto, M. Nagatsu, K. Nakamura, K. Suzuki, and H. Toyoda, Plasma Phys. Control. Fusion **39**, A445 (1997).
91. S. Xu, K. N. Ostrikov, W. Luo, and S. Lee, J. Vac. Sci. Technol. A, **18**, 2185 (2000).
92. K. N. Ostrikov, S. Xu, and M. Y. Yu, J. Appl. Phys. **88**, 2268 (2000).
93. S. Xu, K. N. Ostrikov, Y. Li, E. L. Tsakadze, and I. R. Jones, Phys. Plasmas **8**, 2549 (2001).
94. K. N. Ostrikov, S. Xu, and A. B. M. Shafiul Azam, J. Vac. Sci. Technol. A **20**, 251 (2002).
95. K. N. Ostrikov, I. B. Denysenko, E. L. Tsakadze, S. Xu, and R. G. Storer, J. Appl. Phys. **92**, 4935 (2002).

96 K. Ostrikov, E. Tsakadze, N. Jiang, Z. Tsakadze, J. Long, R. Storer, and S. Xu, IEEE Trans. Plasma Sci. **30**, 128 (2002).

97 K. Ostrikov, E. Tsakadze, S. Xu, S. V. Vladimirov, and R. Storer, Phys. Plasmas **10**, 1146 (2003).

98 M. Tuszewski, Phys. Plasmas **5**, 1198 (1998).

99 H. Sugai, K. Nakamura, and K. Suzuki, Jpn. J. Appl. Phys. Part 1 **33**, 2189 (1994).

100 K. Suzuki, K. Nakamura, H. Ohkubo, and H. Sugai, Plasma Sources Sci. Technol. **7**, 13 (1998).

101 M. Tuszewski, IEEE Trans. Plasma Sci. **27**, 68 (1999).

102 I. El-Fayoumi and I. R. Jones, Plasma Sources Sci. Technol. **7**, 162 (1998).

103 I. El-Fayoumi and I. R. Jones, Plasma Sources Sci. Technol. **7**, 179 (1998).

104 J. Hopwood, Plasma Sources Sci. Technol. **1**, 1009 (1992).

105 J. H. Keller, Plasma Sources Sci. Technol. **5**, 166 (1996).

106 J. H. Keller, Plasma Phys. Control. Fusion **39**, A437 (1997).

107 V. A. Godyak, R. B. Piejak, B. M. Alexandrovich, and V. I. Kolobov, Phys. Plasmas **6**, 1804 (1999).

108 Y. Wu and M. A. Lieberman, Appl. Phys. Lett. **72**, 777 (1998).

109 C. Chakrabarty, Ph.D. thesis, The Flinders University of South Australia (1996).

110 U. Kortshagen, N. D. Gibson, and J. E. Lawler, J. Phys. D.: Appl. Phys. **29**, 1224 (1996).

111 R. B. Piejak and V. A. Godyak, Appl. Phys. Lett. **76**, 2188 (2000).

112 N. A. Azarenkov and K. N. Ostrikov, Phys. Reports **308**, 333 (1999).

113 T. H. Chung, H. J. Yoon, and D. C. Seo, J. Appl. Phys. **86**, 3536 (1999).

114 E. Tatarova, F. M. Dias, C. M. Ferreira, and A. Ricard, J. Appl. Phys. **85**, 49 (1999).

115 E. Tatarova, F. M. Dias, C. M. Ferreira, V. Guerra, J. Loureiro, E. Stoykova, I. Ghanashev, and I. Zhelyazkov, J. Phys. D: Appl. Phys. **30**, 2663 (1997).

116 B. Gordiets, C. M. Ferreira, J. Nahorny, D. Pagnon, M. Touzeau, and M. Vialle, J. Phys. D: Appl. Phys. **29**, 1021 (1996).

117 M. J. Baldwin, G. A. Collins, M. P. Fewell, S. C. Haydon, S. Kumar, K. T. Short, and J. Tendys, Jpn. J. Appl. Phys. **36**, 4941 (1997).

118 J. R. Tynan, J. Appl. Phys. **86**, 5356 (1999).

119 I. P. Herman, *Optical Diagnostics for Thin Film Processing* (Academic Press, New York, 1996).

120 *CRC Handbook of Chemistry and Physics*, edited by D. R. Lide, 78th edition (CRC Press, New York, 1997).

121 R. W. B. Pearse and A. G. Gaydon, *The Identification of Molecular Spectra*, 4th edition (Wiley, New York, 1976).

122 I. M. El-Fayoumi, I. R. Jones, and M. M. Turner, Plasma Sources Sci. Technol. **7**, 3082 (1998).

123 K. Suzuki, K. Nakamura, H. Ohkubo, and H. Sugai, Plasma Sources Sci. Technol. **7**, 13 (1998).

124 G. Cunge, B. Crowley, D. Vender, and M. M. Turner, Plasma Sources Sci. Technol. **8**, 576 (1999).

125 M. M. Turner and M. A. Lieberman, Plasma Sources Sci. Technol. **8**, 313 (1999).

126 E. L. Tsakadze, K. N. Ostrikov, Z. L. Tsakadze, N. Jiang, R. Ahmad, and S. Xu, Int. J. Mod. Phys. B **16**, 1143 (2002).

127 A. Garscadden and R. Nagpal, Plasma Sources Sci. Technol. **4**, 268 (1995).

128 S. Ashida, C. Lee, and M. A. Lieberman, J. Vac. Sci. Technol. A **13**, 2498 (1995).

129 B. Gordiets, C. M. Ferreira, M. J. Pinheiro, and A. Ricard, Plasma Sources Sci. Technol. **7**, 363 (1998).

130 V. Guerra, M. Pinheiro, B. Gordiets, J. Loureiro, and C. M. Ferreira, Plasma Sources Sci. Technol. **6**, 220 (1997).

131 R. Nagpal and A. Garscadden, Chem. Phys. Lett. **232**, 211 (1994).

132 B. Gordiets, C. M. Ferreira, V. Guerra, J. Loureiro, J. Nahorny, D. Pagnon, M. Touzeau, and M. Vialle, IEEE Trans. Plasma Sci. **23**, 750 (1995).

133 V. Guerra and J. Loureiro, Plasma Sources Sci. Technol. **6**, 361 (1997).

134 V. Vahedi, M. A. Lieberman, G. DiPeso, T. D. Rognlien, and D. Hewett, J. Appl. Phys. **78**, 1446 (1995).

135 V. E. Golant, A. P. Zhilinskii, and I. E. Sakharov, *Fundamentals of Plasma Physics* (Wiley, New York, 1980).

136 R. A. Stewart, P. Vitello, D. B. Graves, E. F. Jaeger, and L. A. Berry, Plasma Sources Sci. Technol. **4**, 36 (1995).

137 N. A. Azarenkov, I. B. Denysenko, A. V. Gapon, and T. W. Johnston, Phys. Plasmas **8**, 1467 (2001).

138 I. B. Denysenko, A. V. Gapon, N. A. Azarenkov, K. N. Ostrikov, and M. Y. Yu, Phys. Rev. E **65**, 046419 (2002).

139 H. M. Wu, B. W. Yu, A. Krishnan, M. Li, Y. Yang, J. P. Yan, and D. P. Yuan, IEEE Trans. Plasma Sci. **25**, 776 (1997).

140 I. Peres, M. Fortin, and J. Margot, Phys. Plasmas **3**, 1754 (1996).

141 L. M. Biberman, V. S. Vorob'ev, and I. T. Yakubov, *Kinetics of Non-Equilibrium Low-Temperature Plasma* (Nauka, Moscow, 1982), in Russian.

142 *Plasma Etching: An Introduction*, edited by D. M. Manos and D. L. Flamm (Academic Press, New York, 1989).

143 *Plasma-Surface Interactions and Processing of Materials*, edited by O. Auciello et al. (Kluwer, Boston, 1990).

144 K. N. Ostrikov, S. Xu, and S. Lee, Phys. Scripta **62**, 189 (2000).

145 E. L. Tsakadze, K. N. Ostrikov, S. Xu, R. G. Storer, and H. Sugai, J. Appl. Phys. **91**, 1804 (2002).

146 E. L. Tsakadze, K. Ostrikov, Z. L. Tsakadze, S. V. Vladimirov, and S. Xu, Phys. Plasmas **11**, 3915 (2004).

147 E. L. Tsakadze, K. Ostrikov, Z. L. Tsakadze, and S. Xu, J. Appl. Phys. **97**, 013301 (2005).

148 J. Hopwood, C. R. Guarnieri, S. J. Whitehair, and J. J. Cuomo, J. Vac. Sci. Technol. A **11**, 152 (1993).

149 Y. Wu and M. A. Lieberman, Plasma Sources Sci. Technol. **9**, 210 (2000).

150 S. S. Kim, H. Y. Chang, and C. S. Chang, Appl. Phys. Lett. **77**, 492 (2000).

151 S. Xu, K. Ostrikov, and E. L. Tsakadze, *An Apparatus and Method for Generating Uniform Plasmas*, International Patent Application PCT/SG2004/000210 (filed July 2003).

152 J. D. Jackson, *Classical Electrodynamics* (Wiley, New York, 1975).

153 R. B. Piejak and V. A. Godyak, Appl. Phys. Lett. **76**, 2188 (2000).

154 Z. L. Tsakadze, K. Ostrikov, E. L. Tsakadze, and S. Xu, J. Vac. Sci. Technol. A **23**, 440 (2005).

155 E. L. Tsakadze, K. N. Ostrikov, S. Xu, J. Long, N. Jiang, Z. L. Tsakadze, M. Y. Yu, and R. Storer, Phys. Scripta **64**, 360 (2001).

156 M. Chan, S. Xu, N. Jiang, J. Long, and C. H. Diong, Int. J. Mod. Phys. B **16**, 254 (2002).

157 S. Xu, K. Ostrikov, J. Long, and S. Y. Huang, Vacuum **80**, 621 (2006).

158 Y. Shiratori, H. Hiraoka, Y. Takeuchi, and M. Yamamoto, Appl. Phys. Lett. **82**, 2485 (2003).

159 G. Gapellini, M. de Seta, C. Spinella, and F. Evangelisti, Appl. Phys. Lett. **82**, 1772 (2003).

160 Z. L. Tsakadze, K. Ostrikov, and S. Xu, Surf. Coat. Technol. **191/1**, 49 (2005).

161 Z. L. Tsakadze, K. Ostrikov, R. Storer, and S. Xu, J. Metastable Nanocryst. Mater. **23–25**, 297 (2005).

162 Z. L. Tsakadze, K. Ostrikov, J. D. Long, and S. Xu, Diam. Relat. Mater. **13**, 1923 (2004).

163 L. Nilsson, O. Groening, O. Kuettel, P. Groening, and L. Schlapbach, J. Vac. Sci. Technol. **20**, 326 (2002).

164 A. Ilie, A. C. Ferrari, T. Yagi, S. E. Rodil, J. Robertson, E. Barborini, and P. Milani, J. Appl. Phys. **90**, 2024 (2001).

165 J. Perrin, M. Shiratani, P. Kae-Nune, H. Videlot, J. Jolly, and J. Guillion, J. Vac. Sci. Technol. **16**, 278 (1998).

166 I. B. Denysenko, S. Xu, P. P. Rutkevych, J. D. Long, N. A. Azarenkov, and K. Ostrikov, J. Appl. Phys. **95**, 2713 (2004).

167 K. Ostrikov, Z. Tsakadze, I. Denysenko, P. P. Rutkevych, J. D. Long, and S. Xu, Contr. Plasma Phys. **45**, 514 (2005).

168 K. B. K. Teo, M. Chhowalla, G. A. J. Amaratunga, W. I. Milne, D. G. Hasko, G. Pirio, P. Legagneux, F. Wyczisk, and D. Pribat, Appl. Phys. Lett. **79**, 1534 (2001).

169 D. Reznik, C. H. Olk, D. A. Neumann, and J. R. D. Copley, Phys. Rev. B **52**, 116 (1995).

170 J. W. Ager, D. K. Veirs, and G. M. Rosenblatt, Phys. Rev. B **43**, 8 (1991).

171 V. Ligatchev, Phys. B **337**, 333 (2003).

172 Y. J. Li, Z. Sun, S. P. Lau, G. Y. Chen, and B. K. Tay, Appl. Phys. Lett. **79**, 11 (2001).

173 J. D. Long, S. Xu, S. Y. Huang, P. P. Rutkevych, M. Xu, and C. H. Diong, IEEE Trans. Plasma Sci. **33**, 240 (2005).

174 K. Ostrikov, J. D. Long, P. P. Rutkevych, and S. Xu, Vacuum **80**, 1126 (2006).

175 A. C. Ferrari and J. Robertson, Phys. Rev. B **64**, 075414 (2001).

176 C. Thomsen and S. Reich, Phys. Rev. Lett. **85**, 5214 (2000).

177 M. Ge and K. Sattler, Chem. Phys. Lett. **220**, 192 (1994).

178 A. Krishnan, E. Dujardin, M. Treacy, J. Hugdahl, S. Lynum, and T. Ebbesen, Nature **388**, 451 (1997).

179 M. Endo, K. Takeuchi, K. Kobori, K. Takahashi, H. W. Kronto, and A. Sarkar, Carbon **33**, 873 (1995).

180 K. Sattler, Carbon **33**, 915 (1995).

181 H. L. Chua and S. Xu, *Ab initio Density Functional Theory Simulations of Single-Crystalline Carbon Nanotip Structures*, Internal Report 2678/2005, National Institute of Education, Singapore.

182 P. Hohenberg and W. Kohn, Phys. Rev. **136**, B864 (1964).

183 W. Kohn and L. J. Sham, Phys. Rev. **140**, A1133 (1965).

184 S. Lunqvist and N. H. March, *The Theory of the Homogenous Electron Gas* (Plenum, New York, 1983).

185 http://www.accelrys.com/mstudio/

186 B. Delley, J. Chem. Phys. **113**, 7756 (2000).

187 R. A. Serway and R. J. Beichner, *Physics for Scientists and Engineers with Modern Physics*, 4th edition (Saunders, Philadelphia, PA, 1996).

188 V. P. Veedu, A. Cao, X. Li, K. Ma, C. Soldano, S. Kar, P. M. Ajayan, and M. N. Ghasemi-Nejhad, Nature Mater. **5**, 457 (2006).

189 O. A. Louchev, Y. Sato, and H. Kanda, Appl. Phys. Lett. **80**, 2752 (2002).

190 M. Terrones, N. Grobert, J. Olivares, J. P. Zhang, H. Terrones, K. Kordatos, W. K. Hsu, J. P. Hare, P. D. Townsend, K. Prassides, A. K. Cheetham, H. W. Kroto, and D. R. M. Walton, Nature **388**, 52 (1997).

191 M. Keidar, Y. Raitses, A. Knapp, and A. M. Waas, Carbon **44**, 1022 (2006).

192 S.-H. Jeong, H.-Y. Hwang, K.-H. Lee, and Y. Jeong, Appl. Phys. Lett. **78**, 2052 (2001).

193 K. Bradley, J.-C. P. Gabriel, A. Star, and G. Grüner, Appl. Phys. Lett. **83**, 3821 (2003).

194 Z. Yu, C. Rutherglen, and P. J. Burke, Appl. Phys. Lett. **88**, 233115 (2006).

195 J. Zhu, H. Peng, F. Rodriguez-Macias, L. J. Margrave, N. V. Khabashesku, M. A. Imam, K. Lozano, and V. E. Barrera, Adv. Funct. Mater. **14**, 643 (2004).

196 L.-H. Wong, Y. Zhao, G. Chen, and A. T. Chwang, Appl. Phys. Lett. **88**, 183107 (2006).

197 E. Dujardin, V. Derycke, M. F. Goffman, R. Lefèvre, and J. P. Bourgoin, Appl. Phys. Lett. **87**, 193107 (2005).

198 Y.-T. Kim, Y. Ito, K. Tadai, T. Mitani, U.-S. Kim, H.-S. Kim, B.-W. Cho, Appl. Phys. Lett. **87**, 234106 (2005).

199 S. Shenogin, A. Bodapati, L. Xue, R. Ozisik, and P. Keblinski, Appl. Phys. Lett. **85**, 2229 (2004).

200 P. W. Chiu, G. S. Duesberg, U. Dettlaff-Weglikowska, and S. Roth, Appl. Phys. Lett. **80**, 3811 (2002).

201 E. A. Whitsitt and A. R. Barron, Nano Lett. **3**, 775 (2003).

202 V. Stolojan, S. R. P. Silva, M. J. Goringe, R. L. D. Whitby, W. K. Hsu, D. R. M. Walton, and H. W. Kroto, Appl. Phys. Lett. **86**, 063112 (2005).

203 B. N. Khare, M. Meyyappan, J. Kralj, P. Wilhite, M. Sisay, H. Imanaka, J. Koehne, and C. W. Bauschlicher, J. Appl. Phys. Lett. **81**, 5237 (2002).

204 J. Zhao, J. P. Lu, J. Han, and C.-K. Yang, Appl. Phys. Lett. **82**, 3746 (2003).

205 N. O. V. Plank, L. Jiang, and R. Cheung, Appl. Phys. Lett. **83**, 2426 (2003).

206 Y. Yue, Z. Zhang, and C. Liu, Appl. Phys. Lett. **88**, 263115 (2006).

207 E. T. Michelson, C. B. Huffman, A. G. Rinzler, R. E. Smalley, R. H. Hauge, and J. L. Margrave, Chem. Phys. Lett. **296**, 188 (1998).

208 B. N. Khare, M. Meyyappan, A. M. Cassell, C. V. Nguyen, and J. Han, Nano Lett. **2**, 73 (2002).

209 L. H. Chan, K. H. Hong, D. Q. Xiao, W. J. Hsieh, S. H. Lai, H. C. Shih, T. C. Lin, F. S. Shieu, K. J. Chen, and H. C. Cheng, Appl. Phys. Lett. **82**, 4334 (2003).

210 N. O. V. Plank, L. Jiang, and R. Cheung, Appl. Phys. Lett. **83**, 2426 (2003).

211 V. Krasheninnikov, K. Nardlund, J. Keinonen, and F. Banhart, Phys. Rev. B **66**, 245403 (2002).

212 P. He, D. Shi, J. Lian, L. M. Wang, R. C. Ewing, W. van Ooij, W. Z. Li, and Z. F. Ren, Appl. Phys. Lett. **86**, 043107 (2005).

213 D. Shi, S. X. Wang, W. van Ooij, L. M. Wang, J. Zhao, and Z. Yu, Appl. Phys. Lett. **78**, 1234 (2001).

214 D. Shi, P. He, S. X. Wang, W. van Ooij, L. M. Wang, J. Zhao, and Z. Yu, J. Mater. Res. **17**, 981 (2002).

215 D. Shi, J. Lian, P. He, L. M. Wang, W. van Ooij, M. Shulz, Y. J. Liu, and D. B. Mast, Appl. Phys. Lett. **83**, 5301 (2003).

216 M. Kuball, Surf. Interf. Anal. **31**, 987 (2001).

217 S. C. Jain, M. Willander, J. Narayan, and R. van Overstraeten, J. Appl. Phys. **87**, 965 (2000).

218 H. Morkoc, *Nitride Semiconductors and Devices* (Springer, Berlin, 1999).

219 K. Jagannadham, A. K. Sharma, Q. Wei, R. Kalyanraman, and J. Narayan, J. Vac. Sci. Technol. A **16**, 2804 (1998).

220 A. R. Goni, H. Siegle, K. Syassen, C. Thomsen, and J.-M. Wagner, Phys. Rev. B **64**, 035205 (2001).

221 A. L. Alvarez, F. Calle, E. Monroy, J. L. Pau, M. A. Sanchez-Garcia, E. Calleja, E. Munoz, F. Omnes, P. Gibart, and P. R. Hageman, J. Appl. Phys. **92**, 223 (2002).

222 G. W. Auner, F. Jin, V. M. Naik, and R. Naik, J. Appl. Phys. **85**, 7879 (1999).

223 A. Sarua, M. Kuball, and J. E. van Nostrand, Appl. Phys. Lett. **81**, 1426 (2002).

224 T. Prokofyeva, M. Seon, J. Vanbuskirk, M. Holtz, S. A. Nikishin, N. N. Faleev, H. Temkin, and S. Zollner, Phys. Rev. B **63**, 125313 (2001).

225 H. Harris, N. Biswas, H. Temkin, S. Gangopadhyay, and M. Strathman, J. Appl. Phys. **90**, 5825 (2001).

226 C. T. M. Ribeiro, F. Alvarez, and A. R. Zanatta, Appl. Phys. Lett. **81**, 1005 (2002).

227 C. Men, Z. Xu, Z. An, P. K. Chu, Q. Wan, X. Xie, and C. Lin, Appl. Surf. Sci. **199**, 287 (2002).

228 R. S. Naik, R. Reif, J. J. Lutsky, and C. G. Sodini, J. Electrochem. Soc. **146**, 691 (1999).

229 C.-C. Cheng, Y.-C. Chen, R.-C. Horng, H.-J. Wang, W.-R. Chen, and E.-K. Lai, J. Vac. Sci. Technol. A **16**, 3335 (1998).

230 C.-C. Cheng, Y.-C. Chen, H.-J. Wang, and W.-R. Chen, J. Vac. Sci. Technol. A **14**, 2238 (1996).

231 R. D. Vispute, J. Narayan, H. Wu, and K. Jagannadham, J. Appl. Phys. **77**, 4724 (1995).

232 B. G. Streetman and S. Banerjee, *Solid State Electronic Devices* (Prentice-Hall, New York, 2000).

233 F. Engelmark, G. Fucntes, I. V. Katardijev, A. Harsta, U. Smith, and S. Berg, J. Vac. Sci. Technol. A **18**, 1609 (2000).

234 C. Mirpuri, J. D. Long, and S. Xu, *RF Magnetron Sputtering Deposition of AlN Quantum Dots*, Internal Report 2687/2005, National Institute of Education, Singapore.

235 M. O. Aboelfotoh, R. S. Kern, S. Tanaka, R. F. Davis, and C. I. Harris, Appl. Phys. Lett. **69**, 2873 (1996).

236 F. Semond, B. Damilano, S. Vézian, N. Grandjean, M. Leroux, and J. Massies, Appl. Phys. Lett. **75**, 82 (1999).

237 C. J. Doss and R. Zallen, Phys. Rev. B **48**, 15626 (1993).

238 S. Y. Huang, S. Y. Xu, J. D. Long, Z. Sun, X. Z. Wang, Y. W. Chen, T. Chen, C. Ni, Z. J. Zhang, L. L. Wang, X. D. Li, P. S. Guo, and W. X. Que, Physica E **31**, 200 (2006).

239 M. D. Kim, S. K. Noh, S. C. Hong, and T. W. Kim, Appl. Phys. Lett. **82**, 553 (2003).

240 G. Ortner, M. Bayer, A. Larionov, V. B. Timofeev, A. Forchel, Y. B. Lyanda-Geller, T. L. Reineche, P. Hawrylak, S. Fafard, and Z. Wasilewski, Phys. Rev. Lett. **90**, 086404 (2003).

241 P. Ballet, J. B. Smathers, H. Yang, C. L. Warkman, and G. J. Salamo, Appl. Phys. Lett. **77**, 3406 (2000).

242 N. Y. Jin-Phillipp and F. Phillipp, J. Appl. Phys. **88**, 710 (2000).

243 A. D. Andreev and E. P. O'Reilly, Appl. Phys. Lett. **79**, 521 (2001).

244 E. Martinez-Guerrero, F. Chabuel, B. Daudin, J. L. Rouvière, and H. Mariette, Appl. Phys. Lett. **81**, 5117 (2002).

245 J. -F. Carlin and M. Ilegems, Appl. Phys. Lett. **83**, 668 (2003).

246 S. Yamaguchi, M. Kariya, S. Nitta, H. Kato, T. Takeuchi, C. Wetzel, H. Amano, and L. Akasaki, J. Cryst. Growth **195**, 309 (1998).

247 M. J. Lukitsch, Y. V. Danylyuk, V. M. Naik, C. Huang, G. W. Auner, L. Rimai, and R. Naik, Appl. Phys. Lett. **79**, 632 (2001).

248 N. Nakarona, H. Ishikawa, T. Egawa, T. Jimbo, and M. Umeno, J. Cryst. Growth **237**, 961 (2002).

249 A. B. Preobrajenski, K. Barucki, and T. Chasse, Phys. Rev. Lett. **85**, 4337 (2000).

250 H. C. Jeon, Y. S. Jeong, T. W. Kang, T. W. Kim, K. J. Chung, K. J. Chung, W. Jhe, and S. A. Song, Adv. Mater. **14**, 1725 (2002).

251 O. Briot, B. Maleyre, and S. Ruffenach, Appl. Phys. Lett. **83**, 2919 (2003).

252 S. Huang, Z. Dai, F. Qu, L. Zhang, and X. Zhu, Nanotechnology **13**, 691 (2002).

253 T. Peng, J. Piprek, G. Qiu, J. O. Olowolafe, K. M. Unruh, C. P. Swann, and E. F. Schubert, Appl. Phys. Lett. **71**, 2439 (1997).

254 S. J. Weiner, P. A. Kollman, D. T. Nguyen, and D. A. Case, J. Comput. Chem. **7**, 230 (1986).

255 Y. P. Guo, J. C. Zheng, A. T. S. Wee, C. H. A. Huan, K. Li, J. S. Pan, Z. C. Feng, and S. J. Chua, Chem. Phys. Lett. **339**, 319 (2001).

256 S. Charpentier, A. Kassiba, J. Emery, and M. Cauchetier, J. Phys.: Condens. Matter. **11**, 4887 (1999).

257 I. V. Kityk, A. Kassiba, K. Tuesu, S. Charpentier, Y. Ling, and M. Makowska-Januski, Mater. Sci. Eng. B **77**, 147 (2000).

258 S. Bandyopadhyay and H. S. Nalwa, *Quantum Dots and Nanowires* (American Scientific Publisher, New York, 2003).

259 R. Roucka, J. Tolle, A. V. G. Chizmeshya, P. A. Crozier, C. D. Poweleit, D. J. Smith, I. S. T. Tsong, and J. Kouvetakis, Phys. Rev. Lett. **88**, 206102 (2002).

260 H. Morkoc, S. Strite, G. B. Gao, M. E. Lin, B. Sverdlov, and M. Burns, J. Appl. Phys. **76**, 1363 (1994).

261 C. M. Zetterling, *Process Technology for Silicon Carbide Devices* (Institute of Electrical Engineering, London, 2002).

262 D. M. Teter, MRS Bull. **23**, 22 (1998).

263 L. B. Rowland, R. S. Kern, S. Tanaka, and R. F. Davis, Appl. Phys. Lett. **62**, 3333 (1993).

264 S. Tanaka, R. S. Kern, and R. F. Davis, J. Appl. Phys. **66**, 37 (1995).

265 G. S. Solomon, M. Pelton, and Y. Yamamoto. Phys. Rev. Lett. **86**, 3903 (2001).

266 Y. P. Guo, J. C. Zheng, A. T. S. Wee, C. H. A. Huan, K. Li, J. S. Pan, Z. C. Feng, and S. J. Chua, Chem. Phys. Lett. **339**, 319 (2001).

267 Y. Glinka, S. H. Lin, L. P. Hwang, Y. T. Chen, and N. H. Tolk, Phys. Rev. B **64**, 085421 (2001).

268 A. Kassiba, M. Makowska-Janusik, J. Boucle, J. F. Bardeau, A. Bulou, and N. Herlin-Boime, Phys. Rev. B **66**, 155317 (2002)

269 T. Rajagopalan, X. Wang, B. Lahlouh, C. Ramkumar, P. Dutta, and S. Gangopadhyay, J. Appl. Phys. **94**, 5252 (2003).

270 S. Kerdiles, A. Berthelot, F. Gourbilleau, and R. Rizk, Appl. Phys. Lett. **76**, 2373 (2000).

271 L. F. Marsal, J. Pallares, X. Correig, A. Orpella, D. Bardés, and R. Alcubilla. J. Appl. Phys. 85, 1216 (1999)

272 M. B. Yu, Rusli, S. F. Yoon, Z. M. Chen, J. Ahn, Q. Zhang, K. Chew, and J. Cui, J. Appl. Phys. **87**, 8155 (2000).

273 G. Soto, E. C. Samano, R. Machorro, and L. Cota, J. Vac. Sci. Technol. A **16**, 1311 (1998).

274 J. Chen, A. J. Steckl, and M. J. Loboda, J. Vac. Sci. Technol. B **16**, 1305 (1998).

275 K. Jagannadham, A. K. Sharma, Q. Wei, R. Kalyanraman, and J. Narayan, J. Vac. Sci. Technol. A **16**, 2804 (1998).

276 M. O. Aboelfotoh, R. S. Kern, S. Tanaka, R. F. Davis, and C. I. Harris, Appl. Phys. Lett. **69**, 2873 (1996).

277 W. Seifert, N. Carlsson, J. Johansson, M.-E. Pistol, and L. Samuelson, J. Cryst. Growth **170**, 39 (1997).

278 D. J. Eaglesham and M. Cerullo, Phys. Rev. Lett. **64**, 1943 (1990).

279 N. Carlsson, K. Georgsson, L. Montelius, L. Samuelson, W. Seifert, and R. Wallenberg, J. Cryst. Growth **156**, 23 (1995).

280 F. M. Ross, J. Tersoff, and R. M. Tromp, Phys. Rev. Lett. **80**, 984 (1998).

281 A. Elshabini and F. D. Barlow, *Thin Film Technology Handbook* (McGraw-Hill, New York, 1997).

282 V. Ng, S. Y. Huang, J. D. Long, and S. Xu, *Nanofabrication of SiC Quantum Dots on Si with AlN Buffer Interlayers*, Internal Report 2687/2005, National Institute of Education, Singapore.

283 F. Ren and J. C. Zolper, *Wide Energy Bandgap Electronic Devices* (World Scientific, River Edge, NJ, 2003).

284 H. Y. Liu, I. R. Sellers, M. Hopkinson, C. N. Harrison, D. J. Mowbray, and M. S. Skolnick, Appl. Phys. Lett. **83**, 3716 (2003).

285 K. Hiramatsu, S. Itoh, H. Amano, I. Akasaki, N. Kuwano, T. Shiraishi, and K. Oki, J. Cryst. Growth **115**, 628 (1991).

286 G. Burns, *Solid State Physics* (Academic, San Diego, 1985), Chapter 18.

287 Q. J. Cheng, S. Xu, J. D. Long, and K. Ostrikov, Appl. Phys. Lett. **90**, 173112 (2007).

288 *Silicon Carbide and Related Materials*, Materials Science Forum, Vols. 433–436, edited by P. Bergman and E. Jantzen (Trans Tech Publications, Zurich, Switzerland, 2003).

289 F. A. Reboredo, L. Pizzagalli, and G. Galli, Nano Lett. **4**, 801 (2004).

290 H. Colder, R. Rizk, M. Morales, P. Marie, J. Vicens, and I. Vickridge, J. Appl. Phys. **98**, 024313 (2005).

291 Y. Sun, T. Miyasato, and J. K. Wigmore, J. Appl. Phys. **86**, 3076 (1999).

292 V. Cimalla, A. A. Schmidt, T. Stauden, K. Zekentes, O. Ambacher, and J. Pezoldt, J. Vac. Sci. Technol. B **22**, L20 (2004).

293 J. Y. Fan, X. L. Wu, H. X. Li, H. W. Liu, G. G. Siu, and P. K. Chu, Appl. Phys. Lett. **88**, 041909 (2006).

294 A. Fissel, K. Pfennighaus, and W. Richter, Appl. Phys. Lett. **71**, 2981 (1997).

295 A. Fissel, K. Pfennighaus, and W. Richter, Thin Solid Films **88**, 318 (1998).

296 F. Rosei, J. Phys.: Condens. Matter **16**, S1373 (2004).

297 W. K. Choi, T. Y. Ong, L. S. Tan, F. C. Loh, and K. L. Tan, J. Appl. Phys. **83**, 4968 (1998).

298 T. Berlind, N. Hellgren, M. P. Johansson, and L. Hultman, Surf. Coat. Technol. **141**, 145 (2001).

299 K. Yamamoto, Y. Koga, and S. Fujiwara, Diam. Relat. Mater. **10**, 1921 (2001).

300 F. Liao, S. L.Girshick, W. M. Mook, W. W. Gerberich, and M. R. Zachariah, Appl. Phys. Lett. **86**, 171913 (2005).

301 Z. Hu, X. Liao, H. Diao, G. Kong, X. Zeng, and Y. Xu, J. Cryst. Growth **7**, 264 (2004).

302 W. Yu, W. Lu, L. Han, and G. Fu, J. Phys. D: Appl. Phys. **37**, 3304 (2004).

303 S. Y. Huang, J. D. Long, M. Xu, and S. Xu, *Reactive Plasma Assembly of Large Area Amorphous SiO$_2$ Nanowires*, unpublished.

304 D. D. D. Ma, C. S. Lee, F. C. K. Au, S. Y. Tong, and S. T. Lee, Science **299**, 1874 (2003).

305 D. Zhang, A. Alkhateeb, H. Han, H. Mahmood, and D. N. Mcllroy, Nano Lett. **3**, 983 (2003).

306 J.-F. Lin, J. P. Bird, Z. He, P. A. Bennett, and D. J. Smith, Appl. Phys. Lett. **85**, 281 (2004).

307 Y. J. Chang, B. H. Kang, G. T. Kim, S. J. Park, and J. S. Ha, Appl. Phys. Lett. **84**, 5314 (2004).

308 Z. W. Pan, Z. R. Dai, and Z. L. Wang, Science **291**, 1947 (2001).

309 I. Solomon, Appl. Surf. Sci. **184**, 3 (2001).

310 H.-F. Zhang, C.-M. Wang, E. C. Buck, and L.-S. Wang, Nano Lett. **3**, 577 (2003).

311 K.-H. Lee, H. S. Yang, K. H. Baik, J. Bang, R. R. Vanfleet, and W. Sigmund, Chem. Phys. Lett. **383**, 380 (2004).

312 D. J. Zhang and R. Q. Zhang, Chem. Phys. Lett. **394**, 437 (2004).

313 D. N. Mcllroy, D. Zhang, and Y. Kranov, Appl. Phys. Lett. **79**, 1540 (2001).

314 M. Xu, S. Xu, S. Y. Huang, J. W. Chai, V. M. Ng, J. D. Long, and P. Yang, Physica E **35**, 81 (2006).

315 *Silicon Based Microphotonics: from Basic to Applications*, edited by O. Bisi, S. U. Campisano, L. Pavesi, and F. Priolo (IOS Press, Amsterdam, 1999).

316. L. T. Canham, Appl. Phys. Lett. **57**, 1045 (1990).
317. H. Takagi, H. Ogawa, Y. Yanazaki, and T. Nakagiri, Appl. Phys. Lett. **56**, 2397 (1990).
318. T. S. Iwayama, S. Nakao, and K. Saitoh, Appl. Phys. Lett. **65**, 1814 (1994).
319. K. D. Hirschman, L. Tsybeskov, S. P. Duttagupta and P. M. Fauchet, Nature **384**, 338 (1996).
320. S. Hayashi and K. Yamamoto, J. Lumin. **70**, 352 (1996).
321. L. Pavesi, L. D. Negro, C. Mazzoleni, G. Franzo, and J. P. Prolo, Nature **408**, 440 (2000).
322. M. Zacharias and P. Streitenberger, Phys. Rev. B **62**, 8391 (2000).
323. M. Zacharias, J. Heitmann, R. Scholz, U. Kahler, M. Schmidt, and J. Blasing, Appl. Phys. Lett. **80**, 661 (2002).
324. L. S. Liao, X. M. Bao, Z. F. Yang, and N. B. Min, Appl. Phys. Lett. **66**, 2382 (1995).
325. M. Molinari, H. Rinnert, and M. Vergnat, Appl. Phys. Lett. **77**, 3499 (2000).
326. M. Molinari, H. Rinnert, and M. Vergnat, Appl. Phys. Lett. **79**, 2172 (2001).
327. L. C. Chen, C. K. Chen, S. L. Wei, D. M. Bhusari, K. H. Chen, Y. F. Chen, Y. C. Jong, and Y. S. Huang, Appl. Phys. Lett. **72**, 2463 (1998).
328. H. Ueda, O. Kitakami, Y. Shimada, Y. Goto, and M. Yamamoto, Jpn. J. Appl. Phys. **33**, 6173 (1994).
329. V. M. Ng, M. Xu, S. Y. Huang, J. D. Long, and S. Xu, Thin Solid Films **506**, 283 (2006).
330. A. R. Wilkinson and R. G. Elliman, Appl. Phys. Lett. **83**, 5512 (2003).
331. G. G. Qin and Y. J. Li, Phys. Rev. B **68**, 085309 (2003).
332. S. Y. Huang, J. D. Long, M. Xu, and S. Xu, *Self-assembled Si quantum dots on amorphous AlN films synthesized by RF magnetron sputtering*, (unpublished).
333. A. T. Tike, F. C. Simmel, R. H. Bick, H. Lorenz, and J. P. Kotthaus, Progr. Quant. Electron. **25**, 97 (2001).
334. S. Tiwari, F. Rhana, H. Hanafi, A. Harstein, E. F. Crabbe, and K. Chan, Appl. Phys. Lett. **68**, 1377 (1996).
335. H. Akazawa, Appl. Phys. Lett. **82**, 1464 (2003).
336. M. Fukuda, K. Nakagawa, S. Miyazaki, and M. Hirose, Appl. Phys. Lett. **70**, 2291 (1997).
337. A. B. Preobrajenski, K. Baruchi, and B. Chasse, Phys. Rev. Lett. **85**, 4337 (2000).
338. S. Y. Huang, S. Xu, J. D. Long, Z. Sun, and T. Chen, *Plasma reactive synthesis single layer ordered SiC quantum dots with self-assembled growth mode*, (unpublished).
339. B. Yang, F. Liu, and M. G. Lagally, Phys. Rev. Lett. **92**, 025502 (2004).
340. S. Guha, A. Madhukar, and K. C. Rajkumar, Appl. Phys. Lett. **57**, 2110 (1990).
341. D. J. Eaglesham and M. Cerullo, Phys. Rev. Lett. **64**, 1943 (1990).
342. Y. W. Mo, D. E. Savage, B. S. Swartzentruber, and M. G. Lagally, Phys. Rev. Lett. **65**, 1020 (1990).
343. C. Teichert, J. C. Bean, and M. G. Lagally, Appl. Phys. A **67**, 675 (1998).
344. T. Kitajima, B. Liu, and S. R. Leone, Appl. Phys. Lett. **80**, 497 (2002).
345. L. Vescan, K. Grimm, and C. Dieker, J. Vac. Sci. Technol. B **16**, 1549 (1998).
346. G. S. Solomon, Appl. Phys. Lett. **84**, 2073 (2004).
347. F. Poster, A. Bhattacharya, S. Weeke, and W. Richter, J. Cryst. Growth **248**, 317 (2003).
348. C. L. Zhang, Z. G. Wang, F. A. Zhao, B. Xu, and P. Jin, J. Crystal Growth **265**, 64 (2004).
349. J. Liang, H. Chik, A. Yin, and J. Xu, J. Appl. Phys. **91**, 2544 (2002).
350. R. Noetzel and K. H. Plong, J. Cryst. Growth **227–228**, 8 (2001).
351. H. J. Kim, Y. J. Park, Y. M. Park, E. K. Kim, and T. W. Kim, Appl. Phys. Lett. **78**, 3253 (2001).
352. R. Leon, S. Chaparro, S. R. Johnson, C. Navarro, X. Jin, Y. H. Zhang, J. Siegert, S. Marcinkevicius, X. Z. Liao, and J. Zou, J. Appl. Phys. **91**, 5826 (2002).
353. K. Zhang, Ch. Heyn, and W. Hansen, Appl. Phys. Lett. **76**, 2229 (2000).
354. S. Huang, Z. Dai, F. Qu, L. Zhang, and X. Zhu, Nanotechnology **13**, 691 (2002).

355 B. J. Spencer and J. Tersoff, Phys. Rev. Lett. **79**, 4858 (1997).

356 M. Xu, S. Y Huang, V. Ng, S. Xu, and J. D. Long, *SiCAlN nanorods synthesized by plasma-assisted RF magnetron sputtering deposition*, (unpublished).

357 V. Ng, M. Xu, S. Y Huang, S. Xu, and J. D. Long, *Transition of SiCN nanorods to nanowires: control by the plasma process parameters*, (unpublished).

358 S. Xu, J. D. Long, L. Sim, C. H. Diong, and K. Ostrikov, Plasma Proc. Polym. **2**, 373 (2005).

359 B. D. Ratner, *Biomaterials Science: An Introduction to Materials in Medicine* (Academic Press, San Diego, 1996).

360 D. F. Williams, *The Williams Dictionary of Biomaterials* (Liverpool University Press, Liverpool, UK, 1999).

361 R. Z. Le Geros and J. P. Le Geros, *Calcium Phosphate Bioceramics: Past, Present and Future*, in *Bioceramics*, edited by B. Ben-Nissan, D. Sher, and W. Walsh, Intern. Soc. for Ceramics in Medicine (Trans Tech Publications, Sydney, 2003), vol. 15, pp. 3–10.

362 L. Sun, C. C. Berndt, K. A. Gross, and A. Kucuk, J. Biomed. Mater. Res.: Appl. Biomater. **58**, 570 (2001).

363 K. Yamagishi, K. Onuma, T. Suzuki, F. Okada, J. Tagami, M. Otsuki, and P. Senawangse, Nature **433**, 819 (2005).

364 K. A. Gross, C. C. Berndt, and H. Herman, J. Biomed. Mater. Res. **39**, 407 (1998).

365 American Society for Testing and Materials (ASTM), *Standard Specification for Composition of Hydroxyapatite for Surgical Implants*, F1185-03, ASTM, 2003. See also http://www.astm.org

366 Y. C. Tsui, C. Doyle, and T. W. Clyne, Biomaterials **19**, 2031 (1998).

367 C. P. Klein, P. Patsa, J. G. C. Wolke, J. Blieck-Hogervorst, and K. Groot, J. Biomed. Mater. Res. **28**, 909 (1994).

368 C. C. Berndt, G. N. Haddad, A. J. D. Farmer, and K. A. Gross, Mater Forum **14**, 161 (1990).

369 K. De Groot, R. Geesink, C. P. Klein, and P. Serekian, J. Biomed. Mater. Res. **21**, 1375 (1987).

370 J. H. C. Lin, M. L. Liu, and C. P. Ju, J. Mater. Sci.: Mater. Med. **5**, 279 (1994).

371 F. Brossa, A. Cigada, R. Chiesa, L. Paraccini, and C. Consonni, Biomed. Mater. Eng. **3**, 127 (1993).

372 C. M. Cotell, Appl. Surf. Sci. **69**, 140 (1993).

373 L. Cleries, E. Martinez, J. M. Fernandez-Prads, G. Sardin, J. Esteve, and J. L. Morenza, Biomaterials **21**, 967 (2000).

374 J. L. Ong and L. C. Lucas, Biomaterials **15**, 337 (1994).

375 P. Ducheyne, W. V. Raemdonck, J. C. Heughebaert, and M. Heughebaert, Biomaterials **6**, 97 (1986).

376 W. V. Raemdonck, P. Ducheyne, and P. D. Meester, J. Amer. Ceram. Soc. **63**, 381 (1984).

377 W. R. Lacefield, Ann. Rep. N.Y. Acad. Sci. **523**, 72 (1988).

378 T. T. Li, J. H. Lee, T. Kobayashi, and H. Aoki, J. Mater. Sci.: Mater. Med. **7**, 97 (1996).

379 G. L. Darimont, B. Gilbert, and R. Cloots, Mater. Lett. **58**, 71 (2003).

380 M. J. Filliagi, N. A. Coombs, and R. M. Pillar, J. Biomed. Mater. Res. **25**, 1211 (1991).

381 S. Xu, J. D. Long, K. N. Ostrikov, J. H. Lu, and C. H. Diong, IEEE Trans. Plasma Sci. **30**, 118 (2002).

382 S. Mohammadi, M. Esposito, J. Hall, L. Emanuelsson, A. Krozer, and P. Thomsen, Int. J. Oral Maxillofac. Implants **19**, 498 (2004).

383 B. Feddes, A. M. Vredenberg, J. G. C. Wolke, and J. A. Jansen, Surf. Coat. Technol. **185**, 346 (2004).

384 K. Ozeki, A. Mishima, T. Yuhta, Y. Fukui, and H. Aoki, Biomed. Mater. Eng. **13**, 451 (2003).

385 A. Boyd, M. Akay, and B. J. Meenan, Surf. Interf. Anal. **35**, 188 (2003).

386 J. L. Ong, K. Bessho, and D. L. Carnes, Int. J. Oral Maxillofac. Implants **27**, 581 (2002).

387 K. Ozeki, T. Yuhta, Y. Fukui, H. Aoki, and I. Nishimura, J. Mater. Sci.: Mater. Medicine **13**, 253 (2002).

388 J. D. Long, S. Xu, J. W. Cai, N. Jiang, J. H. Lu, K. N. Ostrikov, and C. H. Diong, Mater. Sci. Eng. C: Biomimetic Supramol. Syst. **20**, 175 (2002).

389 S. Xu, J. D. Long, K. N. Ostrikov, and H. Y. Foo, Key Eng. Mater. **240**, 307 (2003).

390 J. D. Long, S. Xu, H. Y. Foo, and C. H. Diong, Key Eng. Mater. **240**, 303 (2003).

391 I. S. Lee, H. E. Kim, and S. Y. Kim, Surf. Coat. Technol. **131**, 181 (2000).

392 J. D. Long, K. Ostrikov, S. Xu, and V. Ligatchev, Proc. Mater. Res. Soc. Symp. **740**, I12.17.1 (2003), (Materials Research Society, Boston, 2003).

393 K. Hayashi, T. Inadome, H. Tsumura, Y. Nakashima, and Y. Sugioka, Biomaterials **15**, 1187 (1994).

394 E. J. Evans, Biomaterials **12**, 574 (1991).

395 B. D. Cullity, *Elements of X-ray Diffraction*, 3rd edition (Prentice-Hall, New York, 2001).

396 Y. Yang, K.-H. Kim, and J. L. Ong, Biomaterials **26**, 327 (2005).

397 J. L. Ong and L. C. Lukas, Biomaterials **15**, 337 (1994).

398 K. van Dijk, H. G. Schaeken, J. G. C. Wolke, and J. A. Jansen, Biomaterials **17**, 405 (1996).

399 Y. Yang, K.-H. Kim, C. M. Agrawal, and J. L. Ong, Biomaterials **24**, 5131 (2003).

400 Y. Yang, K.-H. Kim, C. M. Agrawal, and J. L. Ong, J. Dent. Res. **82**, 833 (2003).

401 J. E. G. Hulshoff, T. Hayakawa, K. van Dijk, A. F. M. Leidekkers-Govers, J. P. C. M. van der Waerden, and J. A. Jansen, J. Biomed. Mater. Res. **36**, 75 (1997).

402 F. C. Chang, M. Levy, R. Huie, M. Kane, P. Buckley, T. Z. Kattamis, and G. R. Lakshminarayan, Surf. Coat. Technol. **49**, 87 (1991).

403 J. D. Long, *Triadic Bioactive CaPTi Thin Films Synthesized by Concurrent Sputtering Deposition*, Ph.D. Thesis, Nanyang Technological University, Singapore, 2003.

404 T. Kokubo, S. Ito, and T. Yamamuro, et al., J. Biomed Mater Res. **24**, 331 (1990).

405 H. Zreiqat, C. R. Howlett, A. Zannettino, C. Evans, P. Knabe, G. Schulze-Tanzil, and M. Shakibaei, Key Eng. Mater. **240–242**, 707 (2003).

406 H. M. Kim, H. F. Miyaji, T. Kokubo, and T. J. Nakamura, J. Biomed. Mater. Res. **32**, 409 (1996).

407 S. Y. Huang, S. Xu, J. Long, Z. Sun, X. Z. Wang, Y. W. Chen, T. Chen, C. Ni, Z. J. Zhang, L. L. Wang, X. D. Li, P. S. Guo, and W. X. Que, Surf. Rev. Lett. **13**, 123 (2006).

408 S. S. Iyer and Y. H. Xie, Science **260**, 280 (1993).

409 S. Schmitt-Rink, C. M. Varma, and A. F. J. Levi, Phys. Rev. Lett. **66**, 2782 (1991).

410 N. Q. Vinh, H. Przybylinska, Z. F. Krasil'nik, and T. Gregorkiewicz, Phys. Rev. Lett. **90**, 06601 (2003).

411 R. M. Fujili, M. Yoshida, Y. Kanzawa, S. Hayashi, and K. Yamamoto, Appl. Phys. Lett. **71**, 1198 (1997).

412 J. S. Ha, C. H. Bae, S. H. Nam, S. M. Park, Y. R. Jang, and K. H. Yoo, Appl. Phys. Lett. **82**, 3436 (2003).

413 K. Rerbal, F. Jomard, J.-N. Chazalviel, F. Ozanam, and I. Solomon, Appl. Phys. Lett. **83**, 45 (2003).

414 M. Iwami, Nucl. Instr. Meth. Phys. Res. A **466**, 406 (2001).

415 M. G. Park, W. S. Choi, B. Hong, Y. T. Kim, and D. H. Yoon, J. Vac. Sci. Technol. A **20**, 861 (2002).

416 S. Dogan, A. Teke, D. Huang, C. B. Roberts, J. Parish, B. Ganguly M. Smith, R. E. Myers, and S. E. Saddow, Appl. Phys. Lett. **82**, 3107 (2003).

417 G. Pasold, F. Albrecht, J. Grillenberger, U. Grossner, C. Hülsen, W. Withuhn, and R. Sielemann, J. Appl. Phys. **93**, 2289 (2003).

418 I. V. Kityk, A. Kassiba, K. Tuesu, S. Charpentier, Y. Ling, and M. Makowska-Januski, Mater. Sci. Eng. B **77**, 147 (2000).

419 Y. C. Ee, Z. Chen, S. B. Law, and S. Xu, Thin Solid Films **504**, 218 (2006).

420 C. Marcadal, M. Eizenberg, and L. A. Yoon, J. Eletrochem. Soc. **149**, C52 (2002).

421 J. S. Reid, X. Sun, E. Kolawa, and M.-A. Nicolet, IEEE Electron. Device Lett. **15**, 298 (1994).

422 E. L. Tsakadze, K. N. Ostrikov, Z. L. Tsakadze, N. Jiang, R. Ahmad, and S. Xu, Int. J. Mod. Phys. B **16**, 1143 (2002).

423 T. Fujii, J. Muraki, S. Arulmozhiraja, and M. Kareev, J. Appl. Phys. **88**, 5592 (2000).

424 K. M Yu, M. Cohen, E. E. Haller, W. L, Hansen, A. Y. Liu, and I. C. Wu, Phys. Rev. B **49**, 5034 (1994).

425 Z. Wu, Y. Yu, and X. Liu, Appl. Phys. Lett. **68**, 1291 (1996).

426 N. Jiang, S. Xu, K. N. Ostrikov, J. Chai, Y. Li, K. M. Ling, and S. Lee, Thin Solid Films **357**, 55 (2001).

427 N. Jiang, *Carbon Nitride and Its Modifications Synthesized by Means of RF Plasmas*, PhD thesis, Nanyang Technological University, Singapore, 2002.

428 S. Bhattacharyya, A. Granier, and G. Turban, J. Appl. Phys. **86**, 4668 (1999).

429 C. Mirpuri, S. Xu, J. D. Long, and K. Ostrikov, J. Appl. Phys. **101**, 024312 (2007).

430 R. D. Vispute, J. Narayan, H. Wu, and K. Jagannadham, J. Appl. Phys. **77**, 4724 (1995).

431 V. Ligatchev, Rusli, and Z. Pan, Appl. Phys. Lett. **87**, 242903 (2005).

432 H. Morkoc, *Nitride Semiconductors and Devices* (Springer, New York, 1999).

433 N. Laidani, L. Vanzetti, M. Anderle, A. Basillais, C. Boulmer-Leborgne, and J. Perriere, Surf. Coat. Technol. **122**, 242 (1999).

434 J. F. Moulder, W.F. Stickle, P. E. Sobol, K. D. Bomben, *Handbook of X-ray Photoelectron Spectroscopy*, (Physical Electronics, Minneapolis, 1995)

435 M. Kuball, Surf. Interf. Anal. **31**, 987 (2001).

436 N. F. Mott and E. A. Davis, *Electronic Processes in Non-Crystalline Solids* (Clarendon, Oxford, 1979).

437 S. P. Lim, J. D. Long, S. Xu, and K. Ostrikov, J. Phys. D: Appl. Phys. **40**, 1085 (2007).

438 W. R. Roach, Appl. Phys. Lett. **19**, 453 (1971).

439 C. G. Granqvist, Thin Solid Films **193**, 730 (1994).

440 C. R. Aita, Y.-L. Liu, M. L. Kao, and S. D. Hansen, J. Appl. Phys. **60**, 749 (1986).

441 M. Mozetic, U. Cvelbar, M. K. Sunkara, and S. Vaddiraju, Adv. Mater. **17**, 2138 (2005).

442 K. West and B. Zachau, Solid State Ion. **76**, 1068 (1995).

443 R. Ramirez, B. Casal, L. Utera, and E. Ruiz-Hitzky, J. Phys. Chem. **94**, 8965 (1990).

444 S. F. Cogan, N. M. Nguyen, S. J. Perrotti, and R. D. Rauh, J. Appl. Phys. **66**, 1333 (1989).

445 X. J. Wang, H. D. Li, Y. J. Fei, X. Wang, Y. Y. Xiong, Y. X. Nie, and K. A. Feng, Appl. Surf. Sci. **177**, 8 (2001).

446 G. S. Nadkarni and V. Shirodkar, Thin Solid Films **105**, 115 (1983).

447 J. Cui, D. Da, and W. Jiang, Appl. Surf. Sci. **133**, 225 (1998).

448 H.-T. Yuan, K.-C. Feng, X.-J. Wang, C. Li, C. J. He, and Y.-X. Nie, J. Chinese Phys. **13**, 82 (2004).

449 F. Guinneton, L. Sauques, J.-C. Valmalette, F. Cros, and J. R. Gavarri, Thin Solid Films **446**, 287 (2004).

450 S.-H. Lee, C. H. Cheong, M. J. Seong, P. Liu, C. E. Tracy, A. Mascarenhas, J. R. Pitts, and S. K. Deb, Solid State Ion. **165**, 111 (2003).

451 T. D. Manning and I. P. Parkin, J. Mater. Chem. **14**, 2554 (2004).

452 H. S. Choi, J. S. Ahn, J. H. Jung, T. W. Noh, and D. H. Kim, Phys. Rev. B **54**, 7 (1996).

453 Y. Muraoka and Z. Hiroi, Appl. Phys. Lett. **80**, 4 (2002).

454 L. J. Wang and F. C. N. Hong, Micropor. Mesopor. Mater. **77**, 167 (2005).

455 H. C. Foley, Micropor. Mesopor. Mater. **4**, 407 (1995).

456 Y. Yin and R. E. Collins, Carbon **31**, 1333 (1993).

457 H. Matsuyama, T. Shiraishi, and M. Teramoto, J. Appl. Polym. Sci. **54**, 1665 (1994).

458 M. Yamamoto, J. Sakata, and M. Hirai, J. Appl. Polym. Sci. **29**, 2981 (1984).

459 J. Sakata, M. Yamamoto, and M. Hirai, J. Appl. Polym. Sci. **31**, 1999 (1986).

460 Z. Sun, S. Xu, and K. N. Ostrikov, Diam. Relat. Mater. **11**, 92 (2002).

461 F. Huber, J. Springer, and M. Muhler, J. Appl. Polym. Sci. **63**, 1517 (1997).

462 S. Roualdes, A. van der Lee, R. Berjoan, J. Sanchez, and J. Durand, AICHE J. **45**, 1566 (1999).

463 M. L. Steen, W. C. Flory, N. E. Capps, and E. R. Fisher, Chem. Mater. **13**, 2749 (2001).

464 E. Piskin, J. Biomater. Sci.—Polymer Edition **4**, 45 (1992).

465 Q. T. Le, C. M. Whelan, H. Struyf, H. Bender, T. Conrad, S. H. Brongersma, W. Boullart, S. Vanhaelemeersch, and K.

Maex, Electrochem. Solid State Lett. **7**, F49 (2004).

466 T. Abell and K. Maex, Microelectron. Eng. **76**, 16 (2004).

467 V. Rouessac, P. Ferreira, and J. Durand, Separation Purification Technol. **32**, 37 (2003).

468 J. Weichart and J. Muller, Surf. Coat. Technol. **59**, 342 (1993).

469 K. Beltsios, G. Charalambopoulou, G. Romanos, and N. Kanellopoulos, J. Porous Mater. **6**, 25 (1999).

470 L. Lin and J. T. Guthrie, J. Membrane Sci. **173**, 73 (2000).

471 S. Veprek, S. Mukherjee, P. Karvankova, H.-D. Mannling, J. L. He, K. Moto, J. Prochazka, and A. S. Argon, J. Vac. Sci. Technol. A **21**, 532 (2003).

472 S. Veprek, H.-D. Mannling, M. Jilek, and P. Holubar, Mater. Sci. Eng. A: Struct. Mater. Prop. Microstruct. Process. **366**, 202 (2004).

473 Y. J. T. Lii, in *ULSI Technology*, edited by C. Y. Chang and S. M. Sze (McGraw-Hill, New York), pp. 329–370.

474 Q. Yang, W. Chen, C. Xiao, R. Sammynaiken, and A. Hirose, Carbon **43**, 748 (2005).

475 M. J. Kim, J. S. Lee, S. K. Kim, G. Y. Yeom, J. B. Yoo, and C. Y. Park, Thin Solid Films **475**, 41 (2005).

476 H. W. Huang, C. C. Kao, T. H. Hsueh, C. C. Yu, C. F. Lin, J. T. Chu, H. C. Kuo, and S. C. Wang, Mater. Sci. Eng. B, Solid-State Mater. Adv. Technol. **113**, 125 (2004).

477 K. A. Lister, B. G. Casey, P. S. Dobson, S. Thoms, D. S. Macintyre, C. D. W. Wilkinson, and J. M. R. Weaver, Microelectron. Eng. **73–74**, 319 (2004).

478 L. Menon, K. B. Ram, S. Patibandla, D. Aurongzeb, M. Holtz, J. Yun, V. Kuryatkov, and K. Zhu, J. Electrochem. Soc. **151**, C492 (2004).

479 X. F. Hua, C. Stolz, G. S. Oehrlein, P. Lazzeri, N. Coghe, M. Anderle, C. K. Inoki, M. Anderle, T. S. Kuan, and P. Jiang, J. Vac. Sci. Technol. A **23**, 151 (2005).

480 W. J. Cho, O. Rodriguez, R. Saxena, M. Ojha, R. A. Chanta, J. L. Plawsky, and W. N. Gill, J. Electrochem. Soc. **152**, F26 (2005).

481 T. Kimura, Micropor. Mesopor. Mater. **77**, 97 (2005).

482 G. N. Panin, Y. S. Park, T. W. Kang, T. W. Kim, K. L. Wang, and M. Bao, J. Appl. Phys. **97**, 043527 (2005).

483 J. M. Lee, S. H. Oh, C. W. Lee, H. Ko, S. Park, K. S. Kim, and M. H. Park, Thin Solid Films **475**, 189 (2005).

484 C. Arnault and X. Devaux, J. Mater. Sci. Technol. **20**, 63 (2004).

485 K. S. Cho, N. M. Park, T. Y. Kim, K. H. Kim, G. Y. Sung, and J. H. Shin, Appl. Phys. Lett. **86**, 071909 (2005).

486 Z. W. Pei, A. Y. K. Su, H. L. Hwang, and H. L. Hsiao, Appl. Phys. Lett. **86**, 063503 (2005).

487 M. Lin, K. P. Loh, C. Boothroyd, and A. Y. Du, Appl. Phys. Lett. **85**, 5388 (2004).

488 S. P. Song, M. A. Crimp, V. M. Ayres, C. J. Collard, J. P. Holloway, and M. L. Brake, J. Nanosc. Nanotechnol. **4**, 817 (2004).

489 J. L. H. Chau, M. K. Hsu, C. C. Hsieh, and C. C. Kao, Mater. Lett. **59**, 905 (2005).

490 Y. S. Kwon, J. S. Kim, P. P. Choi, J. H. Song, and D. Dudina, J. Industr. Engn. Chem. **11**, 103 (2005).

491 K. H. Jeong, S. G. Park, and S. W. Rhee, J. Vac. Sci. Technol. B **22**, 2799 (2004).

492 R. F. Service, Science **314**, 45 (2006).

493 H. Kanzow and A. Ding, Phys. Rev. B **60**, 11180 (1999).

494 A. N. Obraztsov, I. Pavlovsky, A. P. Volkov, E. D. Obraztsova, A. L. Chuvilin, and V. L. Kuznetsov, J. Vac. Sci. Technol. B **18**, 1059 (2000).

495 S. H. Tsai, F. K. Chiang, T. G. Tsai, F. S. Shieu, and H. C. Shih, Thin Solid Films **366**, 11 (2000).

496 Lee Y. J., Mater. Lett. **59**, 615 (2005).

Index

a

acetylene   156
AlCN nanocomposite: tailoring composition   246
AlN buffer interlayer   168
AlN nanostructures: fabrication   162
AlN: properties   163
amorphous silicon   20
anisotropic selective etching   274
$Ar+H_2$ plasma   182
$Ar+H_2+C_2H_2$ plasma   156
$Ar+H_2+CH_4$ plasma   124, 137
$Ar+N_2+CH_4$ plasma   245
$Ar+N_2+H_2$ plasma   168
$Ar+O_2$ plasmas   271
argon   X
Atomic Force Microscopy (AFM)   221
atomic force microscopy (AFM)   165

b

Bohm sheath criterion   11
Bohm velocity   11
Boltzmann's relation   10
building units   19, 33, 82, 266, 272

c

calcium phosphate CaP   225
calcium titanate $(CaTiO_3)$ 226
carbon nanotips   121, 126, 132
carbon nanotips, single-crystalline   137
carbon nanotips: field emission properties   134
carbon nanotips: hydrogen termination   144
carbon nanotips: temperature-controlled growth (TCG) regime   130
carbon nanotube microemitters   26
carbon nanotubes   X, 2, 19
carbon nanotubes: doping   153
carbon nanotubes: multiwalled   XI, 15
carbon nanotubes: single-walled   XI, 15
carbon nanotubes: vertical alignment   276
carbon nanowall-like structures (CNWLS)   156
carbon pyramid-like structures (CPLS)   126, 128
carbon-based nanostructures   25
charge neutrality   6
charged cluster theory   270
clustering in a plasma   271
conduction band   147, 159
cross-sectional SEM   255, 262
crystalline Si nanoparticles   22

d

dangling bonds   18
DC substrate bias   X
Debye length   10
Debye shielding   9
defect of mass   1
density functional theory (DFT)   142
discharge mode transitions: $E \leftrightarrow H$ 99
discharge mode transitions: $D \leftrightarrow M$ 111
discharge mode transitions: $E \leftrightarrow H$ 51, 54, 66
discharge mode transitions: self-transition   101
discharges in gas mixtures: control of excited species   72
$DMol^3$ package   142
dusty (complex) plasma   20

e

EDX spectroscopy   256
electron energy distribution/probability functions (EEDF/EEPF)   246
electron field emission   146
electron temperature   5
energy bandgap   147

f

Fermi energy level   146
floating temperature regime   X
Fowler–Nordheim tunneling   146
FTIR spectroscopy   187, 196, 249

g

GaN nanorod LEDs   274
gas discharge   3
gas separation   264
gas temperature   5
germane   29

h

hexamethyldisiloxane (HMDSO) plasma   265
High-Resolution TEM (HRTEM)   137, 154, 189, 205, 243
highest occupied molecular orbitals (HOMO)   147
hydrogen   X
hydroxyapatite   272
hydroxyapatite: biomimetic response   233
hydroxyapatite: chemical formula   210
hydroxyapatite: crystallinity   219
hydroxyapatite: cytocompatibility assessment   233
hydroxyapatite: fabrication methods   212
hydroxyapatite: growth scenario   225
hydroxyapatite: microscratch test   226
hydroxyapatite: process optimization   217
hydroxyapatite: SBF *in vitro* assessment   230
hydroxyapatite: surface morphology   221

i

ICP discharge: equivalent circuit   45
ICP plasma sources   44, 50
ICP-DC magnetron hybrid discharges   114
ICP: discharge hysteresis   54
ICP: EEDF/EEPF measurements   47
ICP: electron temperature   52, 78, 95
ICP: Langmuir probe diagnostics   47, 67, 95
ICP: magnetic probe diagnostic   46
ICP: nonlinear electromagnetic fields   57
ICP: nonlinear Lorentz force   62
ICP: plasma density   52, 78, 95
ICP: plasma potential   52, 95
ICP: RF circuit diagnostic   46
ICP: RF power deposition   87
ICP: second-harmonic generation   57, 61
ICP: two-dimensional fluid model   72
inductively coupled plasma (ICP)   XI, 41, 81, 86
Integrated Plasma-Aided Nanofabrication Facility (IPANF)   XI, 116, 199, 253
IOCPS plasma source   87
IOCPS: operating pressure range   97
IOCPS: plasma uniformity   95
ion bombardment   X, 33, 266
ion fluxes   X, 27
ion temperature   5
ionization   3
ionization degree   5
ITER   1

l

liquid precursor feed system   117
local-density approximation (LDA)   142
low-temperature plasma   32
lowest unoccupied molecular orbitals (LUMO)   147
luminescent Si nanoparticles   24

m

methane   X
molecular sieve membranes   265
multistability of plasma discharges   114

n

nano-/microporous membranes   264
nano-pyramids   XI
nano-scale   IX
nanoassembly   XI
nanoassembly: chemical purity   16
nanoassembly: cohesive energy   144
nanoassembly: controlled BU supply   278
nanoassembly: elemental composition   15
nanoassembly: plasma substrate heating   278
nanoassembly: precursor dissociation in plasmas   278
nanoassembly: reactivity   17
nanoassembly: size   14
nanoassembly: stoichiometry   16
nanoassembly: structural stability   144
nanoassembly: structure   15
nanoassembly:shape   15
nanocantilevers   275
nanocluster charge   271
nanocluster-assembled crystalline TiO_2 270
nanoclusters   18
nanocones   XI
nanocrystalline AlN: columnar structure   255
nanocrystalline AlN: plasma-controlled properties   257
nanocrystalline AlN: stoichiometry   256
nanocrystalline V_2O_5: main features   263
nanocrystalline V_2O_5: plasma-controlled properties   258
nanocrystalline V_2O_5: Raman fingerprints   259
nanocrystals   2
nanofabrication   2, 19
nanofibers   XI

nanoparticle superlattices: SiCN/AlN 192
nanoparticles 2, 14, 20
nanoparticles: carbon 25
nanoparticles: plasma synthesis 278
nanoparticles: titanium dioxide 25
nanopatterns/arrays 2
nanopatterns: size and shape uniformity 18
nanorod-to-nanowire transformation 206
nanorods: SiCAlN 205
nanorods: SiCN 206
nanoscience 12
nanostructure dimensionality XI, 160
nanostructure reshaping 28
nanostructure synthesis X
nanostructured films 2
nanostructured films: AlCN 245
nanostructured materials 20
nanostructured surfaces X
nanostructures 2
nanostructures: dimensionality 29
nanotechnology 2, 12
nanowires: a-SiO$_2$ 189
nanowires: SiCN 30, 206
Ni/Fe/Co catalyst X
Ni/Fe/Co catalysts: thermal fragmentation 123
nonequilibrium plasmas 32
nuclear fusion 1

o
Optical Emission Spectroscopy (OES) 48, 63, 99, 156, 218, 249
optical microscopy 233
Ostwald ripening 166, 183

p
Pauli's exclusion principle 146
PEMSF: discharge hysteresis 115
photoluminescence (PL) 169, 170, 175, 192, 196, 200
photoluminescence (PL): doping induced 239
photovoltaic applications 20
plasma applications 1
Plasma Assisted Magnetron Sputtering Deposition (PAMSD) 203, 213, 238, 245, 259
plasma density 7
plasma discharges: boundary conditions 76
plasma discharges: ionization and excitation rates 75
plasma discharges: particle and power balance 75

Plasma Enhanced Chemical Vapor Deposition (PECVD) 29, 124, 130, 137
plasma environment X
Plasma Nanoscience XII
plasma nanotools XII
plasma requirements for nanofabrication 35
plasma sheath 9
plasma source XI
plasma sources: requirements 3, 42
plasma state 1
Plasma- assisted magnetron sputtering deposition (PAMSD) 263
Plasma-Assisted Magnetron Sputtering Deposition (PAMSD) 162, 169
plasma-assisted RF magnetron sputtering 29
plasma-enhanced magnetron sputtering 109
Plasma-Enhanced Magnetron Sputtering Facility (PEMSF) 108
plasma-enhanced nanolithography 274
plasma: definition 3
plasma: partially and fully ionized 5
polymorphous nanomaterials 22, 242

q
Quadrupole Mass Spectrometry (QMS) 49, 116
quantum confinement structures XI
quantum dots: Al$_x$In$_{1-x}$N/AlN/Si 168, 169
quantum dots: AlN/Si 165
quantum dots: crystallinity 165
quantum dots: GaN/Al$_x$Ga$_{1-x}$N 275
quantum dots: Ge/Si 29
quantum dots: Ge/SiO$_2$ 275
quantum dots: growth modes 175
quantum dots: Si/AlN 199
quantum dots: SiC 29
quantum dots: SiC, effect of AlN buffer layer 177
quantum wells: SiC-AlN 174

r
Raman spectroscopy 127, 139, 167, 259

s
Scanning Electron Microscopy (SEM) 124, 131, 156, 164, 169, 177, 183, 189, 204–206, 220, 224, 230, 233, 254, 262
Scanning electron microscopy (SEM) 137
Scanning Transmission Electron Microscopy (STEM) 170, 180
Scherrer's equation 165, 223
semiconductors: acceptor atoms 150

semiconductors: donor atoms 150
semiconductors: Group III nitride 161
semiconductors: n-type 150
semiconductors: p-type 150
Si nanopillar arrays 274
SiC nanoassemblies: fabrication techniques 173
SiC nanoparticle films: Er doping 239
SiC quantum dots: crystallinity 185
SiC quantum dots: pattern uniformity 184
SiC quantum dots: plasma process control 187
SiC quantum dots: stoichiometry 185
SiC: properties 172
silane plasma 20
Simulated Body Fluid (SBF) 230
size-dependent properties 14
skin effect 89
substrate heating XI
superhard nc-$Al_xTi_{1-x}N$/a-Si_3N_4 nanocomposite 273
surface passivation 18
surface temperature X
surface-to-volume ratio 17

t

thermal plasma spraying 212
thermal plasmas 6, 32
Ti–Si–N–O barrier alloys 242
Ti6Al4V orthopedic alloy 210

$TiCl_4+O_2$ plasma 272
Transmission Electron Microscopy (TEM) 189, 193, 201
tricalcium phosphate (TCP) 232

u

ULSI technology 2
ultrananocrystalline diamond 24
US National Nanotechnology Initiative 12
UV/Vis spectrophotometry 261

v

valence band 147, 159
Van der Waals bonding 148
vertical alignment X, 26, 136

w

work function 146
working units 33

x

X-ray Diffractometry (XRD) 127, 220, 222, 231, 253
X-ray diffractometry (XRD) 239, 254, 262
X-ray Photoelectron Spectroscopy (XPS) 184, 218, 255, 261
X-ray reflectivity (XRR) 194

z

ZnO nanorods 28

Related Titles

Reich, S., Thomsen, C., Maultzsch, J.
Carbon Nanotubes
Basic Concepts and Physical Properties

224 pages with 126 figures
2004
Hardcover
ISBN: 978-3-527-40386-8

Schmid, G. (Ed.)
Nanoparticles
From Theory to Application

444 pages with 257 figures
2004
Hardcover
ISBN: 978-3-527-30507-0

Hodes, G. (Ed.)
Electrochemistry of Nanomaterials

326 pages with 125 figures and 5 tables
2001
Hardcover
ISBN: 978-3-527-29836-5

Wang, Z. L. (Ed.)
Characterization of Nanophase Materials

426 pages with 266 figures and 6 tables
2000
Hardcover
ISBN: 978-3-527-29837-2

Fendler, J. H. (Ed.)
Nanoparticles and Nanostructured Films
Preparation, Characterization and Applications

488 pages with 275 figures and 36 tables
1998
Hardcover
ISBN: 978-3-527-29443-5